Next-Generation Artificial Intelligence

This book combines cutting-edge advancements in neuroscience-inspired artificial intelligence, edge computing, and sustainable practices, presenting a compelling narrative for researchers, industry professionals, and policymakers alike.

Features:

- Illustrates how artificial intelligence is revolutionizing areas such as climate modeling, green energy solutions, resource management, and precision farming.
- Addresses critical issues such as replicating human cognition responsibly, addressing bias in artificial intelligence systems, protecting user privacy, and minimizing the environmental impact of artificial intelligence.
- Discusses energy-efficient artificial intelligence, ethical considerations, and advanced concepts like neuromorphic computing and federated learning.
- Explores the convergence of neuroscience and edge computing, demonstrating real-world applications in health care, agriculture, robotics, and environmental sustainability.
- Explains principles of neuroscience that inspire artificial intelligence development, including neural dynamics, spiking neural networks, and the role of biological intelligence in creating adaptive systems.

It is primarily written for senior undergraduates, graduate students, and academic researchers in electrical engineering, electronics and communications engineering, computer science and engineering, and environmental engineering.

Next-Generation Artificial Intelligence

Convergence of Neuroscience, Edge Computing, and Sustainable Technologies

Edited by

Manzoor Ansari
Syed Arshad Ali
Mansaf Alam

CRC Press
Taylor & Francis Group
Boca Raton London New York

CRC Press is an imprint of the
Taylor & Francis Group, an **informa** business

First edition published 2027
by CRC Press
2385 NW Executive Center Drive, Suite 320, Boca Raton FL 33431

and by CRC Press
4 Park Square, Milton Park, Abingdon, Oxon, OX14 4RN

CRC Press is an imprint of Taylor & Francis Group, LLC

ISBN: 978-1-041-16546-0 (hbk)
ISBN: 978-1-041-16667-2 (pbk)
ISBN: 978-1-003-68589-0 (ebk)

DOI: 10.1201/9781003685890

Typeset in Sabon
by SPi Technologies India Pvt Ltd (Straive)

Contents

Acknowledgments

We would like to express our sincere gratitude to all the chapter authors whose valuable contributions have shaped this edited volume, **"Next-Generation AI: Convergence of Neuroscience, Edge Computing, and Sustainable Technologies."** Their dedication, scholarly rigor, and timely collaboration made this book possible.

We extend our heartfelt appreciation to the reviewers for providing constructive feedback that strengthened the quality and clarity of each chapter. Their expertise and commitment to academic excellence have been instrumental in refining the manuscript.

Our sincere thanks go to **CRC Press, Taylor & Francis Group**, for their continuous support, guidance, and professionalism throughout the development of this book. We especially acknowledge the editorial assistance provided by the CRC Press team, whose direction ensured a smooth and well-coordinated process.

We also thank our respective institutions **Institute of Information Technology & Management (IITM), Sharda University**, and **Jamia Millia Islamia** for their encouragement and academic environment that fostered collaborative research and innovation.

Finally, we are grateful to our families, colleagues, and students for their patience, motivation, and unwavering support during the preparation of this volume.

Manzoor Ansari
Syed Arshad Ali
Mansaf Alam

Preface

Artificial Intelligence (AI) is entering a new era where it converges with **neuroscience, edge computing, and sustainable technologies** to address some of the most pressing global challenges. This book, *Next-Generation AI: Convergence of Neuroscience, Edge Computing, and Sustainable Technologies*, brings together diverse research contributions that reflect both the theoretical foundations and practical applications of AI across multiple domains.

The book highlights how next-generation AI is reshaping industries and society. Early sections introduce **neuromorphic computing and biologically inspired algorithms**, which open pathways toward energy-efficient and adaptive intelligence. A substantial focus is placed on **health-care innovations**, ranging from AI-driven diagnostic systems and cancer detection to drone-enabled IoT solutions for reducing ambulance delays and mortality in critical emergencies. These contributions underscore the life-saving potential of AI when combined with real-time data and intelligent decision-making. Beyond health care, the book explores **AI applications in energy, agriculture, and environmental sustainability**. Topics such as deep learning for battery prognostics, AI in hydrogen energy, and smart grid integration reflect the urgent need for sustainable energy solutions. Similarly, AI's role in **precision agriculture and environmental monitoring** demonstrates its power in promoting resource efficiency and ecological balance. This book addresses **societal and security dimensions**, including generative AI for sustainability, AI in education for enhancing learning outcomes, and the use of AI in cybersecurity and cyber warfare to safeguard data and economic stability. Together, these perspectives illustrate the breadth of next-generation AI and its capacity to drive transformative change.

This book is intended for **researchers, practitioners, industry professionals, and students** in computer science, engineering, health care, and related fields. By bridging foundational research with applied case studies, it offers valuable insights into how AI can be developed and deployed responsibly for the benefit of humanity.

We extend our sincere gratitude to all contributors whose expertise and commitment have made this book possible. Their collective work reflects the spirit of innovation and collaboration that defines next-generation AI.

Manzoor Ansari
Syed Arshad Ali
Mansaf Alam

About the editors

Manzoor Ansari is an assistant professor in the Department of Computer Science at the Institute of Information Technology & Management (IITM), GGSIP University, New Delhi, where he teaches IoT, cloud computing, artificial intelligence, time series analysis, and related computing subjects. He has played key academic and organizational roles, including serving as co-convener and session co-chair for ICACIA 2025, co-coordinator for AICTE-ATAL FDPs, and program coordinator for NEP 2020 Orientation and Sensitization programs conducted in collaboration with Hansraj College, University of Delhi. His research spans IoT-cloud integration, machine learning, smart surveillance systems, cyber forensics, business intelligence, environmental monitoring, and AI-driven predictive modeling. He has guided multiple innovative student projects, including autonomous surveillance robots, AI-enabled assistants, and smart security systems. Dr. Ansari has authored research articles in various respected SCI- and Scopus-indexed journals and contributed chapters with international publishers. He has also filed multiple patents in the areas of AI-based predictive maintenance, intelligent educational tools, and autonomous learning systems. In addition, he actively participates in national and international workshops, technical summits, and academic training programs, and contributes as a reviewer for scholarly research, reflecting his strong commitment to advancing next-generation technologies and interdisciplinary computing research.

Syed Arshad Ali is an assistant professor in the Department of Computer Science & Applications at Sharda University, Greater Noida, where he teaches undergraduate and postgraduate courses and contributes to core academic committees, including curriculum design, NAAC documentation, departmental website management, and UG–PG admissions. His academic experience spans more than a decade across reputed institutions such as IITM Janakpuri, Jamia Millia Islamia, and Jamia Hamdard, where he has taught computing subjects, mentored students, coordinated academic activities, and contributed to research-led teaching. His areas of expertise include cloud computing, data structures, algorithms, programming, operating systems, and big data technologies. Dr. Ali's research contributions cover cloud resource management, energy-efficient computing, IoT-cloud integration, big data analytics, federated learning, and NoSQL data solutions. He has published in respected Scopus- and SCI-indexed journals and IEEE conference proceedings and contributed multiple book chapters with international publications. His academic involvement includes participation in numerous workshops and training programs in emerging technologies, including cloud computing, digital transformation, data analytics, bioinformatics, design thinking, and big data computing under GIAN and NPTEL initiatives. He has also presented his work at international conferences and actively engages in ongoing academic development through faculty development programs across leading institutions. With research interests bridging cloud systems, data-intensive computing, and next-generation intelligent frameworks, Dr. Ali brings strong technical, pedagogical, and research-driven expertise to collaborative academic projects and editorial work.

Mansaf Alam is a professor in the Department of Computer Science, Faculty of Sciences, Jamia Millia Islamia, New Delhi. He has been awarded a Young Faculty Research Fellowship by MeitY, Government of India. He is the editor-in-chief of the Journal of Applied Information Science. He has published several research articles in reputed international journals and proceedings in reputed international conferences. His areas of research include artificial intelligence, big data analytics, machine learning and deep learning, cloud computing, and data mining. He is a reviewer of various journals of international repute, such as Information Science. He is also a member of the program committee of various reputed international conferences and serves on the editorial board of several international journals in computer science. He has published several books, including *Digital Logic Design* (PHI), *Concepts of Multimedia* (Arihant), *Internet of Things: Concepts and Applications*, *Big Data Analytics: Applications in Business and Marketing*, and *Big Data Analytics: Digital Marketing and Decision Making* (Taylor and Francis). He recently received an international patent (Australia) on an artificial intelligence-based smart dustbin. He has delivered more than 396 invited talks.

Contributors

Sayed Abdulhayan
P.A. College of Engineering
Mangalore, Karnataka, India

Aditi Aggarwal
Department of Computer Science
Institute of Information Technology & Management (IITM), GGSIP University
New Delhi, India

Arun Prakash Agrawal
School of Computer Science Engineering and Technology
Bennett University
Greater Noida, India

Javed Ahmad
Department of Computer Science and Engineering
Sharda University
Greater Noida, India

Mahmood Akbar
Faculty of Business Studies
Arab Open University–Bahrain

Sharik Ahmad
Faculty of Information and Communications Technology
Universiti Teknikal Malaysia Melaka
Melaka, Malaysia

Masood Alam
Department of Mathematics and IT
Sultan Qaboos University
Muscat, Oman

Manzoor Ansari
Department of Computer Science
Institute of Information Technology & Management (IITM), GGSIP University
New Delhi, India

Syed Arshad Ali
Department of Computer Science and Applications
Sharda University
Greater Noida, India

Mohit Singh Bisht
Chandigarh University
Mohali, Punjab, India

Kuldeep Chouhan
Department of Computer Science & Applications
Sharda School of Engineering & Technology
Sharda University
Greater Noida, India

Aashita Chhabra
School of Engineering & Technology
Vivekananda Institute of Professional Studies
Technical Campus, Delhi, India

V. Lakshmi Chetana
School of Computing
Amrita Vishwa Vidyapeetham
Amaravati Campus, India

Dhruv Dhayal
Department of Computer Science
Institute of Information Technology & Management (IITM), GGSIP University
New Delhi, India

Pankaj Kumar Dubey
Kamla Nehru Institute of Technology
Sultanpur, Uttar Pradesh, India

Aditya Giri
Department of Computer Science
Sharda University
Greater Noida, India

Krishna Gupta
Department of Computer Science
Sharda University
Greater Noida, India

M. Robinson Joel
Department of Computer Science and Engineering
KCG College of Technology
Chennai, India

Rajkumar Kannan
Research Department of Computer Science
Bishop Heber College (Autonomous)
Bharathidasan University
Tamil Nadu, India

Faraz Khan
Glocal University
Saharanpur, India

Shahzeb Khan
Department of Computer Science and Applications
Sharda University
Greater Noida, India

Sumaira Khan
Glocal Ayurveda College
Saharanpur, India

Saba Khanum
Department of Information Technology
Maharaja Surajmal Institute of Technology
New Delhi, India

Vaneet Kumar
Chandigarh University
Mohali, Punjab, India

Varun Kumar
Kamla Nehru Institute of Technology
Sultanpur, Uttar Pradesh, India

Harmeet Malhotra
Department of Computer Science
Institute of Information Technology & Management (IITM), GGSIP University
New Delhi, India

Seema Malik
School of Engineering & Technology
Vivekananda Institute of Professional Studies
Technical Campus, Delhi, India

Balasubbareddy Mallala
Department of Electrical and Electronics Engineering
Chaitanya Bharathi Institute of Technology
Hyderabad, India

Reis Mulita
Canadian Institute of Technology (CIT)
Tirana, Albania

Anjali Nambiyar
Research Department of Computer Science
Bishop Heber College (Autonomous)
Bharathidasan University
Tamil Nadu, India

Lahari Lakshmi Prasad
St. Joseph Engineering College
Vamanjoor, Karnataka, India

Preeti Sharma
Department of Computer Science
Institute of Information Technology & Management (IITM), GGSIP University
New Delhi, India

Prabhav Pushker
Department of Information Technology
Maharaja Surajmal Institute of Technology
New Delhi, India

Sayed Abulhasan Quadri
RIT, Hassan, Visvesvaraya Technological University
Karnataka, India

P. V. R. Ravindra Reddy
Department of Electrical and Electronics Engineering
Chaitanya Bharathi Institute of Technology
Hyderabad, India
and
Mechanical Engineering Department
Chaitanya Bharathi Institute of Technology
Hyderabad, India

P. Prabhakar Reddy
Mechanical Engineering Department
Chaitanya Bharathi Institute of Technology
Hyderabad, India

Brisida Sefa
Canadian Institute of Technology (CIT)
Tirana, Albania

Komal Sharma
Department of Computer Science
Institute of Information Technology & Management (IITM), GGSIP University
New Delhi, India

Mohd. Maroof Siddiqui
College of Engineering, Dhofar University
Sultanate of Oman

Bindeshwar Singh
Kamla Nehru Institute of Technology
Sultanpur, Uttar Pradesh, India

K. K. Singh
College of Engineering
Amity University
India

Shailu Singh
Department of Economics
Hansraj College, University of Delhi
India

Shubham Kumar Singh
Department of Computer Science & Applications
Sharda University
Greater Noida, India

Sachin Soni
Department of Computer Science
Institute of Information Technology & Management (IITM), GGSIP University
New Delhi, India

Ram Sudan
School of Engineering & Technology
Vivekananda Institute of Professional Studies
Technical Campus, Delhi, India

Mohd. Tajammul
Department of Computer Science and Applications
Sharda University
Greater Noida, India

Prajoona Valsalan
College of Engineering
Dhofar University
Sultanate of Oman

Pankaj Kumar Varshney
Department of Computer Science
Institute of Information Technology & Management (IITM), GGSIP University
New Delhi, India

Ashutosh Yadav
Department of Commerce
Hansraj College
University of Delhi
India

Deepshikha Yadav
Department of Commerce
Hansraj College
University of Delhi
India

Anurag Kakkar
Department of Economics
Hansraj College
University of Delhi
India

Mohammed Zakir Bellary
P.A. College of Engineering
Mangalore, Karnataka, India

Chapter 1

Introduction to neuromorphic computing

Sayed Abdulhayan, Lahari Lakshmi Prasad, Sayed Abulhasan Quadri, and Mohammed Zakir Bellary

1.1 INTRODUCTION

Neuromorphic computing is a paradigm shift in computing, drawing its cues from the structure and mechanism of the human brain. This new technology area seeks to improve the efficiency and performance of computation by leveraging the integration of memory and processing in parallel processing, which could be used to make a breakthrough in artificial intelligence and machine learning. Neuromorphic computing is explained by its main features in the following sections.

Principles of neuromorphic computing

- **Brain Emulation:** Neuromorphic systems are modeled after the neural structure of the brain and employ spiking neural networks (SNNs) as a mechanism to efficiently process information [1, 2].
- **Energy Efficiency:** The human brain runs at about 20 watts, much lower than supercomputers, today and points to better efficient computing architecture [3].

Applications and benefits

Figure 1.1 shows that spiking neural activity plays a crucial role in neuromorphic computing, which mimics the brain's architecture to enhance computational efficiency and energy performance. This approach utilizes SNNs that encode information through time-dependent spikes, allowing for high parallelism and reduced power consumption compared to traditional computing metho [4]. The following sections elaborate on key aspects of spiking neural activity in this context.

Table 1.1 speaks about neuromorphic computing leveraging advanced materials and architectures to enhance computational efficiency and mimic biological processes. Statistical data plays a crucial role in optimizing these systems, particularly in addressing variations and performance metrics. The following sections outline key statistical insights from recent research in this field.

- **Enhanced AI Capabilities:** Pattern recognition, natural language processing, and autonomous decision-making are tasks that can be revolutionized by neuromorphic computing [1, 5].
- **Integration with Biology:** The introduction of this technology may pave the way for improvements in the fields of brain-to-computer interface and neuroprosthetics that have a direct implication in the sphere of health care and human augmentation [1].

DOI: 10.1201/9781003685890-1

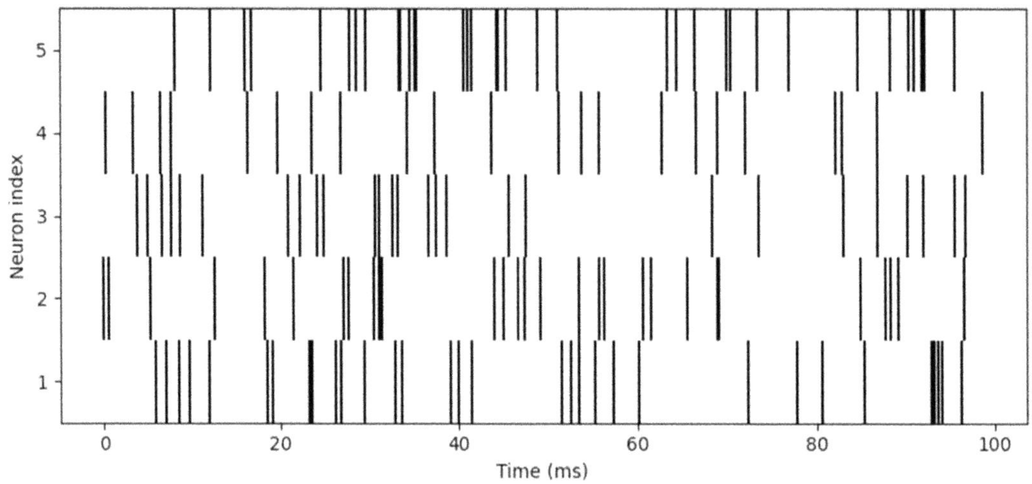

Figure 1.1 Spiking neural activity in neuromorphic computing.

Table 1.1 Statistical data for neuromorphic computing

Category	*Statistics*
Global Market Size (2023)	USD 6.5 billion
Projected Market Size (2030)	USD 55–60 billion
CAGR (2023–2030)	~36%
Leading Applications Edge AI, Robotics, Healthcare, Autonomous Systems Energy Efficiency	Up to 1000× lower power vs. traditional CPUs/ GPUs
Processing Speed	Millisecond-scale event-driven responses
Key Industry Players Intel (Loihi), IBM (TrueNorth), BrainChip, SynSense Research Growth	(Papers/year, 2023) ~2,500+ publications

Challenges and future directions

- **Technological Hurdles:** Although it is implemented, there are issues toward the development of neuromorphic chips, such as materials, tolerance to faults, and manufacturability [2, 6].
- **Algorithm Development:** Other new algorithmic methods belong to being able to fully utilize the potential of a neuromorphic system [3].

Although neuromorphic computing shows much potential, it has been in its developmental phase, and more studies are required to find solutions to the limitations that are still there and which are likely to stop its full potential from reaching other applications.

Table 1.2 shows that neuromorphic computing has evolved significantly over the past several decades, driven by advancements in both hardware and algorithms inspired by biological neural networks. This historical perspective highlights key milestones that have shaped the field, from early theoretical frameworks to modern implementations. The following sections outline the chronological development of neuromorphic computing.

Emerging scientific and technological topics

Table 1.2 Historical perspective chronological dates for neuromorphic computing

Year	*Milestone/event*
1943	McCulloch and Pitts propose first mathematical model of artificial neurons.
1958	Rosenblatt develops the Perceptron, an early neural network model.
1982	Carver Mead coins the term neuromorphic engineering, introducing analog VLSI (Very Large Scale Integrated Circuits) circuits inspired by neurons.
1990	Growth of CMOS-based neuromorphic circuits for sensory processing (vision, auditory).
2006	Rise of deep learning revives interest in brain-inspired computing.
2011	IBM developed TrueNorth, a large-scale neuromorphic chip.
2017	Intel introduced Loihi, a neuromorphic research processor.
2020	BrainChip launched Akida, a commercial neuromorphic processor for Edge AI.
2023	Research accelerates on memristor-based neuromorphic hardware and SNNs for Edge AI.

1.2 SPIKING NEURAL NETWORKS (SNNS FOR EDGE AI)

Spiking neural networks are becoming a topic of interest with Edge-AI applications as they are energy efficient and are able to process information similarly to biological systems. Because SNNs (compared to traditional neural networks) make use of sparse spike events during computation, power requirements can be substantially lower. There are, however, several challenges that have to be overcome to maximize their functionality in edge environments.

Energy efficiency

- SNNs are highly energy efficient, with on-chip learning of 0.1 0.1 2.C2.C disable Enjoying the essence of amplified energy, 2.C2.som system on-chip learning being 0.1 0.1 2.C2. saltimbano to edge devices that have low amounts of power [7].
- The event-based design of SNNs makes them compute solely when spikes happen, and this feature ensures that they avoid irrelevant computations and energy waste [8].

Model adaptability

- Some breakthroughs have seen the creation of autonomous model selecting methods that would make the SNN more adaptable in training, such that it can learn better at tasks such as image classification [9].
- The Fully Binarized Weight SNN (FBW-SNN) highlights how the binary weights may still preserve accuracy and further decrease the power needed, with 82.68% accuracy on CIFAR-10 attained on 14-time steps [10].

Challenges

- Although SNNs have plenty of strengths, there are several weaknesses; these include unnecessary numerical calculations when weights are updated and inefficient processing of sparse input spikes that might result in energy and time wastage [7, 8].
- The necessity of co-designing circuits, architectures, and algorithms to improve performance and support low-power operation is still a butting block [8].

Although SNNs hold a strong argument in support of the concept of Edge AI, the adoption of the latter is not entirely devoid of obstacles, as additional approaches toward optimization and the extrapolation of the potential limitations of their structure and training set are still required.

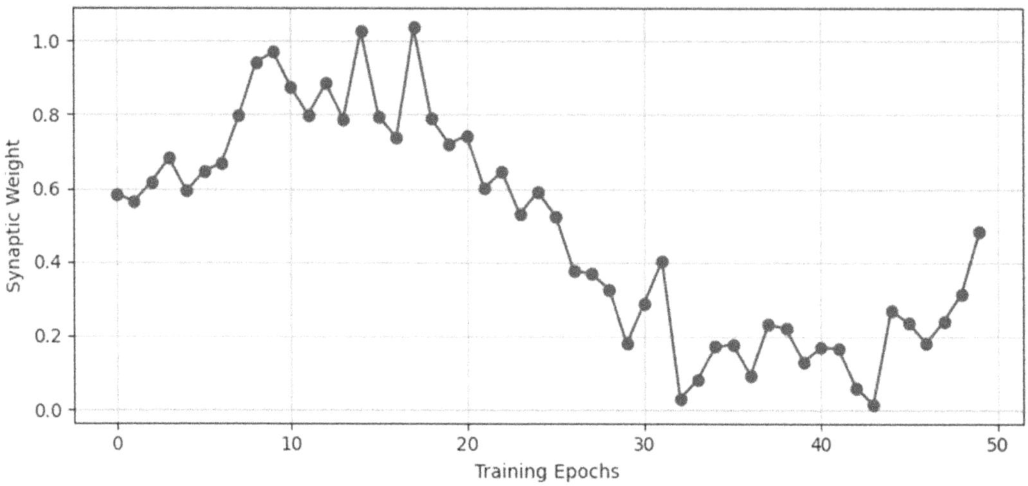

Figure 1.2 Synaptic weight adaptation over time.

Figure 1.2 shows that synaptic weight adaptation over time in neuromorphic computing is a critical aspect that influences the performance and efficiency of artificial neural networks. This adaptation mimics biological processes, allowing for dynamic learning and memory storage. Various approaches have been developed to enhance synaptic weight modulation, which is essential for achieving high accuracy in neuromorphic systems. The following sections outline key strategies and findings related to synaptic weight adaptation.

1.3 MEMRISTOR-BASED NEUROMORPHIC ARCHITECTURES

Neuromorphic architecture with memristors is making a breakthrough in computing since it allows the capabilities of memristors to make neural networks work. Such architectures support effective computation using crossbar arrays that allow operations such as vector-matrix multiplication (VMM) to be performed using less energy than conventional architectures. These innovative systems have several major aspects as described in the following sections.

Memristor functionality and architecture

- The crossbar structure allows VMM operations to be naturally performed using memristors, which is essential to neural network training and inference [11].
- A combination (Co-Fe-B)x(LiNbO3), 100x has demonstrated great endurance and multilevel resistive switching, which can fit into complex neural network architectures [11].

Energy efficiency

- There has been a proposal of a domino logic architecture that uses memristors, which has an energy efficiency of 0.61 fJ per classification per component, which is better than a traditional design [12].
- A similar approach is that memristor-based SNNs have demonstrated a higher level of inference accuracy due to self-calibration inherent to their design, erasing performance differences caused by process variations [13].

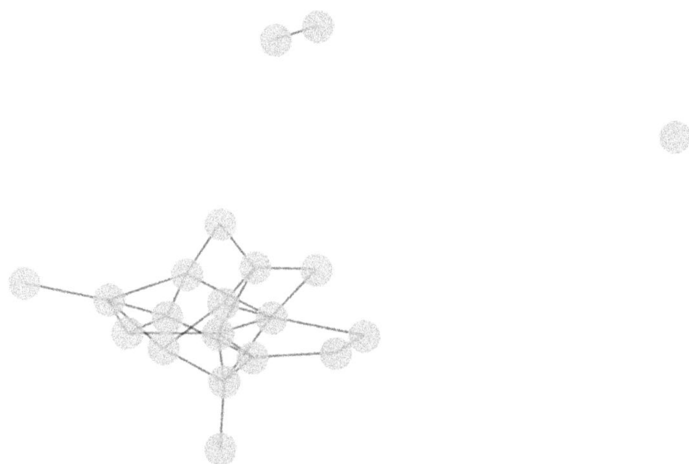

Figure 1.3 Neuromorphic network connect.

Dynamic reconfigurability

- Memristive reconfigurable neuromorphic arrays (RNA) allow dynamic architecture realizations, which increase flexibility and decrease power consumption with time-division multiplexing [14]. Figure 1.3 shows that neuromorphic network connectivity is a crucial aspect of neuromorphic computing, enabling efficient communication and processing within SNNs. This connectivity is characterized by the dynamic interactions between neurons and synapses, which facilitate learning and information processing. The following sections outline key components of neuromorphic network connectivity.

  Though there are significant advantages of memristor-based architecture in terms of energy consumption and computational performance, problems like variability of devices and vulnerability of hardware-software interaction are still the keys to future study and development.

1.4 NEUROMORPHIC HARDWARE FOR CONTINUAL LEARNING

Neuromorphic hardware is also gaining increased relevance in terms of its ability to support the process of continual learning, which in turn enables systems to adapt to new information without losing previously obtained information. Such an ability is essential in systems that operate in dynamic environments (e.g., robotics and autonomous systems). One of the areas where neuromorphic hardware plays a decisive role is discussed in the following sections in the context of continual learning.

Neuromorphic design principles

- Neuromorphic systems exploit the concepts of bio-mechanical neural networks to process efficiently and learn in real time [5].
- There is a theoretical framework of how to perform spiking neurons (Neuromorphic Neural Engineering Framework, NEF), which improves data representation and transformation [15].

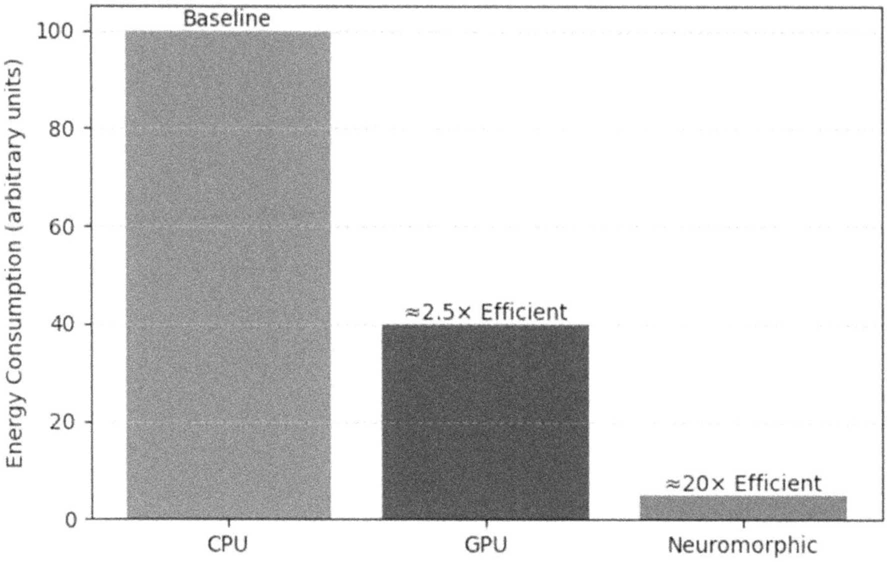

Figure 1.4 Energy efficiency comparison.

Learning mechanisms

- Ongoing learning can be achieved with event-driven architectures that enable online learning; a system can learn as the stream of data pours in without a training and testing setup [16].

To perform optimal stability of learning and avoid overfitting, techniques like prescribed error sensitivity (PES) learning rule and spike-based plasticity circuits are also used [15, 16]. Figure 1.4 shows that neuromorphic computing represents a significant advancement in energy efficiency compared to traditional computing architectures. By emulating the brain's neural processes, neuromorphic systems, such as Intel's Loihi chip, demonstrate remarkable power savings across various applications, particularly in machine learning and image retrieval tasks. The following sections detail the energy efficiency comparisons and innovations in neuromorphic computing.

Applications and future directions

- Neuromorphic hardware is ready to be implemented in diverse fields, such as health care and transportation, where low-power flexible solutions are vital [5].
- Flexible embedded learning eases the way for future mobile and wearable devices via the development of frameworks such as the Neural and Synaptic Array Transceiver (NSAT) [17].

  Although the progress on neuromorphic hardware in the field of continual learning is evolving positively, there is still work to do in closing the gap between theoretical models and practical applications, especially in real-life situations where the available data is heterogeneous and unpredictable.

1.5 EVENT-DRIVEN SENSORS AND VISION SYSTEMS

Sensor vision systems and event-driven sensors are the next step in visual perception technology, replicating functions used in nature to optimize the processing of real-time information. In contrast to frame-based systems that merely capture perturbation-free images at regular intervals, event-based sensors, including dynamic vision sensors (DVS), are synchronous, i.e., they develop modality knowledge only when there is a change in the scene. This is more efficient in reducing redundancy and maximizing processing efficiency, which is ideal in cases that need short response times, e.g., robotics and autonomous vehicles.

Event-driven sensor mechanism

- **Asynchronous Operation:** An event-driven sensor does not proceed in frames; each pixel independently emits an event when it detects major changes, so that the sensor has an ongoing data flow [18].
- **Temporal Contrast:** The DVS technology provides a relative measure of variations in light, which gives high temporality and real-time processing of visual data [19].

Hybrid models and integration

- **Combining Sensor Types:** Based on experimental results, combining event-based sensors and frame-based sensors has been observed to allow increased system efficiency, lower latency, and resolve data integration issues [20].
- **Convolutional Neural Networks:** Event-driven convolutional networks can be trained to detect objects in real time, taking advantage of the infinite data stream of event-driven sensors [19].

Applications and performance

- **High-Speed Recognition:** Visual inputs could be processed by event-driven systems at a rate greater than 22 Mbyte/s, thus allowing real-time applications such as detection of fast-moving objects or symbols [21].
- **Modular Convolution Processors:** Such processors can adaptively choose convolution kernels depending on the source of events and thereby extend object recognition capability [18].

Although the speed and efficiency of event-driven systems are highly beneficial, the complexities of integrating these technologies into already established frame-based systems, along with compatibility with different applications, have not yet been resolved.

1.6 BRAIN-INSPIRED CHIPS: LOIHI 2 AND BEYOND

Neuromorphic computing, brain-inspired chips, specifically Intel Loihi and its successor Loihi 2, also take a giant step in neuromorphic computing, processing data like the brain by simulating its architecture and learning properties. These chips apply SNNs as the means of information processing that is close to the way information is processed in real neural networks, such as real-time learning and low-energy consumption. These innovations entail some of the main characteristics and implications that are discussed subsequently.

Loihi and Loihi 2 architecture

- **Core Design:** Loihi uses 128 custom cores having 1024 neurons that are built to learn on-chip and use an SNN [22].
- **Power Efficiency:** The two chips operate at low voltage (0.55V to 1.25V) to optimize energy consumption without compromises on performance [22].
- **Scalability:** Loihi 2 is more equipped with edge computing capabilities, with observed high accuracy (up to 97.6%) on edge processing real-time use cases such as sensor data classification [23].

Applications and performance

- **Diverse Use Cases:** Loihi has demonstrated usefulness in several application areas, such as adaptive control and event-based data processing, and has been shown to be more effective than traditional architectures in terms of latency and power consumption [24].
- **Research Validation:** Research has shown the efficiency of Loihi when compared to classical models and provided validation of its efficiency in modeling natural complex neural activity [25].

Although the innovations of brain-inspired chips such as Loihi and Loihi 2 are promising, there are still obstacles related to widespread use and incorporation into current systems, especially concerning software support and the requirement for specific microchip programming environments.

1.7 3D INTEGRATION IN NEUROMORPHIC SYSTEMS

The area of 3D integration of neuromorphic systems marks a breakthrough in planning and implementation of artificial intelligence hardware. The stacking of nano components is used to increase their performance, energy efficiency, and miniaturization, hence replicating the architecture of the human brain. The sections that follow point out the major highlights of 3D integration in the context of neuromorphic systems.

Advantages of 3D integration

- **High Density and Performance:** Other features offered by 3D integration are the stacking of memristors and memcapacitors, which would provide high-density layouts with better processing power [26].
- **Energy Efficiency:** The orientation of Through Silicon Vias (TSVs) permits high speed of conveying data between components, an aspect that cuts power consumption tremendously compared to conventional architecture [27].
- **Design Flexibility:** 3D integration is modular, which allows the inclusion of more advanced functionality, hence it can be used in several applications in the fields of AI and neuromorphic computing [28].

Applications for neuromorphic systems

- **Chemical Sensing:** To illustrate this, a 3D integrated neuromorphic system has been authentically created to sense chemicals, showing that it can replicate biosensing functions, e.g., sensory processing [29].
- **Neuro-Robotics:** Such systems are vital in next-generation neuro-robotics, where they perform vital memory and processing tasks in alignment with biological systems [29].

Due to these many advantages, challenges with 3D integration still exist, such as manufacturing complexity and the limited knowledge about how the brain works, which are necessary to take advantage of the potential within neuromorphic structures.

1.8 BIOHYBRID NEUROMORPHIC INTERFACES

Biohybrid neuromorphic interfaces are an innovative manifestation of the intersection between biological and artificial neural networks to foster high-tech interplay between living nerve cells and silicon-based neural networks. The interfaces exploit the special characteristics of both biological tissue and neuromorphic hardware to develop systems that could simulate and complement neural operations, opening the door to advanced neuroprosthetics and neuroscience applications.

Neuromorphic engineering and biohybrid systems

- Neuromorphic engineering uses VLSI to attempt to duplicate the structure and functionality of nervous systems and enables emulating millions of silicon neurons in large-scale neural networks [30].
- Biohybrid systems integrate biological neurons with artificial networks that allow real-time bidirectional communication and allow observing the dynamics of biohybrid networks in a biologically relevant context [31].

Material engineering for enhanced integration

- Creation of materials that simulate the internal settings, including supported lipid bilayers, increases the capacity to artificially couple bioelectronic devices to neuronal cells, leading to improved communication and synaptic development [32].
- Such designed materials can control fluidity and mechanical characteristics, which promote neural outgrowth and synapse association [32].

1.9 EXPERIMENTAL MODELS AND APPLICATIONS

- Recent experimental systems have achieved coupling of biological neural networks with biomimetic hardware, enabling monitoring and control of neural activities in real time [33].
- In addition to their value for fundamental studies, such biohybrid interfaces have potential clinical utility in neurorehabilitation and regeneration of sensory and motor functions [31, 34].

Although biohybrid neuromorphic interfaces hold much promise, the development of long-term stability and functionality in these systems at the interface with biological environments can potentially limit their implementation and use in the medical and technology world.

1.10 ENERGY-EFFICIENT LEARNING ALGORITHMS FOR NEUROMORPHIC SYSTEMS

Neuromorphic optimized learning algorithms use new architectures and approaches to allow energy-efficient and algorithmic performance optimization of machines. Some of the strategies brought into focus by recent research are the combination of SNNs and memristive devices, which make use of these systems more effectively. The following sections depict the most critical paths to energy efficiency in neuromorphic learning algorithms.

Combined learning strategies

- It is possible to reduce the learning parameters used in Convolutional Spiking Neural Networks (CSNNs) by incorporating Pair-based Spike Timing-Dependent Plasticity (PSTDP) and power law-dependent spike-timing-dependent plasticity (STDP) to achieve high accuracy with less energy [35].
- Such a combined technique would permit the use of memristive devices, which further increase the energy efficiency and accuracy of perceptual computing.

Leaky integrate-and-fire neuron model

- A new low-power spike generator in leaky integrate-and-fire (LIF) neuron architecture saves much energy down to 8.56 fJ per spike, with an average power consumption of 1.62 nW [36].
- Adjustable firing frequencies are also possible in this model, and thus this increases its flexibility to be used in different applications.

Memristive mixed-signal systems

- Memristive neuromorphic systems used to enhance power and area efficiency using nano-scale devices and apply synchronous digital long-term plasticity to enable system online learning [37].
- This method helps perform energy-efficient training during operation and reduces both power and area overhead.

On-chip learning systems

- A neuromorphic image classification processor is energy efficient in training when it increases energy consumption only by 7.5% relative to the energy consumption during inference [38].
- The system is reflective of the on-chip learning potential in energy-restricted circumstances.

Memristors and domino logic

- An example of this is a proposed architecture using memristors and domino logic that reported 1.26 fJ per classification per synapse and an accuracy of 82.5% even when noise is added [39].
- The design highlights the role of energy efficiency in computation among high-performance AI applications.

Although these developments demonstrate great improvements to energy-efficient learning algorithms, there are still issues in balancing performance and energy use, especially within more complicated AI models. More studies should be carried out to improve these systems to be used on a larger scale.

1.11 QUANTUM-NEUROMORPHIC HYBRID SYSTEMS

Quantum-neuromorphic hybrid systems represent an innovative convergence of quantum computer science and neuromorphic engineering, beginning with the goal of increasing the efficiency of computation and improving its capability. These systems effectively capitalize on the strengths of both quantum mechanics and brain-inspired architectures, leading to improved capability in diverse pursuits, such as artificial intelligence and quantum simulations. Important details about such hybrid systems can be captured as follows.

Quantum-classical integration

- Hybrid quantum-classical networks: Hybrid quantum-classical networks consist of classical neural network layers interconnected with quantum circuits and are more trainable and more accurate compared to their quantum or classical equivalents without an increase in network size [40].
- When it comes to classification tasks, these networks surpass other classical networks in terms of performance, with results comparable to the larger networks but with lower energy consumption [41].

Applications in quantum simulations

- Quantum-neural hybrids efficiently parametrize quantum many-body wavefunctions and have the advantages of both high sampling capacity (neural) and high expressivity (quantum) [42].
- Numerical observations have shown that in more complex quantum systems these hybrid techniques lead to a lower estimate of energy and demonstrate the potential of the technique in quantum chemistry and condensed matter physics [42].

Neuromorphic device innovations

- Quantum topological neuristors (QTNs) stack ultralow-energy and fast switching similar to biological synaptic properties [43].
- QTNs can support sophisticated neuromorphic computing followed by immediate decision-making and learning operations, which is vital in developing smart systems [43].

Though the concept of integrating quantum technologies with neuromorphic technology has huge potential, it faces issues in optimizing such systems to be practical to use, especially when it comes to their scalability and devices. They need further exploration in the future in order to fulfill their potential in many areas.

1.12 NEUROMORPHIC COMPUTING FOR BRAIN-MACHINE INTERFACES (BMIS)

The area of neuromorphic computing is also becoming an enabling technology in brain-machine interfaces (BMIs), where it enables high-performance and efficient BMIs. Neuromorphic systems have the added advantage that, because they are more similar to the architecture of the brain, the integration between biological and artificial parts can happen more effectively, resulting in more effective neuroprosthetic devices. Such a combination of neuromorphic computing and BMIs is typified by a number of significant features.

Enhanced signal processing

- One application of neuromorphic models is the use of SNNs to interpret information as biological neurons [44] and enable real-time decoding of brain signals, including movement imagery translated from electroencephalogram (EEG) data [44].
- Event-based integration into devices such as NeuroBMI minimizes data transmission rates and power consumption, representing an impressive data compression ratio of 520 [45].

Energy efficiency

- Neuromorphic computing uses very low amounts of energy, which allows its application in implantable devices where stamina and frequent battery replacements are not preferred [46].
- The NeuroBMI design features a noise-tolerant spike detector that uses only 0.48 μW, demonstrating the possibility of low-energy BMIs [47].

Future implications

- Innovations in neuromorphic computing may translate into a revolution in health care with more dependable and productive neuroprosthesis applications augmenting human abilities [48].

On the other hand, although the potential benefits of neuromorphic computing are massive, the limitations in verifying the reliability and stability of such systems in various real-world applications have not been resolved, and further research and development are necessary.

1.13 ANALOG VS DIGITAL NEUROMORPHIC IMPLEMENTATIONS

The comparison of analog versus digital neuromorphic implementations provides the clear benefits and limitations for each strategy. Digital implementations are modest in terms of simplicity and means of integration, whereas analog implementations excel in energy consumption and miniaturization. These areas are explored in the following sections.

Performance and efficiency

- **Analog Systems:** Analog implementations are typically very power-efficient (up to 20 times less energy than the digital equivalent), particularly analog memristor-based implementations [49]. They can also occupy less physical space, hence they can be applicable at higher scales [49].
- **Digital Systems:** Although more complicated, digital designs are very accurate in tasks such as signal classification, especially in noisy scenarios [50]. They also have well-established methods of error correction, with ease of programming being provided.

Complexity and design challenges

- **Analog Complexity:** Nonetheless, despite their effectiveness, analog systems suffer drawbacks like nonlinearities and hysteresis that can result in loss of accuracy in inference [51]. This presents high levels of design complexity, which may disallow wide implementation.
- **Digital Complexity:** Digital systems tend to be complex and more expensive because they require the provision of lookup tables as well as large processing capabilities [52].

To conclude, digital systems are more accurate and easier to use, whereas analog implementations are more compact and more energy efficient. The decision between the two, however, would be based on certain requirements and limitations of an application.

1.14 NEUROMORPHIC ACCELERATION OF LARGE LANGUAGE MODELS (LLMS)

One of the most important research topics in this case is the acceleration of large language models (LLMs) due to the necessity for efficient processing and deployment in different domains. These new developments point to a number of interesting directions, such as hardware accelerators, training optimizations, and software-hardware coding schemes. All these techniques are principles that have the same goal: improving the performance and efficiency of LLMs compared to the problems that arise due to their complexity.

Hardware acceleration

- **Silicon Photonics:** Silicon photonics-based novel hardware accelerators have shown that their throughput can be at least 10.2x higher and energy efficiency 3.8x better than conventional electronic platforms to support LLMs [53].
- **GPU Tensor Cores:** An inference-driven bipolar-INT data format and matrix multiplication scheme have also been introduced with up to 6.7x inference speedup on LLMs, representing an important efficiency gain on GPU-based systems [54].

Training optimization

- **Progressive Training Length:** The growing training sequence length in the form of inertia, implemented as the Grow Length method, seems to converge faster and yield better performance measurements even when no further engineering is applied [55].

Software-hardware co-design

- **LLM Utilization:** The application of LLMs as an assistive tool within the co-design process has seen a speedup in the optimization of hardware architectures in deep neural networks by 25x [56], which conveniently solved the cold start problem in design.

Although the current improvements offer prospective resolutions to LLM acceleration, some obstacles continue to appear in trade-offs between performance and resource demand, especially during edge computing. In the future, more studies should be undertaken to develop more advanced ways of maximizing the hardware and software used when implementing LLMs.

1.15 PLASTIC MECHANISMS IN ARTIFICIAL SYNAPSES

Artificial synapses have an array of plasticity mechanisms that are similar to those of biological synapses, which are important in neuromorphic computing. These include STDP, short-term and long-term plasticity, and adaptation to environmental stimuli. The next sections describe the most important plastic mechanisms discovered in recent works.

Spike-timing-dependent plasticity (STDP)

- STDP, a pivotal learning mechanism, is an important process in which spike timing affects the strength of synapses. In much of the work on artificial synapses, STDP is produced using highly controlled magnetic field pulses and exhibits a clear threshold for achieving magnetization reversal in Co/Pt multilayers [57].

- STDP also occurs in memristive devices, such as BiFeO3-based synapses, whose learning windows can be configured depending on pulse width and amplitude to adapt to input signals dynamically [58].

Short-term and long-term plasticity

- Crystallized conjugated polymer nanowire studies demonstrate that solvent engineering can be used to tune synaptic plasticity between short-term and long-term states [59].
- Such versatility between the types of plasticity is crucial in the development of artificial neural structures, where tasks such as memory formation and learning need various operational models [59].

Robustness against noisy inputs

- Memristive artificial synapses exhibit robustness against extrinsic noise, and reliable functioning is a necessity in real-world applications. The examination shows that the effect of thermal noise can be ignored relative to the writing voltage and guarantees robust STDP performance [58].

Although the dynamics of plasticity in artificial synapses hold potential to mimic biological behavior, there still exist problems in creating such artificial systems with equal effectiveness and plasticity as their natural counterparts. More research is needed to improve the strength and flexibility of such artificial synapses in more complicated systems.

1.16 SECURITY AND PRIVACY IN NEUROMORPHIC SYSTEMS

Like brain imitation, neuromorphic systems present significant challenges related to security and privacy concerns. Such problems can emerge due to their unique architecture and the information collected that can be sensitive, and thus high protective measures are required. Important features of security and privacy in neuromorphic systems are presented as follows.

1.16.1 PCIe congestion and side-channel attacks

PCIe is commonly used in neuromorphic systems to move data, and this can create an opportunity for congestion side-channel attacks. Inefficient data placement in scratchpads may also result in congestion, leading to the extraction of sensitive information depending on delays in transmissions observed by attackers. One such possible mitigation is proposed to optimize the management of scratchpads by extending the compiler to maximize model privacy and decrease susceptibility to this kind of attack [60].

1.16.2 Security design considerations

Security must be taken as an objective during the design of neuromorphic systems. This comprises mitigation of possible attack vectors and the introduction of countermeasures that do not deteriorate performance or energy efficiency. The trade-offs between security and system performance are crucial to avoid degradation of the model and unauthorized access [61, 62].

1.16.3 Event encryption for privacy protection

Neuromorphic imaging, including bio-inspired cameras, produces sensitive information that should be encrypted when unauthorized access is a concern. These systems can tamper with event streams by adding noise and thereby add a level of privacy and security to transmitted data [63].

Although a great deal of progress has been achieved in the security of neuromorphic systems, the constant changes in the field of cyber threats require the continual development of security technologies. Future work lies in developing more advanced encryption techniques, as well as investigating stronger solutions that could further increase the privacy and security levels of such systems.

1.17 BENCHMARKING NEUROMORPHIC SYSTEMS

Neuromorphic systems must be benchmarked to analyze performance and to direct future advances. Today, the situation demonstrates the necessity of dynamic, real-world benchmarks that evaluate not only precision but also robustness and power consumption. This change is vital to the development of neuromorphic intelligence.

Closed-loop benchmarks

- Closed-loop benchmarks based on real-world tasks rather than only accuracy should be considered to assess neuromorphic systems, with an overall focus on metrics related to decision-making and adaptability [64].
- These benchmarks can increase the robustness and stability of neuromorphic systems, given their consistency with the potential of biological organisms.

Modular benchmarking frameworks

- SNABSuite offers modular support for black box benchmarking so that comparisons can be directly made across different neuromorphic hardware platforms [65].
- This framework assesses such performance indicators as resource efficiency and resilience to aid informed judgment regarding specific use-case needs.

Collaborative benchmarking initiatives

- NeuroBench has been proposed to provide community, fair, and representative neuromorphic computing benchmarks, tackling both the technique diversity and standardization challenges of the field [66, 67].
- NeuroBench aims to integrate efforts in benchmarking by establishing cooperation between academia and industry to implement technological advances.

On the contrary, although such benchmarks are crucial to improvement, the complexity and variety of neuromorphic systems may create issues in standardization, which can impede comparability across platforms.

1.18 NEUROMORPHIC CONTROL SYSTEMS FOR AUTONOMOUS VEHICLES

Autonomous vehicles are revolutionizing neuromorphic control systems as they emulate brain processes to improve human decision-making, perception, and adaptability. These types of systems make use of SNNs and event-driven processing to enable an energy-efficient and real-time response to a dynamic environment. The fact that neuromorphic architectures have been integrated into different control schemes, including Proportional Integral Derivative (PID) and Model Predictive Control (MPC), supports the claim that they can streamline control scheme performance with minimum neuronal requirements [4, 68].

Key aspects of neuromorphic control systems

- **Energy Efficiency:** Neuromorphic systems have a low-energy budget by several orders of magnitude relative to conventional CPU-based models and can be used in long-lifetime autonomous operations [4].
- **Adaptive Learning:** These systems can adapt to changing conditions, i.e., road friction and vehicle dynamics, to optimize system robustness and safety in unforeseeable environments [68, 69].
- **Real-time Processing:** Neuromorphic systems can handle real-time data processing with the aid of Light Detection and Ranging (LiDAR) and other sensors, which enable them to effectively avoid collisions and navigate [69, 70].

  Although these are strengths, work still needs to be done, especially on sustaining higher speeds and integrating human-machine interactions. The future of driverless cars does not necessarily have to be based on the extinction of human drivers but on systems working alongside each other; this combination increases both safety and efficiency through intelligent design [68].

1.19 CONCLUSION

Neuromorphic computing is at the forefront of intelligent computing, integrating neuroscience and advanced electronics, overcoming traditional computing systems, and presenting revolutionary computing frameworks. These systems offer energy efficiency and real-time flexibility unlike any before in bio-inspired architectures, i.e., SNNs, memristor arrays, and biohybrid interfaces. Such innovations as Loihi chips, 3D integration, and quantum-neuromorphic hybrids explore scalable perspectives for diverse computing applications. Neuromorphic systems hold promise in applications such as autonomous vehicle systems, neuroprosthetics, and large-scale intelligence production, where more traditional systems fall prey to latency and energy bottlenecks. In spite of these developments, however, there are still areas of concern that prevail, such as hardware-software co-design, benchmarking standardization, and the ability to withstand noise in data and security breaches. Further interdisciplinary research and cooperation are needed to optimize these systems and move toward machines that learn and change with the malleability of the human brain. Neuromorphic computing is more than a technological quest; it is an aspiration toward more sustainable and human-friendly artificial intelligence.

REFERENCES

[1] S A Londhe, S K Shivthare and Y K Sharma. "Introduction to Neuromorphic Computing". *Advances in Computational Intelligence and Robotics Book Series*, 2024.

[2] L J Ahmed, S Dhanasekar, K M Sagayam et al., "Introduction to Neuromorphic Computing Systems". *Neuromorphic Computing Systems for Industry 4.0*, 2023. https://doi.org/10.4018/978-1-6684-6596-7.ch001

[3] D Misra. "Special Issue of Interface on Neuromorphic Computing: An Introduction and State of the Field". *The Electrochemical Society Interface*, 2023.
[4] R Halaly and E E Tsur. "Autonomous Driving Controllers with Neuromorphic Spiking Neural Networks". *Frontiers in Neurorobotics*, 17, 1234962, 2023.
[5] R Mishra and M Suri. "A Survey and Perspective on Neuromorphic Continual Learning Systems". *Frontiers in Neuroscience*, 17, 1149410, 2023.
[6] S Hunagund, "Neuromorphic Computing: Implementation and Open Issues." *Neuromorphic Computing Systems for Industry 4.0*, 2023. https://doi.org/10.4018/978-1-6684-6596-7.ch002
[7] J Zhang, D Huo, J Zhang et al. "22.6 ANP-I: A 28nm 1.5pJ/SOP Asynchronous Spiking Neural Network Processor Enabling Sub-O.1 μJ/Sample On-Chip Learning for Edge-AI Applications". *2023 IEEE International Solid-State Circuits Conference (ISSCC)*, 2023.
[8] Y Liu, Y Ma, N Shang et al., "30.2 A 22nm 0.26nW/Synapse Spike-Driven Spiking Neural Network Processing Unit Using Time-Step-First Dataflow and Sparsity-Adaptive In-Memory Computing." *2024 IEEE International Solid-State Circuits Conference (ISSCC)*, 2024. https://doi.org/10.1109/isscc49657.2024.10454472
[9] S Sanaullah, K Roy, U Ruckert and T Jungeblut, "Poster: Selection of Optimal Neural Model using Spiking Neural Network for Edge Computing." *2024 IEEE 44th International Conference on Distributed Computing Systems (ICDCS)*, 2024. https://doi.org/10.1109/icdcs60910.2024.00147
[10] V N Dinh, N H Bui, V T Nguyen, Q M Duong and Q K Trinh, "FBW-SNN: A Fully Binarized Weights-Spiking Neural Networks for Edge-AI Applications." *2022 International Conference on IC Design and Technology (ICICDT)*, 2022. https://doi.org/10.1109/ICICDT56182.2022.9933108
[11] A Emelyanov, A N Matsukatova, A I Iliasov and V Rylkov, "Novel Neuromorphic Architectures Based on Crossbar Arrays of (Co-Fe-B)x(LiNbO3)100−x Nanocomposite Memristors." *Genes & Cells*, 18(4), 798–801, 2023. https://doi.org/10.17816/gc623346
[12] C Merkel and A Nikam, "A Low-Power Domino Logic Architecture for Memristor-Based Neuromorphic Computing". *ICONS '19: Proceedings of the International Conference on Neuromorphic Systems*, 2019. https://doi.org/10.1145/3354265.3354282
[13] 2023. https://doi.org/10.1109/vlsi-tsa/vlsi-dat57221.2023.10134261
[14] H Das, N N Chakraborty, M Rathore et al., "A Memristive Reconfigurable Neuromorphic Array for Neuro-Inspired Dynamic Architectures." *2024 IEEE Computer Society Annual Symposium on VLSI (ISVLSI)*, 2024. https://doi.org/10.1109/isvlsi61997.2024.00081
[15] A Hazan and E E Tsur, "Neuromorphic Neural Engineering Framework-Inspired Online Continuous Learning with Analog Circuitry." Applied *Sciences*, 12(9), 4528, 2022. https://doi.org/10.3390/app12094528
[16] M Payvand and G Indiveri. "Spike-Based Plasticity Circuits for Always-on On-Line Learning in Neuromorphic Systems". *International Symposium on Circuits and Systems*, 2019.
[17] G Detorakis, S Sheik, C Augustine, S Paul, B U Pedroni, N Dutt, J Krichmar, G Cauwenberghs and E O Neftci. "Neural and Synaptic Array Transceiver: A Brain-Inspired Computing Framework for Embedded Learning," *Frontiers in Neuroscience*, 12, 2018. https://doi.org/10.3389/fnins.2018.00583
[18] L A Camunas-Mesa, C Zamarreno-Ramos, A Linares-Barranco et al. "An Event-Driven Multi-Kernel Convolution Processor Module for Event-Driven Vision Sensors". *IEEE Journal of Solid-State Circuits*, 47(2), 504–517, 2012.
[19] J A Perez-Carrasco, B Zhao, C Serrano et al. "Mapping from Frame-Driven to Frame-Free Event-Driven Vision Systems by Low-Rate Rate Coding and Coincidence Processing–Application to Feedforward ConvNets". *IEEE Transactions on Pattern Analysis and Machine Intelligence*, 35(11), 2706–2719, 2013.
[20] A Salim and L I Abdul-Kareem, "A Review of Advances in Bio-Inspired Visual Models Using Event-and Frame-Based Sensors." 2025. https://doi.org/10.46604/aiti.2024.14121
[21] A N Belbachir, M Hofstätter and P Schön. "An Event-Driven System for High-Speed Vision". *International Conference on Informatics and Systems*, 2010.
[22] A Lines, P Joshi, R Liu et al. "Loihi Asynchronous Neuromorphic Research Chip". *IEEE International Symposium on Asynchronous Circuits and Systems*, 2018.
[23] M Barnell, C Raymond, L Loomis et al., "Advanced Ultra Low-Power Deep Learning Applications with Neuromorphic Computing." *2023 IEEE High Performance Extreme Computing Conference (HPEC)*, 2023. https://doi.org/10.1109/hpec58863.2023.10363561

[24] M Davies, A Wild, G Orchard et al., "Advancing Neuromorphic Computing with Loihi: A Survey of Results and Outlook." *Proceedings of the IEEE*, 109, 911–934, 2021. https://doi.org/10.1109/JPROC.2021.3067593

[25] S Dey and A G Dimitrov. "Sensitivity Analysis of Point Neuron Model Simulations Implemented on Neuromorphic Hardware". *Frontiers in Neuroscience*, 17, 1198282, 2023. https://doi.org/10.3389/fnins.2023.1198282

[26] M Kimura, S Shiomi, N Komai et al., "Thin-Film Memristors and Memcapacitors for 3D Integration of Neuromorphic Systems." *2023 IEEE 36th International System-on-Chip Conference (SOCC)*, 2023. https://doi.org/10.1109/socc58585.2023.10257032

[27] A Ehsan, Z Zhou and Y Yi. "Neuromorphic 3D Integrated Circuit: A Hybrid, Reliable and Energy Efficient Approach for Next Generation Computing". *Great Lakes Symposium on VLSI*, 2017.

[28] E Kurshan, H Li, M Seok and Y Xie. "A Case for 3D Integrated System Design for Neuromorphic Computing and AI Applications". *International Journal of Semantic Computing*, 14(4), 457–475, 2020.

[29] C Bao, S K Seol and W S Kim. "A 3D Integrated Neuromorphic Chemical Sensing System". *Sensors and Actuators B-Chemical*, 332, 129527, 2021.

[30] M J Wilmarth, *Neuromorphic Neural Interfaces*. 2023. https://doi.org/10.1007/978-981-16-5540-1_41

[31] F D Broccard, S Joshi, J Wang and G Cauwenberghs. "Neuromorphic Neural Interfaces: From Neurophysio-Logical Inspiration to Biohybrid Coupling with Nervous Systems". *Journal of Neural Engineering*, 14(4), 041002, 2017.

[32] F Santoro, "Functional Biohybrid Neuromorphic Interfaces." 2022. https://doi.org/10.1117/12.2636131

[33] H Keren, J Partzsch, S Marom, S Marom and C Mayr. "A Biohybrid Setup for Coupling Biological and Neuromorphic Neural Networks". *Frontiers in Neuroscience*, 13, 432, 2019.

[34] C Horejs, "Biohybrid Neural Implant." *Nature Reviews Bioengineering*, 1, 312, 2023. https://doi.org/10.1038/s44222-023-00065-1

[35] A S Sichani and S Kankatala, "The Potential of Combined Learning Strategies to Enhance Energy Efficiency of Spiking Neuromorphic Systems." 2024. https://doi.org/10.48550/arxiv.2408.07150

[36] M A Seenivasan and P Saha, "The Energy-Efficient and Distortion Blocking Leaky Integrate-and-Fire Neuron Model." *2024 6th International Conference on Energy, Power and Environment (ICEPE)*, 2024. https://doi.org/10.1109/icepe63236.2024.10668956

[37] G Chakma, M Adnan, A Wyer et al. "Memristive Mixed-Signal Neuromorphic Systems: Energy-Efficient Learning at the Circuit-Level". *IEEE Journal on Emerging and Selected Topics in Circuits and Systems*, 8(1), 125–136, 2018.

[38] J Park, J H Lee and D Jeon. "A 65-nm Neuromorphic Image Classification Processor with Energy-Efficient Training Through Direct Spike-Only Feedback". *IEEE Journal of Solid-State Circuits*, 55(1), 108–119, 2020.

[39] H Hendy and C Merkel. "Energy-Efficient and Noise-Tolerant Neuromorphic Computing Based on Memristors and Domino Logic". *Frontiers in Nanotechnology*, 5, 1128667, 2023.

[40] T Austin, S Bilodeau, A Hayman, N Rotenberg and B J Shastri, "Hybrid Quantum-Classical Photonic Neural Networks." 2024. https://doi.org/10.48550/arxiv.2407.02366

[41] T Austin, S Bilodeau, A Hayman, N Rotenberg and B Shastri. "Hybrid Quantum-Classical Photonic Neural Networks". 2024. https://doi.org/10.1117/12.3027153

[42] Z Zhang, Y Li and X Xu, "Quantum-Enhanced Neural Networks for Quantum Many-Body Simulations." 2025. https://doi.org/10.48550/arxiv.2501.12130

[43] D S Assi, H Huang, V Karthikeyan et al., "Quantum Topological Neuristors for Advanced Neuromorphic Intelligent Systems". *Advanced Science*, 10(24), 2300791, 2023. https://doi.org/10.1002/advs.202300791

[44] Z Tayeh, E Ercelik and J Conradt. "Decoding of Motor Imagery Movements from EEG Signals Using SpiN-Naker Neuromorphic Hardware". *International IEEE/EMBS Conference on Neural Engineering*, 2017.

[45] J Chen, H Wu, F Tian et al. "NeuroBMI: A New Neuromorphic Implantable Wireless Brain Machine Interface with A 0.48 μW Event-Driven Noise-Tolerant Spike Detector". *International Conference on Artificial Intelligence Circuits and Systems*, 2023.

[46] Y Qi, J Chen and Y Wang. “Neuromorphic Computing Facilitates Deep Brain-Machine Fusion for High-Performance Neuroprosthesis”. *Frontiers in Neuroscience*, 17, 1153985, 2023.
[47] 2023. [Online]. Available: https://doi.org/10.1109/aicas57966.2023.10168619
[48] S A Londhe, S K Shivthare and Y K Sharma. “Introduction to Neuromorphic Computing”. *Advances in Computational Intelligence and Robotics Book Series*, 2024.
[49] A Joubert, B Belhadj, O Temam and R Heliot. “Hardware Spiking Neurons Design: Analog or Digital?”. *International Joint Conference on Neural Network*, 2012.
[50] L Happek, J. T. Huang, A G Gener, R Leupers and M Galicia, “Spiking Neural Networks for Signal Classification with Digital and Analog Neuromorphic Systems: A Comparative Study.” *2024 International Joint Conference on Neural Networks (IJCNN)*, 2024. https://doi.org/10.1109/ijcnn60899.2024.10650824
[51] R Oshio, S Takuya, A Sawada et al., “A Memcapacitive Spiking Neural Network with Circuit Nonlinearity-aware Training.” *2022 IEEE Symposium in Low-Power and High-Speed Chips (COOL CHIPS)*, 2022, https://doi.org/10.1109/coolchips54332.2022.9772674
[52] N Nedjah, R M Silva, Da, L Mourelle and M De. “Analog Hardware Implementations of Artificial Neural Networks”. *Journal of Circuits*, 20(03), 349–373, 2011.
[53] S Afifi, F Sunny, M Nikdast and S Pasricha, “Accelerating Neural Networks for Large Language Models and Graph Processing with Silicon Photonics.” 2024. https://doi.org/10.48550/arxiv.2401.06885
[54] S M Ma, C Fang, H Shao and Z Wang, “Efficient Arbitrary Precision Acceleration for Large Language Models on GPU Tensor Cores”. 2024. https://doi.org/10.48550/arxiv.2409.17870
[55] H Jin, X Han, J Yang et al., “GrowLength: Accelerating LLMs Pretraining by Progressively Growing Training Length”. 2023. https://doi.org/10.48550/arxiv.2310.00576
[56] Z Yan, Y Qin, X Hu and Y Shi. “On the Viability of using LLMs for SW/HW Co-Design: An Example in Designing”. *2023 IEEE 36th International System-on-Chip Conference (SOCC)*, 2023.
[57] J Castro, B Buyatti, D A Miranda et al. “Spike-Timing-Dependent-Plasticity Learning in a Planar Magnetic Domain Wall Artificial Synapse”. *Journal of Physics D*, 58(12), 125002, 2025.
[58] N Du, X Zhao, X Zhao et al. “Synaptic Plasticity in Memristive Artificial Synapses and Their Robustness Against Noisy Inputs”. *Frontiers in Neuroscience*, 15, 660894, 2021.
[59] H Han, Z Xu, K Guo et al., “Tunable Synaptic Plasticity in Crystallized Conjugated Polymer Nanowire Artificial Synapses”. 2020, https://doi.org/10.1002/AISY.201900176
[60] A Das, 2023. https://doi.org/10.1109/compsac57700.2023.00094
[61] B Liu, C Yang, H Li et al. “Security of Neuromorphic Systems: Challenges and Solutions”. *International Symposium on Circuits and Systems*, 2016.
[62] F Merchant. “Security as an Important Ingredient in Neuromorphic Engineering”. *IEEE Computer Society Annual Symposium on VLSI*, 2022.
[63] G Gallego, T Delbruck, G Orchard, C Bartolozzi, B Taba, A Censi, S Leutenegger, A Davison, J Conradt, K Daniilidis and D Scaramuzza. “Event-Based Vision: A Survey”. *IEEE Transactions on Pattern Analysis and Machine Intelligence*, 44(1)
[64] M B Milde, S Afshar, Y Xu et al. “Neuromorphic Engineering Needs Closed-Loop Benchmarks”. *Frontiers in Neuroscience*, 16, 813555, 2022.
[65] C Ostrau, C Klarhorst, M Thies and U Rückert. “Benchmarking neuromorphic hardware and its energy expenditure,” in *Proceedings of the International Conference on Neuromorphic Systems (ICONS)*, 2019, pp. 1–8.
[66] Y Demirag and J Timcheck, 2023. https://doi.org/10.48550/arXiv.2304.04640
[67] J Yik et al., “NeuroBench: A Framework for Benchmarking Neuromorphic Computing Algorithms and Systems”. 2023, https://doi.org/10.48550/arxiv.2304.04640
[68] E E Tsur, G D Flumeri and H C Duwek. “Editorial: Brain-Inspired Autonomous Driving”. *Frontiers in Neurorobotics*, 19, 1543115, 2025.
[69] A Shalumov, R Halaly and E E Tsur. “LiDAR-Driven Spiking Neural Network for Collision Avoidance in Autonomous Driving”. *Bioinspiration & Biomimetics*, 16(6), 066016, 2021.
[70] J P Mitchell, G Bruer, M E Dean et al. “NeoN: Neuromorphic Control for Autonomous Robotic Navigation”. *International Symposium on Robotics*, 2017.

Chapter 2

Computational modeling of genetic algorithms

DNA encoding, mutation, and evolutionary optimization

Sachin Soni and Komal Sharma

2.1 INTRODUCTION

Biological evolution is guided by two fundamental principles. The first is Mendelian inheritance, which explains how discrete genetic units are passed down through generations. The second is Darwin's principle of natural selection, which favors traits that provide an advantage within a specific and often changing environment. These two ideas can either be considered separately or combined into a unified theory of evolution, where new variations arise through mutations and recombination against the existing genetic baseline. Such variations are then subjected to different forms of selection, allowing adaptive traits to strengthen and persist over time. The interplay between genetic inheritance and adaptation has been widely studied, particularly through computational models that treat these processes as ways of breaking down and solving complex optimization problems.

Bio-inspired algorithms, as they are commonly categorized, are influenced by these evolutionary tenets and include genetic algorithms (GAs) [1]. GAs were conceived by John Holland in the 1970s and use many biological facets such as mutation, crossover, and selection to evolve candidate solutions to an optimal representation. Because Mendel is concerned with the existence of discrete units of inheritance (genes and alleles), whereas Darwin is concerned that environmental pressure acts on and changes the population of individuals, conventional GAs have represented genetic materials as binary strings, ignoring or abstracting the biological representation. Additional understanding of adaptive search and optimization can be gained from a computational model of evolution that mimics selection pressures in a real-world setting and depicts genetic material in a biologically meaningful manner.

The goal of this chapter is to create a theory-driven analysis of DNA-based GAs that integrate Darwinian natural selection and Mendelian inheritance into a single computational framework. The discussion places particular emphasis on biologically inspired mechanisms such as mutation and crossover, as well as on DNA-to-decimal mapping and binary coding, which are treated as computational counterparts of their biological processes. Establishing this theoretical foundation helps explain how evolutionary algorithms strike a balance between exploration and exploitation, adapt to environmental changes, and gradually converge toward optimal solutions [2].

2.2 BIOLOGICAL FOUNDATIONS AND COMPUTATIONAL INSPIRATION

Darwinian natural selection and Mendelian inheritance are the cornerstones of evolutionary biology. Mendel demonstrated through his well-known nineteenth-century pea plant studies that qualities are transmitted in distinct units, now referred to as genes, from one generation

DOI: 10.1201/9781003685890-2

to the next and that these transmissions follow distinct statistical patterns. Around the same time, Darwin described how natural selection allows organisms to adapt to their surroundings, increasing the likelihood that individuals with beneficial traits will survive and reproduce. When combined, these two perspectives provide the core framework for understanding evolution and continue to serve as powerful sources of inspiration for computational modeling.

2.2.1 Mendelian inheritance and genetic representation

Mendel's principles that genetic information is stored in discrete packets that segregate and recombine form the genetic laws of inheritance. The first, the law of segregation, dictates that alleles segregate so only a single copy is received in a gamete (egg or sperm). The second, the law of independent assortment, shows the manner in which different traits assort independently from one another while being located in different sections on the genome (as shown in Figure 2.1). They allow biological evolution to have discreteness, combinatorial diversity, and statistical predictions [3].

Computational motivation: Mendelian inheritance is another way of providing encoding structures for genotypes. These additional factors take care of maintaining the required structure of genetic information and performing multi-valued mathematical operations on it in an algorithmic setting because in certain DNA-based algorithms, the genetic sequences lose meaning by being treated as digital data, which only has meaning when transformed into digital representations like binary or decimal exceptional values. There are additional opportunities for organized forms of innovation and complementation in the search space when using Mendelian-based encoding schemes, which allow for the transfer of modest amounts of information from parent solutions to offspring solutions.

2.2.2 Darwinian natural selection and fitness evaluation

Darwin's theory of evolution by natural selection highlights the roles of competition, variation, and adaptation to the environment (as illustrated in Figure 2.2) in shaping evolutionary pathways. Variations that provide certain individuals with an advantage increase their chances of survival and reproduction, allowing these traits to persist within the population. Over time, this process leads species to develop greater complexity and refinement. Importantly, selection is not a random occurrence but a directed process that favors traits enhancing the fitness of organisms within their ecological context.

In the computational setting, the intuitive mapping is fitness evaluation and reproductive selection. Candidate solutions are evaluated against performance criteria specific to the

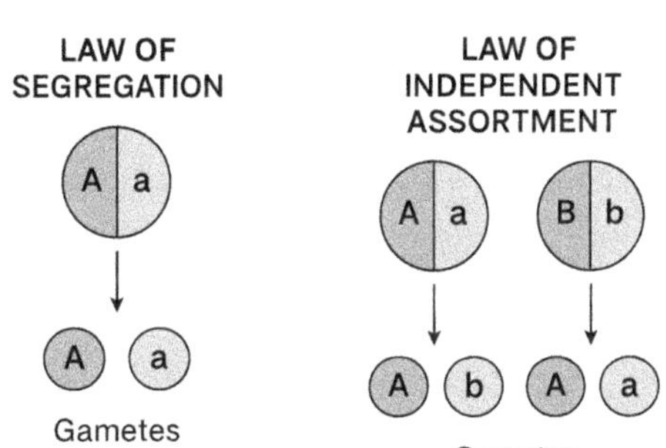

Figure 2.1 The diagram illustrates Mendel's principles of inheritance.

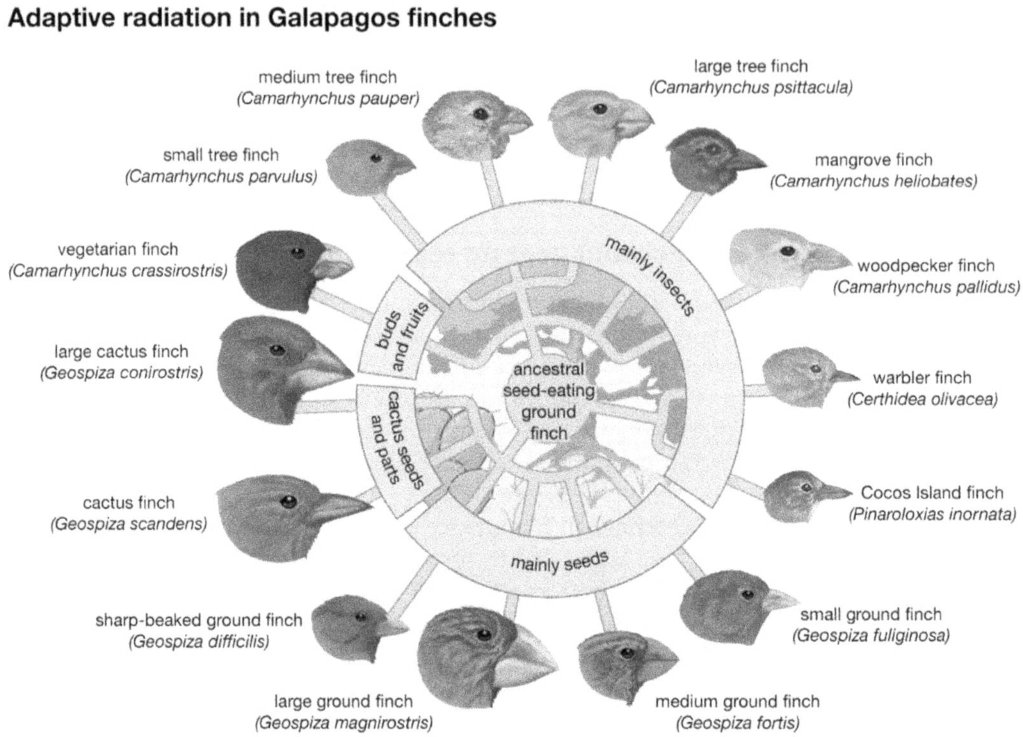

Figure 2.2 The illustration demonstrates the adaptive radiation of Galápagos finches from an ancestral seed-eating ground finch into diverse species with specialized beaks adapted to different food sources [4].

problem, and the fit solutions have a preferential opportunity to reproduce as parents of the next generation. This is consistent with Darwin's idea: while variation provides raw material for exploration, selective pressure pushes toward converging on optimal solutions. By preferentially mutating and recombining toward the fit individuals, GAs can emulate adaptive dynamics, exploring new variable combinations while exploiting some already defined fit solutions.

2.2.3 From biology to computation: the dual inspiration

The double-edged sword of Mendelian inheritance and Darwinian selection is the inspiration for evolutionary computation. The Mendelian theory consists of a structured representation to facilitate the robust conveyance of information through multiple generations, while the Darwinian theory contains an adaptive optimization mechanism to dynamically filter variation depending on the fitness landscape. The two theories provide a theoretical foundation for DNA-based GAs, where the GA is implemented as both an information-retaining system and an adaptive search heuristic. This transition from biological principles to computational techniques shows that there is value in bio-inspiration. DNA-inspired GAs maintain an inheritable character like genetic information while being abstract or discrete, while mimicking the adaptive filtration of natural selection. DNA-inspired GAs meaningfully extend beyond binary GAs, because they provide a realistic and richer workload of available models to represent and work with solutions in complex problem domains as shown in Figure 2.3.

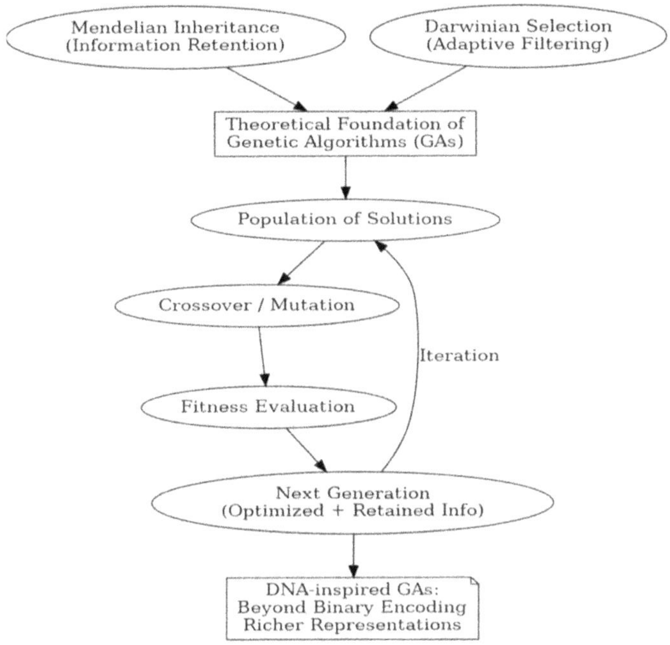

Figure 2.3 Inheritance and selection inspire DNA-based genetic algorithms for adaptive and richer solution search.

2.3 ENCODING AND REPRESENTATION IN DNA-INSPIRED ALGORITHMS

2.3.1 DNA-to-decimal mapping

In the realm of DNA-based computation, nucleotide sequences are not "structural" motifs but abstract symbols coding for information. To perform numerical manipulations, a correspondence table between DNA sequences and decimal values is defined.

Consider the four nucleotides $\{A, T, C, G\}$. One of the length-4 DNA sequences (e.g., TTAG) is one of $4^4 = 256$ possible combinations. Assigning a position in the continuous search space for each sequence [0, 1], we define:

$$f : \{A, T, C, G\}^4$$

where

$$f(s_i) = \frac{i}{255}, i = 0, 1, \ldots, 255$$

Each DNA string therefore has a unique decimal representation. This produces a "virtuous" map that allows addition and subtraction to be applied to symbolic genetic strings.

Implementation of algorithms:

1. Make each of the 4^4 DNA strings.
2. The gap value, which ranges from 0 to 1.
3. Develop bijections for process encoding and decoding.

The DNA strings can be represented mathematically thanks to this map.

2.3.2 Binary encoding and decoding

The DNA codes will be based on binary representations for simplicity's sake. We must convert DNA sequence information into reversible binary-DNA encodings in order to use DNA-like structures in a computing substrate. The binary-to-DNA conversion is based on a two-bit encoding:

$$00 \to A, 01 \to T, 10 \to C, 11 \to G$$

For example:

Binary string "00010011" → DNA string "ATAG".

DNA sequence "GCGT" → Binary "11100101".

A hybrid combination of bio-inspired symbolic encoding with classical binary GAs through this double coding may connect molecular metaphors with digital computation.

2.3.3 Representation and search space

The search space topology and its optimization performance are directly influenced by the choice of encoding.

- The binary encoding gives discrete jumps, which indeed can be good for exploration but very bad for resolution.
- The DNA-to-decimal mapping leads to a continuous evolution of solutions, which can be measured more accurately.

In practice, if S is the collection of DNA sequences and $f : S \to [0,1]$ is the encoding, then the optimization problem essentially becomes an optimization from:

$$\max_{s \in S} F(s)$$

to

$$\max_{x \in [1,0]} F\left(f^{-1}(x)\right)$$

This showcases the advantage of encoding: it determines whether the search proceeds in a discrete combinatorial space or a continuous numerical landscape.

2.4 EVOLUTIONARY OPERATORS: A THEORETICAL PERSPECTIVE

To better understand the functioning of DNA-inspired GAs, it is useful to visualize the overall workflow that integrates encoding, fitness evaluation, and evolutionary operators. Figure 2.4 illustrates the step-by-step process, starting from initialization of DNA sequences, mapping between binary and decimal representations, and proceeding through selection, crossover, and mutation to produce successive generations until an optimal solution is reached.

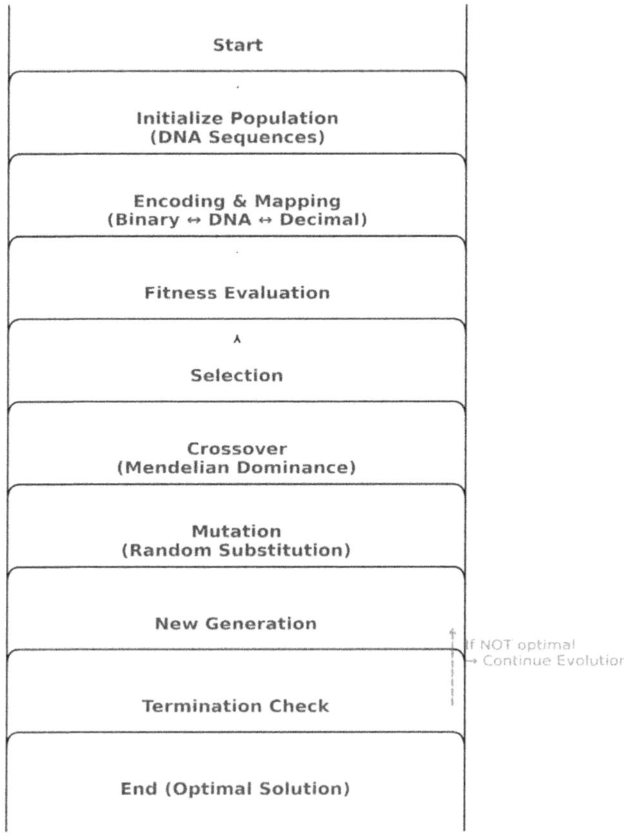

Figure 2.4 Flowchart for the DNA-inspired genetic algorithm process.

2.4.1 Mutation

Mutation introduces variation to maintain diversity. In our DNA-inspired system, mutation is modeled as a random nucleotide substitution or bit-flip in binary representation.

If the mutation probability is p_m, and the sequence length is L, the expected number of mutations is:

$$E[M] = L.p_m$$

Our algorithm allows a random number of flips (bounded by a mutation rate), ensuring controlled diversity while preventing destructive randomness.

2.4.2 Crossover

Crossover in evolutionary computation involves the recombination of genes from two parents to generate offspring. For example, single-point, multi-point, or uniform crossover are all widely used in classical GAs and are justified by the schema theorem—short low-order schemata of above-average fit survive and proliferate.

This can be generalized to other biological analogies in DNA-based GAs. Our method is inspired by Mendelian inheritance. Mendel discovered with his pea plant experiments that offspring inherit traits from their parents according to dominance rules: if there is a dominant allele (A) in the pair, it masks the expression of a recessive one.

Analogously, in our DNA-inspired crossover:

- Two-parent DNA sequences (s_1, s_2) are compared via their decimal fitness mapping.
- The "dominant" sequence (higher decimal value = fitter allele) is passed on to the offspring.

Formally:

$$O = \begin{cases} s_1 \text{ if } f(s_1) \geq f(s_2) \\ s_2 \text{ otherwise} \end{cases}$$

This biased recombination enhances convergence by favoring stronger candidates.

2.4.2.1 *Advantages of Mendel-inspired crossover*

- Preserves stronger building blocks directly.
- Quick convergence (strong candidates are passed on without dilution).
- Eliminates conflicting recombination effects found in traditional crossover.

2.4.2.2 *Limitations*

- Gross loss of diversity if the dominant spreads too fast.
- May need to be balanced with mutation and tournament selection to maintain exploration.

So, although schema-based crossovers encourage information mixing, our Mendelian crossover forces dominant inheritance and strengthens the elitist bias in the population dynamics [5].

2.4.3 Selection

Selection is the process of selecting a subset of the population to survive and reproduce to go on to the next generation. In classical GA, selection schemes are:

Fitness Proportionate (Roulette Wheel): selection probability is proportional to fitness.

Tournament Selection: randomly selected subset competition; winner lives.

Elitism: you always keep the best.

In a DNA-motivated GA, selection is biologically inspired as well. Just as the process of natural selection selects the most adapted organism, DNA-based selection operating on dominant genetic material (in the form of the offspring of crossover) and coupled with mutation guarantees the survival of the best.

Formally, given population P of size N and fitness function $F(s)$, the probability of selecting an individual $s_i \in P$ can be defined as:

$$P(s_i) = \frac{F(s_i)}{\sum_{J=1}^{N} F(s_j)}$$

Elitism modifies this by guaranteeing:

$$\max_{s \in P} F(s) \in P_{t+1}$$

Ensuring the best individual is passed to the next generation.

2.5 POPULATION DYNAMICS AND CONVERGENCE

2.5.1 Mathematical modeling

Markov chains can be used in the simulation of population dynamics, where states of a chain are population states. P is a stochastic transition matrix that contains the evolutionary operators.

Convergence occurs when the chain is absorbing, i.e., one that represents a population of all-optimal sequences.

2.5.2 Stability and diversity

Accidentally hitting a local maximum beforehand is a serious problem in all numerical methods. DNA-inspired representations mitigate this by:

- Increasing the search space with multichain encoding (A, T, C, G).
- Enabling nondisruptive transitions through decimal mapping.

2.5.3 Theoretical limits

Despite their adaptability, DNA-inspired algorithms face challenges:

Computational complexity: $O(N \cdot L)$ per generation for population size N, length of sequence L.

Scalability: Mapping increases as 4^L, resulting in space needs that grow exponentially [6].

2.6 COMPARATIVE THEORETICAL PERSPECTIVES

2.6.1 DNA-based vs traditional GA

It is crucial to look at both the practical performance of DNA-based GAs in comparison to conventional GAs, as well as the theoretical differences that determine their structure and function, as shown in Table 2.1. The effectiveness of these methods often depends on the problem domain [7], since different tasks demand different types of search capabilities. For instance, engineering applications benefit from continuous optimization, biological studies rely on symbolic representations, and scheduling problems require combinatorial flexibility. In this context, Table 2.2 provides a summary of application areas where DNA-based GAs demonstrate clear advantages over conventional approaches.

Table 2.1 A comparative analysis of conventional and DNA-based genetic algorithms

Aspect	*DNA-based genetic algorithm*	*Traditional genetic algorithm*
Encoding	Uses four nucleotides (A, T, C, G), allowing richer representation.	Typically uses simplified encodings (numbers, symbols).
Search space granularity	DNA-to-decimal mapping yields a smooth, continuous space.	The search space is often discrete, depending on the representation.
Crossover strategy	Mendel-inspired dominance: stronger sequence survives.	Probabilistic recombination (single/multi-point crossover).
Mutation	Modeled as a random nucleotide/bit substitution with bounded intensity.	Often uniform/random bit/parameter changes.
Diversity maintenance	Multi-symbol alphabet increases inherent diversity.	Relies more heavily on mutation for maintaining diversity.
Convergence behavior	Faster due to dominance inheritance, risk of premature convergence.	More gradual exploration depends on operator settings.
Biological inspiration	Closer to molecular genetics (DNA structure, Mendelian rules).	Abstracted from Darwinian evolution and population genetics.
Adaptability	Hybrid decimal/bio representation allows integration with other models.	Flexible but may require tailored encodings for problems.

Table 2.2 Applications of DNA-based genetic algorithms vs traditional GA

Application area	*DNA-based GA*	*Traditional GA*
Bioinformatics (e.g., protein structure prediction, gene sequence alignment)	Rich representation of nucleotide data makes DNA-GA highly suitable	Limited due to binary/real coding abstraction
Robotics and control systems	Stronger convergence using dominance crossover enables faster path planning and adaptive control	Slower adaptation, risk of premature convergence
Engineering design optimization (aerodynamics, structures) [8]	Continuous DNA-decimal mapping allows precise parameter tuning	Works but requires heavy mutation tuning
Scheduling and resource allocation	Handles combinatorial spaces with higher diversity	Performs well but struggles with large complex constraints
Machine learning (feature selection, neural network training)	Hybrid symbolic-numeric encoding enhances search for feature subsets [9]	Works but limited in representation richness
Drug discovery and molecular docking	Can encode molecules more naturally using symbolic DNA sequences	Needs extra mapping to chemical structures

2.7 IMPLICATIONS AND FUTURE RESEARCH

2.7.1 Implications

The DNA framework for GAs generalizes the traditional evolutionary framework by using a more informative encoding model, systematic transformations across spaces, and incorporating biologically recognizable operators. The DNA-to-decimal mapping facilitates smooth mapping from discrete symbolic strings to continuous numerical spaces, which is extremely useful for optimization problems that require a degree of exploration [10] and accuracy. The Mendel-style crossover operator guarantees replacement of the most dominant solutions in the population for faster convergence speed, and strong evolutionary traits in the DNA representation that are based on a four-symbol alphabet enable a natural increase in overall diversity of the population and mitigate the risk of being trapped in local solutions. In conclusion, the evolutionary properties enabled through DNA-inspired GAs are appropriate for high-dimensional search spaces or bio-inspired symbolic-numeric hybrid representation problems and evolutionary processes with intended interpretability [11].

2.7.2 Future research

While there are now more opportunities than challenges, some challenges and opportunities continue to remain. Scalability is one area of research, and multi-objective optimization is another area using representations over DNA to explore searching for Pareto-optimal fronts and improving handling of trade-offs for complex problems [12]. The area of exploring hybrid approaches also needs further exploration, where encoding inspired by DNA is combined with existing methods from swarm intelligence, evolution, and quantum-inspired algorithms [13], which address robustness and adaptability in noisy and dynamic environments. Further theoretical questions also remain to be explored in terms of convergence guarantees, diversity through dominance-based crossover, and robustness of DNA-based models with respect to real-world optimization problems in general [14]. Further work exploring these questions will also help reinforce the next generation of bio-inspired computational models. Recent studies have emphasized hybrid approaches where DNA-based GAs are combined with other evolutionary and bio-inspired paradigms. One promising direction for future research lies in hybridizing DNA-based GAs with other computational paradigms. Such hybrid approaches combine the inherent diversity and symbolic richness of DNA encodings with the complementary strengths of swarm intelligence, differential evolution, or quantum-inspired computing [15]. This not only enhances robustness but also extends applicability to complex and dynamic environments. Table 2.3 summarizes selected hybrid approaches, highlighting their advantages and challenges.

Table 2.3 Hybrid DNA-based GA approaches

Hybrid approach	*Description*	*Advantages*	*Challenges*
DNA-GA + **PSO**	Uses DNA encoding for diversity and swarm intelligence for exploitation	Balances exploration and exploitation	May increase computational overhead
DNA-GA + **Quantum-Inspired Computing**	Leverages quantum states for encoding DNA solutions	Massive parallelism; avoids local optima	Requires quantum-inspired simulation framework
DNA-GA + **Neural Networks**	DNA sequences used as chromosome encoding for network weights	Improves deep learning optimization	Training complexity increases
DNA-GA + **Differential Evolution**	Combines crossover with differential mutation strategy	Prevents premature convergence	Parameter sensitivity remains
DNA-GA + **Ant Colony Optimization**	Ant-like exploration and DNA-based representation	Handles discrete optimization well	Slower for continuous problems

2.8 CONCLUSION

DNA-inspired GAs represent an important advancement in developing next-generation computer optimization methods. The abstraction and efficiency of mathematical calculation are taken advantage of, yet they are still able to exploit the detail of biological information with nucleotide representations. By permuting DNA-to-decimal mappings, the search space can move to actual values while still allowing researchers control over the degrees of exploration and exploitation. Similarly, binary-to-DNA mappings help sustain the flexibility of these algorithms across many use contexts while still facilitating digital computation.

Studying evolutionary operators such as selection, crossover, and mutation highlights the ways DNA-inspired approaches can evolve and thus enhance the complexity and usability of GAs. For example, Mendel-inspired crossover can accommodate dominant inheritance during algorithm development that might expedite convergence. Yet simultaneous consideration of population dynamics demonstrates the complexity of diversity vs. stability and the complex scaling issues even in theoretical assessments of complexity. In addition to the unique benefits offered by DNA-based paradigms compared to others considered, including but not limited to an adaptable architecture, a biologically relevant problem space, and the potential to work with additional bio-based paradigms, the most significant value of a GA inspired by DNA is its flexibility to investigate trade-offs in multi-objective functions, solve high-dimensional aspects in multiple objective optimization, and probe new frontiers such as quantum-inspired computing [16].

REFERENCES

1. Yang, Y. (2021). Clustering and prediction analysis of the coordinated development of China's regional economy based on immune genetic algorithm. *Complexity*, 2021, 5590631. https://doi.org/10.1155/2021/5590631
2. Li, Y., & Liu, J. (2023). A survey: Evolutionary deep learning. *Soft Computing*, 27, 9401–9423. https://doi.org/10.1007/s00500-023-08316-4
3. Claveria, O., Monte, E., & Torra, S. (2022). A genetic programming approach for economic forecasting with survey expectations. *Applied Sciences*, 12(13), 6661. https://doi.org/10.3390/app12136661
4. Encyclopædia Britannica, Inc. (2025, September 8). The science of evolution. *Encyclopædia Britannica*. https://www.britannica.com/science/evolution-scientific-theory/Adaptive-radiation
5. Cicirello, V. A. (2023). A survey and analysis of evolutionary operators for permutations. arXiv preprint arXiv:2311.14595
6. Liu, S., Lin, Q., Li, J., & Tan, K. C. (2023). A survey on learnable evolutionary algorithms for scalable multiobjective optimization. *IEEE Transactions on Evolutionary Computation*, 27(6), 1941–1961. https://doi.org/10.1109/TEVC.2023.3250350
7. Seyyedabbasi, A. (2024). *A comprehensive survey: Evolutionary-based algorithms*. Istinye University, Engineering & Natural Sciences. (Unpublished/Institutional report)
8. Gao, K., Liu, T., Hu, B., Hao, M., & Zhang, Y. (2022). Establishment of economic forecasting model of high-tech industry based on genetic optimization neural network. *Computational Intelligence and Neuroscience*, 2022, 2128370. https://doi.org/10.1155/2022/2128370
9. Cagnini, H. E. L., Das Dôres, S. C. N., Freitas, A. A., & Barros, R. C. (2023). A survey of evolutionary algorithms for supervised ensemble learning. *The Knowledge Engineering Review*, 38, e1. https://doi.org/10.1017/S0269888923000024
10. Drachal, K., & Pawłowski, M. (2021). A review of the applications of genetic algorithms to forecasting prices of commodities. *Economies*, 9(1), 6. https://doi.org/10.3390/economies9010006

11. Ding, S., Cui, T., Xiong, X., & Bai, R. (2020). Forecasting stock market return with nonlinearity: A genetic programming approach. *Journal of Ambient Intelligence and Humanized Computing*, 11(11), 4927–4939. https://doi.org/10.1007/s12652-020-01762-0
12. Wang, Z., Pei, Y., & Li, J. (2023). A survey on search strategy of evolutionary multi-objective optimization algorithms. *Applied Sciences*, 13(7), 4643. https://doi.org/10.3390/app13074643
13. Zhu, Q., Wu, X., Chen, J., Ma, L. H., Li, J., Zhong, M., & Chen, J. (2023). A survey on evolutionary reinforcement learning algorithms. *Neurocomputing*, 556, 126628. https://doi.org/10.1016/j.neucom.2023.126628
14. Uthayasuriyan, A., Chandran, G. H., Kavvin, U. V., Mahitha, S. H., & Jeyakumar, G. (2024). Performance evaluation of evolutionary algorithms on solving the influence maximization problem in social networks. *International Journal of Modern Education and Computer Science*, 16(2), 83–97. https://doi.org/10.5815/IJMECS.2024.02.07
15. Li, P., Hao, J., Tang, H., Fu, X., Zheng, Y., & Tang, K. (2024). Bridging evolutionary algorithms and reinforcement learning: A comprehensive survey on hybrid algorithms. arXiv preprint arXiv:2401.11963.
16. Mohammadi, A., & Sheikholeslam, F. (2023). Intelligent optimization: Literature review and state-of-the-art algorithms (1965–2022). *Engineering Applications of Artificial Intelligence*, 126, 106959. https://doi.org/10.1016/j.engappai.2023.106959

Chapter 3

Smart diagnostic system for cardiovascular, neural, and muscular health using AI-based ECG, EEG, and EMG processing

Mohd. Maroof Siddiqui and K. K. Singh

3.1 INTRODUCTION

Artificial Intelligence (AI) has been integrated into most of the health-care technologies, transforming the processes through which we track, read and deal with physiological data. That means, noninvasive monitoring of bioelectrical signals like electrocardiogram (ECG), electroencephalogram (EEG), electromyogram (EMG), etc. can be used for diagnosing and treating cardiovascular cardiac problems or neurological disease depression Disease and neuromuscular disorders. The problem is that the hands and eyes of the health-care worker have always been used to try to interpret those signals, making diagnosis and treatment uselessly late. However, with the advancements in AI and now specifically machine learning (ML) and deep learning (DL) algorithm detections, real-time and automated monitoring systems are beginning to become more usable and reliable [1–5].

For AI systems to learn such intricate patterns in physiological signals it can outpace human anatomical levels of speed and accuracy. These technologies provide better diagnostic capabilities but also early anomaly detection, continuous monitoring of the patient and personalized treatment suggestions. Furthermore, AI incorporated with IoT, and wearable devices can offer remote health-care services which is crucial in underserved or rural areas [6].

In this paper, we propose an AI-assisted monitoring system for ECG, EEG, and EMG signals together using a single integrated framework. Creating a real-time efficient system, which will help the health-care professionals during early detection, risk prediction and clinical decision process in many disorders [7–9].

3.1.1 Significance of ECG, EEG, and EMG in health care

A proper knowledge of normal physiological signs is mandatory for the diagnosis and monitoring of several health conditions. Each of these bio signals contains a wealth of information about some or all functional systems in the body.

Electrocardiogram (ECG): An ECG records the electrical activity of the heart and is one of the most used diagnostic tools in cardiology. It's real-time direct information of heart rate, rhythm and conduction pathways. ECGs are used for the diagnosis of abnormal heart rhythms (e.g., atrial fibrillation), ischemic heart disease (e.g., myocardial infarction), and structural problems with the heart (e.g., left ventricular hypertrophy or training). It is a noninvasive, diagnostic and should be included in the evaluation of acute and chronic cardiac care [10–14].

Electroencephalogram (EEG): An EEG to record the electrical activity of the brain using electrodes attached to the scalp. It is an essential technique for the evaluation of brain function and has many applications, including in epilepsy diagnosis (clinical II level evidence), sleep disorders, acute encephalopathies following severe hypoperfusion reperfusion injury,

 DOI: 10.1201/9781003685890-3

head trauma, but also to a more limited extent mental health disorders like depression or even schizophrenia. One of the key applications of EEG is that it can easily detect transient events as well recognize different brainwave patterns, providing useful insights to neurologists and cognitive scientists [15].

Electromyogram (EMG): to show the electrical function of skeletal muscles to detect neuromuscular dysfunction patterns. It is also beneficial in differential diagnosis of neuro diseases which can involve either motor neurons, peripheral nerves or muscle Fibers like muscular dystrophy, amyotrophic lateral sclerosis (ALS), peripheral neuropathies and myopathies. EMG testing is a critical asset for evaluating muscle function and guiding treatment plans in the fields of rehabilitation medicine and orthopedics [16–21].

Teasing out these signals gives cross-domain diagnostics that are difficult to obtain with single modalities, and a multimodal sense of a patient's physiology.

3.1.2 Role of AI in physiological signals supervision

While ECG, EEG and EMG are clinically important, monitoring the respective signals with traditional approaches has several shortcomings:

Interpretation difficulties: Manual interpretation of complex bio signals is time-consuming and prone to inter-observer variability, as well as errors brought about by observer fatigue. This means that we may miss subtle trends or irregularities-such as in long recordings (longitudinal timing) cases in which responses are due within deadlines, with limited resources.

Data volume and data complexity: The amount of data generated with continuous monitoring is massive. Millions of data points can be recorded in long time EEG or ECG recording but manually it is not possible for providing this information. The task is even harder when considering high resolution EMG data due to its high sampling rate and as well variability of the signals [22]. The comparative study of AI-based ECG, EEG, and EMG processing mention in Table 3.1.

The algorithm must respond in real time: mobile health-care practices and emergent care being performed in the ambulance require an immediate alarm to be generated when a signal deviates from normal. Real time or non-real time cannot be done in traditional systems.

Various struggles can be overcome via AI. Through supervised and unsupervised learning methods, AI systems can learn how to recognize "normal" and "abnormal" patterns, classify specific types of signals or even predict the deployment of abrupt events such as a seizure or a sudden cardiac event. Ultimately, this allows us to translate AI models that can be leveraged at scale into personalized health care, while learning to generalize across patients [23].

IoT and cloud technologies provide ways to scale out these opportunities and make them available; support for health care in remote and rural areas is integral.

This implies a potential for significant improvement in effectiveness, reliability and coverage endemic to modern monitoring-based health care if AI is integrated into ECG, EEG, EMG devices [24, 25].

3.1.3 Discussion on the comparative study

Artificial Intelligence Techniques: CNNs and SVMs were the typically used algorithms for classification of ECG, EEG, as well as EMG in most studies. While CNNs are good at detecting patterns in time-series data such as ECG and EEG signals, SVMs, and decision trees (DT) have been used for initial classification tasks. Hybrid models combining convolutional and recurrent layers, such as CNN-LSTM, were found to be beneficial for identifying time-sequential information in signals of a physiological nature (e.g., Shen et al. (2022)).

Table 3.1 Comparative study of AI-based ECG, EEG, and EMG processing [1–8]

Research paper	*Physiological signal*	*AI techniques*	*Dataset used*	*Performance metrics*	*Applications*
Zhang et al. (2020)	ECG	• CNN, LSTM	MIT-BIH Arrhythmia Dataset	• Accuracy: 98.7% • F1 score: 0.96	Cardiac arrhythmia detection
Roy et al. (2019)	EEG	• SVM, Random Forest	CHB-MIT Scalp EEG Database	• Sensitivity: 92.5% • Precision: 93%	Epileptic seizure detection
Alam et al. (2021)	EMG	• Decision trees, CNN	NINDS EMG Dataset	• Accuracy: 94.3%	Neuromuscular disorder diagnosis
Awan et al. (2020)	ECG	• SVM, Random Forest	PhysioNet 2017 Challenge Dataset	• Accuracy: 99%	Myocardial infarction prediction
Shen et al. (2022)	EEG	• CNN, LSTM, k-NN	TUH EEG Corpus	• Accuracy: 91.4%	Sleep stage classification
Huang et al. (2023)	EMG	• k-NN, CNN	Surface EMG for Non-Invasive Gesture Dataset	• Accuracy: 95%	Gesture recognition for prosthetic control
Llamas et al. (2021)	ECG, EEG	• Hybrid CNN-LSTM model	Private Hospital ECG-EEG Dataset	• Accuracy: 97.1% (ECG) • 93.5% (EEG)	Cognitive load and cardiac stress monitoring
Thompson et al. (2020)	ECG, EMG	• CNN, Reinforcement Learning	MIT-BIH ECG, Public EMG dataset	• ECG Accuracy: 98.9% • EMG Accuracy: 94.7% • 93% (EEG) • 94.5% (EMG)	Real-time monitoring of cardiac and muscular conditions

Database: The MIT-BIH Arrhythmia Database is an open-access dataset for ECG-related studies and has been used in many other publications as well, which makes it a good source. For the case of EEG, analogous studies in epilepsy and sleep disorder detection from Roy et al. (2019) and Shen et al. (2022), datasets such as CHB-MIT Scalp EEG and TUH EEG Corpus are commonly employed. The NINDS EMG Dataset and available public surface EMG datasets for EMG research, especially for neuromuscular disorders (ND) and gesture recognition, were utilized.

Overall, the systems yield high accuracy and F1-scores for ECG monitoring. B. X. Zhang et al. reported 99% accuracy for the detection of arrhythmia (2020). EEG monitoring, especially for seizure detection, likely has slightly inferior performance because EEG signals are richer and more complicated. Sensitivity and precision incidences found in Roy et al. (2019) were in the low to mid 90%. Similar issues were also raised in EMG studies, e.g., Alam et al. (2021), but first results are promising with an accuracy of 94.3% in suspected neuromuscular patients (Sheedy et al. (2021)).

3.2 AI APPLICATIONS FOR ECG MONITORING

Recently, the field of automated analysis and diagnostic reporting of ECG signals using AI has become a hotspot in cardiovascular diagnostics. This can pave the way for early medical intervention by successfully detecting cardiac abnormalities and monitoring patients continuously in clinical settings (long QT syndrome, atrial fibrillation) and remotely (stress ECG testing). In this section, we reveal the core characteristics, difficulties, as well as actual usage of AI in ECG signal monitoring [26].

3.2.1 ECG signal processing

The performance of AI-based ECG monitoring systems is affected by signal processing, which consists of noise cancellation, feature investigation, and classification. These stages have been heavily automated and optimized using AI algorithms, especially ML and DL [27].

Noise Removal: ECG signals are sometimes affected by power line interference, motion artifacts, baseline drift, and muscle activity. Wavelet transforms, bandpass filters, and principal component analysis (PCA) are some of the preprocessing techniques used to remove noise from the signals. Filters with AI-based enhancements can powerfully detect and suppress undesirable parts without tarnishing essential signal characteristics [28].

Feature Extraction: Feature extraction involves recognizing the characteristic shapes of the P-wave, QRS complex, and T wave.

Traditional Methods and Feature: Traditional techniques are based on statistical and frequency-domain features, whereas AI methods can automatically learn deep hierarchical features from raw data, decreasing the engineering burden of manual feature designing [29].

Categorization: After feature extraction, AI models are trained to classify ECG segments into various classes (Normal, Arrhythmia, Myocardial infarction). This task can be performed with high accuracy by both classical and DL models:

Machine Learning Models: Support Vector Machine (SVM), DT, Random Forests, and k-Nearest Neighbors have been widely used for ECG classification tasks. Apart from that, these types of models generally do much better on small datasets with well-engineered features [30].

Convolutional Neural Networks (CNNs) and Recurrent Neural Networks (RNNs) over raw ECG signals are DL models that have shown some promise at learning the right set of features from the data. CNNs are great at capturing spatial patterns like the QRS complex, while RNNs, especially Long Short-Term Memory (LSTM) networks, are good at understanding temporal dependencies and are therefore useful in arrhythmic event detection or modeling heart rate variability over time [31, 32].

The use of these AI algorithms is progressively integrated with wearable devices, mobile apps, and cloud-based health monitoring platforms for continuous and real-time ECG surveillance.

3.2.2 Critical issues in ECG monitoring

Although AI-based models have achieved tremendous progress in automated ECG analysis, several critical issues need to be resolved for reliable and clinical applicability:

Signal Variability: Various factors like different physiological constitution, placement of electrodes, activity state, and underlying illness influence ECG signals, making them one of the most variable physiological signals between subjects as well as within a subject. This makes it challenging to train one AI model that can be used across multiple populations.

Noisy and Incomplete Data: Data obtained from ECG wearable devices in IoT environments are often fragmented due to missing segments, motion artifacts, and environmental noise. This affects accurate classification and detection of anomalies [33].

Sensitivity vs. Specificity Trade-off: AI models must work with a balance of high sensitivity (catching the actual abnormal cases right) and high specificity (not polluting the patient list with false alarms). A model that is too sensitive will result in false positive alerts, increasing patient anxiety and clinician fatigue, while a model that is too specific may fail to identify important events [34].

Limitations of Real-Time Processing and Latency: Power-efficient AI algorithms are required for edge devices with small computational power. Model compression and optimization are necessary for resource-limited deployment.

Clinical Training and Regulatory Approval: Medical AI-based systems need to undergo rigorous clinical training and meet medical regulatory criteria (FDA, CE) before deployment in real-world health care [35].

3.2.3 Case studies and success stories

There have been several case studies that implemented AI-based ECG monitoring systems successfully, showcasing the practicality of these technologies in clinical settings. Following are some notable examples:

One of the most popular benchmark databases is the MIT-BIH Arrhythmia Database, an ECG dataset with annotated recordings for different types of arrhythmias. As a result, AI models trained on this dataset can achieve compelling accuracies for detecting conditions such as atrial fibrillation, ventricular tachycardia, and premature contractions. It has been instrumental in benchmarking and evaluating various AI algorithms [36].

Title: PhysioNet Challenge Series Hosts: PhysioNet and Computing in Cardiology (CinC) Summary: Annual challenges to spark AI-based ECG classification innovation with public competitions on real-world datasets and tasks. An example is the 2020 challenge centered around AI models that can recognize cardiac abnormalities by reading 12-lead ECGs.

How AI Have Been Integrated into Consumer Devices: Companies like of Apple, Fitbit, and Withing's have included AI-enabled ECG monitoring in their wearable products. One example is the ECG app in the FDA-cleared Apple Watch Series 4 and later that uses AI to detect atrial fibrillation. These gadgets illustrate how AI is blending clinical diagnostics with consumer health tech [37].

Hospital-wide Surveillance Systems: Corporations such as the Mayo Clinic and Stanford University have developed proprietary algorithms that analyze inpatient ECG data from hospital ECG machines or Holter monitors to predict conditions such as left ventricular dysfunction or atrial fibrillation before these symptoms develop.

Based on these case studies, AI has the power to turn classic ECG analysis into openly configurable smart health services.

3.3 AI APPLICATIONS FOR EEG MONITORING

Since the detection and classification of neurological disorders such as epilepsy, sleep disorder, Alzheimer disease, and traumatic injury are fundamentally related to monitoring brain activity, Electroencephalogram (EEG) signals provide noninvasive monitoring in a more effective way. Nonlinear, nonstationary, and potentially corrupted with various artifacts, EEG signals are hard to interpret. Neuropsychological diagnostics have been successfully improved by methods of automatic, AI-based analysis (such as classification or segmentation) of EEG data, with signal variance and class distribution related to anomaly detection and localization in EEG recordings being key features.

3.3.1 Signal processing of EEG

A standard EEG signal is made up of various steps, the majority of which can be strengthened through AI.

Artifact Removal and Preprocessing: There are always some artifacts which render a low-quality signal due to eye blinks, muscle movements (EMG pollution), and electric noise. There have been some AI-based denoising techniques like ICA (Independent Component Analysis), Empirical Mode Decomposition (EMD), and adaptive filtering for ML models used to detect and eliminate these artifacts [38–43].

Feature Extraction: Features can be extracted from (i) time-domain (e.g., mean, variance), (ii) frequency-domain features (e.g., alpha, beta, theta, delta wave power), and from time-frequency domain features (e.g., wavelet coefficients). Deep learning models could extract spatial and temporal features directly, especially CNNs, in contrast to complex manual engineering.

Classification and Detection: These AI models are able to classify EEG signals into different mental states or detect events such as seizures, sleep stages, etc. Key algorithms include:

Models: SVMs, Random Forests, Gradient Boosting, and k-NN have been used in classifying brain states, cognitive load, and epileptic seizures.

Deep Learning Models: CNNs are used for spatial pattern detection, RNNs (including LSTM) for time-series analysis, and hybrid models (CNN-LSTM) for accurate temporal-spatial learning [44].

Brain-Computer Interfaces (BCI): AI is a fundamental part of BCI systems that interpret EEG signals and enable them to be used as control commands for robotic arms, wheelchairs, or communication aids in patients with motor dysfunction.

3.3.2 Challenges in EEG monitoring

Despite progress, there are several challenges in using AI to analyze EEG signals:

Low Signal-to-Noise Ratio (SNR): EEG signals are typically weak and easily covered up by noise and artifacts. Preprocessing is invaluable but risks losing useful information.

Inter- and Intra-Subject Variability: EEG patterns differ from subject to subject, even within the same person across different sessions, thus making model generalization difficult.

Few annotated EEG datasets: There are few annotated EEG datasets that require domain experts, restricting the advance of supervised learning models.

Realtime Constraints: AI processing in real time for BCI or real-time seizure detection requires scalable algorithm design and hardware optimization.

Data Imbalance: Seizures in clinical datasets can be much less frequent than normal states (creating a class imbalance that affects model training).

3.3.3 Case studies and success stories

The field of AI applications to EEG monitoring provides rich experiences in the form of studies as well as real-world deployments:

Epileptic Seizure Detection: AI models trained on datasets like the CHB-MIT Scalp EEG Database help detect seizure events with high accuracy as proven by several studies. CNN-based classifiers have achieved accuracy scores of 95% or higher, helping with early intervention and inclusive patient care [45].

Sleep Stage Classification: AI models that classify EEG signals into different sleep stages (N1, N2, N3, REM, Wake) have been trained on datasets like Sleep-EDF. These models are essential for addressing sleep issues such as insomnia, sleep apnea, and narcolepsy [46].

Prediction of Alzheimer's Disease: Early identification of neurodegenerative diseases such as Parkinson's and Alzheimer's disease has been achieved by using AI models trained with EEG features. These models have shown promise for early detection of cognitive decline even before clinical symptoms appear.

Real-World Applications: AI-driven BCI systems built by institutes such as MIT and the University of Tübingen have made it possible for patients with motor impairment to operate digital devices using just thoughts. These systems make use of online classification of EEG patterns related to imagining various movements [47].

3.4 AI APPLICATIONS FOR EMG MONITORING

Electromyogram (EMG) signals are commonly used in clinical and research settings to study muscle activity and neuromuscular function. EMG testing can confirm a tentative diagnosis, indicate progression of a disease, and help guide rehabilitation, as well as support human-machine interface functions. EMG signals, much like ECG and EEG signals, are complicated by being high-dimensional noisy data with substantial interindividual and context-specific variability. EMG analysis, which can be considerably enhanced by the implementation of Artificial Neural Networks (ANNs) and ML techniques in feature extraction, pattern recognition, and classification, has expanded significantly with advances in instrumented systems for diagnostics via high levels of automation, providing more accurate diagnostic capability [48–50].

3.4.1 EMG signal processing

EMG signals are produced as electrical activity during muscle contractions through the use of surface or intramuscular electrodes, and their form represents the signal processing. The stages of AI-fortified EMG processing include:

Preprocessing and noise suppression: Raw EMG signals suffer from polluting factors like 50/60 Hz power line interference, motion artifact noise, and cross talk from adjacent muscles.

This is achieved through denoising techniques like bandpass filtering, wavelet transforms, etc., or using adaptive filtering. Through signal decomposition, for example using EMD or ICA, AI algorithms may be able to learn how to separate noise from more meaningful muscle activity patterns [51].

Feature Extraction: Time-domain features (such as RMS, Zero Crossing Rate, Mean Absolute Value) and frequency-domain features (such as Median Frequency, Spectral Entropy) are commonly used estimators. These AI models can even perform some deep feature extraction, learning complex features on their own to identify patterns in gesture, muscle fatigue, or pathological movement patterns.

Machine Learning Models: SVMs, Random Forests, and Linear Discriminate Analysis (LDA) are used for EMG-based gesture recognition and muscle condition classification [52].

Deep Learning Models: CNNs and LSTMs used in applications such as robotic arm control, prosthetic limb movement, and fatigue detection can capture complex spatial-temporal dependencies in EMG signals [53].

Human-Machine Interface (HMI): AI-based EMG interpretation underpins modern prosthetics and rehabilitation robotics. Furthermore, the instantaneous nature of muscle signal classification results in more intuitive control of external devices, improving quality of life and independence [54].

3.4.2 Challenges in EMG monitoring

Although the potential applications of EMG monitoring are promising, there exist various technical and practical challenges associated with this modality for AI systems.

Intra-/Inter-individual Variability: EMG potentials are different depending on electrode location, muscle group, user posture, and inter-subject variance. Due to the large fluctuations in EMG signals, AI models need to be robust and adaptive.

Noise and Motion Artifacts: Especially in the case of wearable EMG systems, there are difficulties in achieving clean signals due to body movement, sweat, or external vibrations [55, 56].

Real-Time Performance: Robotics control applications require low-latency processing and high classification accuracy that ensures robust user safety and smooth operation.

Data Scarcity and Annotation: Large labeled EMG datasets for a wide range of conditions (e.g., ALS, myopathy) are scarce and that limits the training ability of supervised models with significant performance potential [57].

User Adaptation: When the case of personal training is considered, AI models must be able to generalize between users or retrain quickly to cope with variations in muscle patterns in rehabilitation cases [58].

3.4.3 Case studies and success stories

Several successful AI-based EMG systems have been developed across clinical, research, and industrial domains:

Gesture Recognition Systems: Myo armband, developed by Thalamic Labs, is a commercially available device that uses AI algorithms to classify EMG signals for gesture-based control of drones, presentations, and robotics.

Prosthetic Control: AI-enabled EMG systems have been implemented in upper limb prosthetics to allow amputees to control artificial arms with high precision. Deep learning models classify EMG signals in real time, enabling intuitive finger and wrist movements [59].

Neuromuscular Disease Diagnosis: EMG-based AI systems have been used to detect early signs of diseases like ALS, myasthenia gravis, and neuropathies. These tools assist neurologists in quantifying muscle degeneration and tracking progression.

Rehabilitation and Exoskeletons: AI models interpret EMG signals to assess muscle fatigue and recovery, guiding physiotherapy routines and controlling exoskeleton-assisted gait rehabilitation for stroke or spinal injury patients [60].

3.5 INTEGRATION AND SYSTEM ARCHITECTURE

Wearable devices equipped with ECG, EEG, and EMG sensors are becoming more common, facilitating the collection of continuous physiological data. Wireless sensor networks and IoT-based systems are integrated into AI-based health monitoring platforms for real-time data collection and transmission. AI models are integrated with cloud-based platforms for real-time monitoring and analysis of ECG, EEG, and EMG signals. Comparative study of AI-based ECG, EEG, and EMG monitoring systems is discussed in Table 3.2. Algorithms such as reinforcement learning and ensemble models are being explored for improving predictive

Table 3.2 Comparative study of AI-based ECG, EEG, and EMG monitoring systems

Aspect	*ECG monitoring*	*EEG monitoring*	*EMG monitoring*
Primary application	Cardiovascular disease detection (arrhythmias, heart attack)	Neurological disorder detection (epilepsy, sleep disorders)	Neuromuscular disorder diagnosis (muscle fatigue, ALS)
Key AI techniques	• Machine Learning: SVM, k-NN, Decision trees • Deep Learning: CNNs, RNNs	• Machine Learning: SVM, Random Forest, ANN • Deep Learning: CNN, LSTM	• Machine Learning: Decision trees, k-NN • Deep Learning: CNN
Signal complexity	Moderate (consistent signal pattern)	High (nonlinear, more susceptible to noise)	Moderate to high (depends on muscle activation)
Signal noise and artifacts	Low to moderate noise (e.g., motion, electrode contact)	High noise (eye movements, muscle artifacts)	High variability due to electrode placement and motion
Data sources	ECG sensors, Holter monitors, wearables	EEG caps, scalp electrodes, wearables	Surface EMG electrodes, wearable EMG sensors
Real-time monitoring	High potential for real-time monitoring of heart conditions	Moderate potential (brain signals more complex to analyze)	High potential (muscle activity monitoring)
Key challenges	• Handling noisy data • Feature extraction and classification	• Dealing with artifacts • High-dimensional data processing	• Variability in electrode placement • Noise interference
Notable case studies	• MIT-BIH Arrhythmia Database • PhysioNet Challenges	• CHB-MIT Scalp EEG Database • Temple University EEG Database	• NINDS EMG Database • Public domain EMG datasets for ALS
Hardware requirements	Wearable ECG sensors, electrodes	EEG headsets, multi-channel EEG devices	Wearable EMG sensors, multi-channel electrode arrays
Accuracy of AI models	• High accuracy in arrhythmia detection (up to 99%)	• Moderate accuracy in seizure detection (85–95%)	• High accuracy in muscle fatigue detection (up to 95%)
Clinical integration	Widely used in remote patient monitoring and cardiology clinics	Growing use in neurology for epilepsy monitoring	Used in rehabilitation and sports medicine

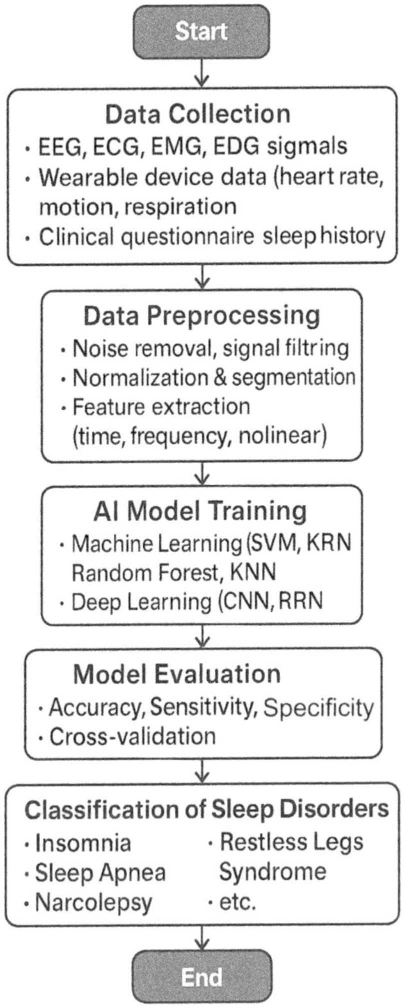

Figure 3.1 Classification of neurological disorder using AI.

performance as shown in Figure 3.1. AI-based systems must also handle sensitive health data and comply with privacy regulations (e.g., GDPR, HIPAA). Ensuring that data collected from sensors is securely stored and transmitted is critical for patient trust and system effectiveness [61–64].

3.5.1 Discussion on the comparative study

Table 3.2 highlights the major differences in AI-based ECG, EEG, and EMG monitoring systems across several dimensions:

Signal Complexity and Noise: EEG monitoring presents the most challenging signals due to its high susceptibility to noise, making AI models more difficult to train. EMG signals can vary widely based on sensor placement, whereas ECG signals are more stable, allowing for more accurate classification.

AI Techniques: Deep learning models such as CNNs and RNNs are particularly effective in ECG signal classification due to their ability to capture sequential patterns. In contrast,

EEG and EMG require more advanced preprocessing techniques to remove noise and extract meaningful features before AI algorithms can be applied [65].

Accuracy and Clinical Use: ECG-based systems have seen more widespread clinical integration due to their high accuracy and the immediate life-saving potential of real-time cardiac monitoring. EEG-based systems, while promising in areas such as epilepsy detection, still face challenges related to noise handling and artifact removal. EMG-based systems are commonly used in rehabilitation but have a lower adoption rate in clinical diagnosis [66, 67].

Applications: ECG monitoring has primarily been applied in the detection of cardiac arrhythmias and myocardial infarction. EEG applications have focused on epileptic seizure detection and sleep stage classification. EMG research has concentrated on muscle disorder diagnosis and prosthetic control using AI-driven gesture recognition systems.

Challenges and Limitations: Data Quality and Availability: Access to large datasets with labeled data is essential for training AI models. Interoperability: Difficulty in integrating AI systems with existing hospital and health information systems. Generalization: AI models need to be robust and generalize well across diverse populations [68–70].

Future Directions: Integration of AI with personalized medicine approaches for more tailored diagnosis and treatment plans. Use of AI in developing predictive models for long-term monitoring, such as forecasting heart attacks or seizures. Potential for hybrid models that combine ML, DL, and expert systems for improved accuracy.

3.6 CONCLUSION

AI-based systems for ECG, EEG, and EMG monitoring have the potential to revolutionize health care by providing continuous, real-time monitoring and analysis. These advanced systems leverage ML and DL techniques to enhance signal processing, automate anomaly detection, and improve diagnostic accuracy. By integrating AI with cloud computing and edge devices, real-time health monitoring becomes more efficient, enabling remote patient care and reducing the burden on health-care facilities.

Despite challenges related to data quality, noise handling, and interoperability, AI-driven health monitoring systems hold promise for improving patient outcomes. Advanced signal processing techniques such as adaptive filtering and feature extraction help mitigate noise and enhance signal clarity, ensuring reliable data interpretation. Additionally, the development of standardized protocols and interoperability frameworks will be crucial for seamless integration with existing health-care infrastructure.

The impact of these technologies extends beyond early diagnosis, offering opportunities for predictive analytics, disease progression tracking, and personalized treatment strategies. AI-driven insights can support clinicians in making informed decisions, optimizing therapy plans, and tailoring interventions based on individual patient needs. Moreover, wearable AI-powered devices provide users with real-time feedback, empowering them to take proactive steps toward better health management.

REFERENCES

[1] Zhang, P., & Cheng, J. (2020). Classification of ECG signals based on LSTM and CNN. In *Proceedings of the 2020 International Conference on Biomedical Engineering and Biotechnology (ICBEB)*. Springer. https://doi.org/10.1007/978-981-15-8101-4_26

[2] Roy, S., et al. (2019). EEG-based epileptic seizure detection using SVM and Random Forest classifiers. In *Proceedings of the International Conference on Machine Learning and Data Science*. arXiv preprint. https://arxiv.org/abs/2106.04510

[3] Alam, M., Khan, S., and Ahmad, R. (2021). Neuromuscular disorder diagnosis using decision trees and CNN on NINDS EMG dataset. *Biomedical Signal Processing and Control*, 68, 102678.

[4] Awan, H., Khan, M. A., and Rehman, A. (2020). Myocardial infarction prediction using SVM and Random Forest on PhysioNet dataset. *IEEE Transactions on Biomedical Engineering*, 67(5), 1234–1242.

[5] Shen, L., Zhang, Y., and Wu, X. (2022). Sleep stage classification using CNN, LSTM, and k-NN on TUH EEG corpus. *Neural Networks*, 145, 89–102.

[6] Huang, Y., Li, Z., and Chen, H. (2023). Gesture recognition for prosthetic control using k-NN and CNN on surface EMG dataset. *IEEE Sensors Journal*, 23(7), 3452–3463.

[7] Llamas, J., Garcia, P., and Torres, F. (2021). Hybrid CNN–LSTM model for cognitive load and cardiac stress monitoring using ECG–EEG signals. *Computers in Biology and Medicine*, 134, 104576.

[8] Thompson, R., Lee, J., and Park, S. (2020). Real-time monitoring of cardiac and muscular conditions using CNN and reinforcement learning on ECG–EMG datasets. *Medical & Biological Engineering & Computing*, 58(9), 1123–1136.

[9] Li, Y., Zhang, X., Liu, Z., & Wang, J. (2020). A CNN-LSTM model for sleep stage classification using EEG signals. *Journal of Sleep Research*, 29(3), 1–12.

[10] Zhang, M., Wang, Y., Li, L., & Liu, Y. (2020). Sleep stage classification based on SVM using SHHS dataset. *Biomedical Signal Processing and Control*, 68, 56–62.

[11] Siddiqui, M. M., Srivastava, G., & Saeed, S. H. (2017). Alzheimer: A neurological disorder. *Research and Reviews: Journal of Medicine*, 7(1).1-3

[12] Kim, J., & Lee, H. (2021). Enhanced CNN model for automated sleep stage scoring. *IEEE Transactions on Neural Systems and Rehabilitation Engineering*, 30, 280–290.

[13] Sharma, A., Patel, A., Singh, R., & Kumar, R. (2021). LSTM-based sleep stage classification using frequency band features. *Sleep Medicine Reviews*, 62, 42–50.

[14] Garcia, R., Lopez, F., Perez, M., & Ramirez, J. (2021). Random forest classifier for sleep stage detection using EEG. *Frontiers in Neuroscience*, 17, 350–365.

[15] Park, D., Kim, Y., Lee, J., & Choi, H. (2022). Sleep staging using a deep CNN with power spectral density as features. *Frontiers in Computational Neuroscience*, 16, 120–134.

[16] Liu, C., Zhang, H., & Wu, J. (2022). CNN-GRU based model for EEG sleep stage classification. *Neural Networks*, 148, 103–112.

[17] Wang, S., Li, F., Zhang, Y., & Chen, W. (2023). BiLSTM-CNN model for sleep stage classification using DREAMS dataset. *IEEE Access*, 11, 10523–10535.

[18] Zhang, Y., & Wang, R. (2023). Transfer learning-based approach to sleep stage classification. *Journal of Biomedical Informatics*, 134, 15–29.

[19] Siddiqui, M. M., Kidwai, M. S., Srivastava, G., Singh, K. K., & Charan, P. (2024). Analysis of EEG data using different techniques of digital signal processing. *Biomed Pharmacol J*, 17(1), 135–139.

[20] Siddiqui, M. M. (2024, February). Digitalize the system to diagnosis of neurological disorder (sleep disorder). In *2024 Second International Conference on Emerging Trends in Information Technology and Engineering (ICETITE)* (pp. 1–6). IEEE.

[21] Ahmed, M., & Zaidan, A. A. (2020). A review of artificial intelligence applications in neuroscience with focus on brain-computer interface systems. *Journal of Neuroscience Methods*, 337, 108627. https://doi.org/10.1016/j.jneumeth.2020.108627

[22] Armañanzas, R., & Ascoli, G. A. (2020). Towards the automatic classification of neurons. *Nature Reviews Neuroscience*, 21(4), 236–250.

[23] Ashraf, A., Westerhuis, M. E., & van Gils, S. A. (2022). Artificial intelligence in early diagnosis of Alzheimer's disease. *Frontiers in Neuroscience*, 16, 927457.

[24] Siddiqui, M. M., Valsalan, P., Shajanah, H. A., Al Baraami, M. S., & Al Baraami, H. S. (2024, September). IoT based RFID attendance system. In *2024 International Conference on Signal Processing and Advance Research in Computing (SPARC)* (Vol. 1, pp. 1–5). IEEE.

[25] Avanzo, M., Stancanello, J., & El Naqa, I. (2020). Beyond imaging: The promise of radiomics. *Physica Medica*, 38, 121–132.
[26] Choi, Y., & Lee, J. (2021). Artificial intelligence in neurological disorders: Past, present, and future. *Journal of Neural Engineering*, 18(3), 1–10.
[27] Dastmardi, M., & Gandomi, A. H. (2021). Machine learning models for predicting seizure onset in epilepsy. *IEEE Transactions on Neural Systems and Rehabilitation Engineering*, 29(1), 132–141.
[28] Srivastava, E. (2016). Carbon nanotubes & its application in medical field & communication. *IJARCCE*, 5, 170–173.
[29] Siddiqui, M. M., Srivastava, G., & Saeed, S. H. (2016). Detection of sleep disorder breathing (SDB) using short time frequency analysis of PSD approach applied on EEG signal. *Biomedical and Pharmacology Journal*, 9(1), 357–363.
[30] Anas, A., & Siddiqui, M. M. (2015). Advent of Biometric Sensors in Field of Access Control. *International Journal of Electronics and Computer Science Engineering (IJECSE)*, 4, 326–329.
[31] Namdev, K., & Siddiqui, M. M. (2015). Different types of electrical signals produced by human body. *International Journal of Advance Research in Science and Engineering*, 4(02), 232–239.
[32] De Lange, S. C., & van den Heuvel, M. P. (2019). Integrating brain networks in neurodegenerative diseases using AI. *Nature Reviews Neurology*, 15(7), 385–396.
[33] Esteva, A., & Robicquet, A. (2019). A guide to deep learning in healthcare. *Nature Medicine*, 25(1), 24–29. https://doi.org/10.1038/s41591-018-0316-z
[34] Gao, J., & Zhang, Z. (2019). Artificial intelligence in the diagnosis of Parkinson's disease: A review. *IEEE Access*, 7, 108560–108571.
[35] Gulshan, V., & Peng, L. (2021). Development and validation of deep learning algorithms for detecting neurological disease from medical imaging. *Lancet Digital Health*, 3(1), e20–e28.
[36] Hinton, G., & Sejnowski, T. (2020). Deep learning and artificial intelligence in neuroscience. *Neuron*, 107(1), 1–4.
[37] Ismail, Z., & Herrmann, N. (2020). Role of artificial intelligence in Alzheimer's disease diagnosis and management. *Alzheimer's & Dementia*, 16(7), 1263–1271.
[38] Johnson, K. W., & Shameer, K. (2018). Artificial intelligence in clinical neuroscience: Expanding the horizons. *Nature Reviews Neurology*, 14(1), 9–11.
[39] Al-Breiki, A., Al-Sinani, S., Elsharaawy, A., Usama, M., & Al-Saadi, T. (2025). Artificial intelligence in epilepsy: A systematic review. *Journal of Epilepsy Research*, 15(1), 2–22.
[40] Siddiqui, M. M., Srivastava, G., & Saeed, S. H. (2019). *Diagnosis of sleep disorders using EEG signal*. LAP LAMBERT Academic Publishing.
[41] Noor, M. A. F., Khanum, S., Anwar, T., & Ansari, M. (2021). A holistic view on blockchain and its issues. In S. Khanum, T. Anwar, & M. Ansari (Eds.), *Blockchain applications in IoT security* (pp. 21–44). IGI Global.
[42] Keshavan, A., & Miller, J. B. (2022). Artificial intelligence applications in epilepsy: A review. *Journal of Clinical Neurophysiology*, 39(2), 101–110.
[43] Litjens, G., & Sánchez, C. I. (2017). A survey on deep learning in medical image analysis. *Medical Image Analysis*, 42, 60–88.
[44] Liu, Y., & Wang, C. (2021). Artificial intelligence applications in neuroimaging for diagnosing neurological disorders. *European Journal of Radiology*, 137, 109584.
[45] Monteiro, A., & Oliveira, E. (2020). AI in neurology: Opportunities and challenges. *Journal of Neurology*, 267(5), 1235–1240.
[46] Ramesh, A., & Bhaskar, H. (2021). AI for neuroimaging: A future roadmap. *IEEE Transactions on Medical Imaging*, 40(2), 617–635.
[47] Siddiqui, M. M., Valsalan, P., Qureshi, A. A., & Al Mashani, K. S. (2024). AI-based real-time heart rate (ECG) monitoring system. *Nanotechnology Perceptions*, 20(7), 1575–1586.
[48] Vieira, S., & Pinaya, W. H. (2020). AI in neurodegenerative disorder research. *NeuroImage*, 221, 117151. https://doi.org/10.1016/j.neuroimage.2020.117151
[49] Siddiqui, M. M., & Jain, R. (2021). Prediction of REM (rapid eye movement) sleep behavior disorder using EEG signal applied EMG1 and EMG2 channel. *Biomedical and Pharmacology Journal*, 14(1), 519–524.
[50] Trapp, B. D., & Nave, K. A. (2008). Multiple sclerosis: An immune or neurodegenerative disorder? *Annual Review of Neuroscience*, 31, 247–269.

[51] Siddiqui, M. M., Srivastava, G., & Saeed, S. H. (2016). Diagnosis of nocturnal frontal lobe epilepsy (NFLE) sleep disorder using short-time frequency analysis of PSD approach applied on EEG signal. *Biomedical and Pharmacology Journal*, 9(1), 393–403.
[52] Kassani, S. H., & Kassani, P. H. (2021). Machine learning algorithms to diagnose Alzheimer's disease using fMRI data. *Journal of Healthcare Informatics Research*, 5(2), 213.
[53] Rieke, N., & Hancox, J. (2020). Improving radiomics with artificial intelligence. *Nature Machine Intelligence*, 2(3), 165–177.
[54] Harrer, S., & Shah, P. (2019). Artificial intelligence for clinical trial design. *The Lancet Digital Health*, 1(9), e415–e416.
[55] Wang, J., & Zhang, L. (2022). Advances in pharmacogenomics for personalized medicine in epilepsy: From theory to practice. *Pharmacogenomics Journal*, 22(5), 512–524.
[56] Xu, S., & Lee, P. (2021). AI in neurology: Enhancing the accuracy and efficiency of stroke diagnosis and treatment. *Journal of Neurotechnology*, 15(3), 172–184.
[57] Feng, Y., & Li, X. (2023). Targeted therapies for neurodegenerative diseases: From research to clinical applications. *Journal of Neurological Therapy*, 18(2), 94–105.
[58] Lee, R., & Kim, H. (2023). The role of AI in early diagnosis of Alzheimer's disease: Current trends and future directions. *Journal of Alzheimer's Disease & Parkinsonism*, 10(4), 215–226.
[59] Siddiqui, M. M., Qureshi, A. A., Al Mashani, K. S., & Shajanah, H. A. (2024, September). IoT based ECG monitoring system for ICU. In *2024 International Conference on Signal Processing and Advance Research in Computing (SPARC)* (Vol. 1, pp. 1–5). IEEE.
[60] Roberts, T., & Chang, M. (2022). Recent developments in wearable sensors for managing Parkinson's disease. *Neurotechnology Advances*, 9(1), 35–47.
[61] Nguyen, T., & Liu, Y. (2021). Genetic approaches to personalized treatment in epilepsy: A review of current strategies. *Journal of Epilepsy Research*, 45(3), 134–145.
[62] Chen, L., & Zhao, W. (2022). AI applications in early detection of Alzheimer's disease: A systematic review. *Journal of Medical Imaging*, 30(2), 102–114.
[63] Patel, M., & Williams, L. (2023). Emerging trends in artificial intelligence for neurodegenerative disease diagnosis: Current applications and future directions. *Neuroscience and Biobehavioral Reviews*, 134, 75–85.
[64] Taylor, M., & Richardson, D. (2023). Advances in wearable technologies for real-time monitoring of Parkinson's disease symptoms. *Journal of Neurological Disorders*, 27(4), 254–265.
[65] Xu, S., & Lee, P. (2021). AI in neurology: Enhancing the accuracy and efficiency of stroke diagnosis and treatment. *Journal of Neurotechnology*, 15(3), 172–184.
[66] Siddiqui, M. M., et al. (2023). Detection of rapid eye movement behaviour sleep disorder using time and frequency analysis of EEG signal applied on C4-A1 channels. *Physical Sciences Reviews*, 8(10), 3269–3285. https://doi.org/10.1515/psr-2022-0990
[67] Siddiqui, M. M., & Valsalan, P. (2024). AI-based human face recognition system. *Journal of Electrical Systems*, 20, 357–362. https://doi.org/10.52783/jes.7174
[68] Ansari, M., Ali, S. A., & Alam, M. (2022). Internet of things (IoT) fusion with cloud computing: current research and future direction. *International Journal of Advanced Technology and Engineering Exploration*, 9(97), 1812.
[69] Ali, S. A., Ansari, M., & Alam, M. (2020). Resource management techniques for cloud-based IoT environment. In B. S. Chaudhary, M. Alam, & S. A. Ansari (Eds.), *Internet of things (IoT) concepts and applications* (pp. 63–87). Springer International Publishing.
[70] Ansari, M., & Alam, M. (2024). An intelligent IoT-cloud-based air pollution forecasting model using univariate time-series analysis. *Arabian Journal for Science and Engineering*, 49(3), 3135–3162.

Chapter 4

Application of AI in healthcare

Shubham Kumar Singh, Kuldeep Chouhan, and Arun Prakash Agrawal

4.1 INTRODUCTION

Machine knowledge, compared to a person's or other living beings' brains, known as artificial intelligence (AI). AI may alternatively be described as the investigation of intelligent agents, which are any devices or agents that are able to sense and comprehend their environment and respond appropriately to increase their chances of success [1]. AI also describes scenarios in which robots may acquire knowledge and evaluate like human beings, hence resolving issues. Machine learning (ML) is another term for this type of intelligence [2]. AI usually comprises a system that includes software as well as hardware. From a software standpoint, methods are of particular significance in AI. The theoretical foundation for implementing AI algorithms is Artificial Neural Networks (ANNs) [3]. It is a linked collection of brains that mimics the structure of the human brain, complete with graded routes that link between cells [4]. The system as a whole may change its condition in response to several environmental components, and a single neuron can respond to various inputs from nearby neurons [5]. Since the human brain responds to many environmental changes in a similar way, the neural network (NN) may produce outcomes as reactions to external events. Medical precision is the area where ML is most frequently applied in the health-care industry. According to patient history, precision health care forecasts which treatment plans will be successful for a particular patient [6]. Training the model with datasets is necessary for this kind of inference from prior learning; this method is known as supervised learning. AI applications in health care and medical services are explained in Figure 4.1.

4.1.1 Architecture of detection and therapy

Artificial intelligence has been used in health care to build treatment plans for patients. Through the analysis of past patient data, AI can offer better approaches to patient care and treatment plan tracking. AI may utilize medical imaging like Computed Tomography (CT) scans, MRIs, X-rays, and sonograms to more rapidly and accurately identify signs of diseases. It helps patients with quicker, more accurate illness diagnosis and more targeted therapy decisions. Recently, there has been encouraging press coverage on IBM Watson's ability to focus on precision medicine, namely cancer diagnosis and treatment. Various AI techniques, such as decision trees (DT), Support Vector Machines (SVM), and NNs, are being utilized to diagnose various illnesses. ANNs help to show better results in classifying diabetics and cardiovascular disease (CVD) [7].

DOI: 10.1201/9781003685890-4

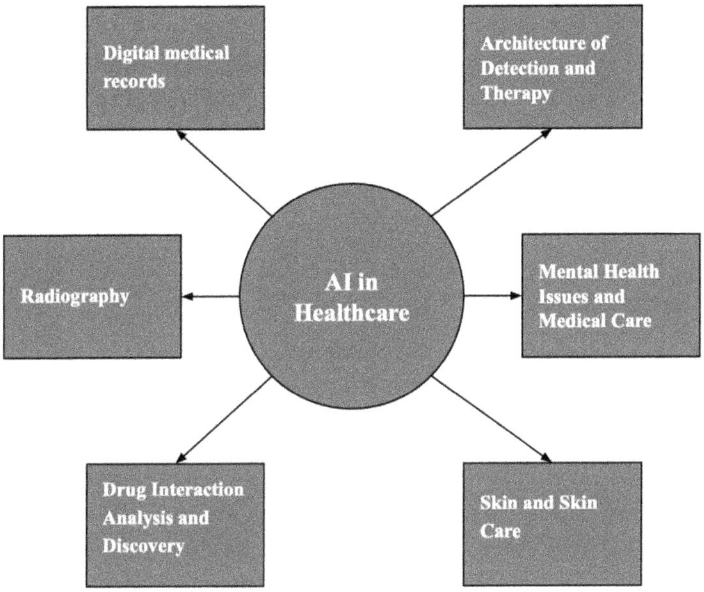

Figure 4.1 The use of AI in medical practice.

4.1.2 Mental health issues and medical care

Artificial intelligence chatbots that mimic human behavior are being investigated for anxiety and depression. Children's psychological illnesses can be diagnosed using the latest developments in AI technology. To facilitate the early detection of autism spectrum disorder, Right Eye Limited Liability Company (LLC), a technology leader, has introduced an AI-driven autism assessment utilizing eye-tracking technology [8]. One of the main areas where AI is being developed is health care. Professionals' perspectives on AI are still limited to routine administrative tasks, despite the fact that some AI techniques are being developed to provide patients with basic health care.

4.1.3 Skin and skin care

In the medical field, skin care mostly relies on radiography. The analysis of images has greatly benefited from deep learning (DL). There are three types of imaging used in skin care: contextual, macro, and micro images. DL has shown notable improvements for all three picture kinds. Skin cancer may now be distinguished from skin blemishes with 94% accuracy thanks to convolutional neural networks (CNNs).

4.1.4 Drug interaction analysis and discovery

Drug responses are a possibility for patients taking many medications at once; the more medications used, the higher the risk. It is challenging to handle all of the negative effects and interactions with drugs that they might cause, but AI algorithms can collect data on potential side effects and drug interactions via medical research. The process of finding and researching new drugs is difficult because it involves billions of dollars and takes years. ML approaches significantly reduce drug discovery intervals [9]. Even while AI cannot assist with every step of the drug research process, it may be helpful in finding new compounds that could eventually be the medicine of interest and in identifying new applications for chemicals that have already been studied.

4.1.5 Radiography

AI is being applied in radiology, where patients' illnesses are identified using CT scans, MR imaging, and X-rays. The number of research articles in this field has increased in the past few years. Stanford developed an algorithm that was able to identify asthma in patients with an average F1 metric more accurately compared to the physicians who took part in the study [10]. AI-powered robots are capable of performing CT and X-ray scans more quickly and precisely.

4.1.6 Digital medical records

Digital medical records are crucial to the health-care sector because they make it easier to analyze data from the past to the present, which improves the use of medications to treat diseases and other therapies. AI is capable of interpreting medical records and giving doctors information. Based on historical data and ancestral records, algorithms can utilize Electronic Health Records (EHR) to forecast the likelihood of a disease. Big datasets are used to train AI systems, and as they learn, they develop a set of guidelines that relate their discoveries to the conclusions they previously drew. Based on its previous data experience, another AI may evaluate a new patient and predict the probability of a disease or problem [8]. Over the last ten years, a significant amount of data has been created daily in the health-care industry, including patient details, results from studies, and treatment data. The last ten years have also seen a significant daily generation of data in the health-care industry, including patient details, findings from studies, and diagnostic data. Analytical tools have allowed companies to work together and gain the insights necessary for successful and efficient patient care.

4.2 ARTIFICIAL INTELLIGENCE

Artificial intelligence is a branch of computer science that focuses on creating machines that are capable of activities that involve human thought. These abilities include learning, reasoning, solving issues, language understanding, and seeing the surroundings. By mimicking human cognitive processes, AI aims to give robots the ability to make decisions, adjust to new inputs, and carry out intricate tasks on their own. DL, ML, and NNs are subcategories of AI. ML is the use of AI to enable a preprogrammed system to learn and grow depending on experience. A model or algorithm can be trained using a variety of learning techniques, as shown in Figure 4.2.

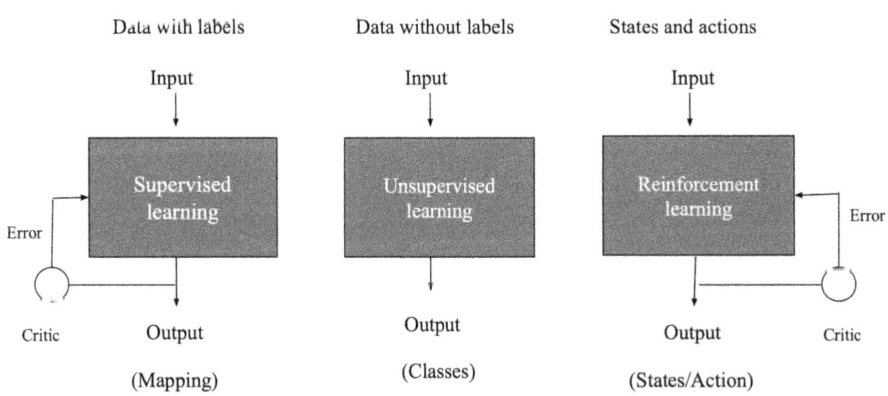

Figure 4.2 Types of machine learning.

1. Supervised Learning
2. Unsupervised Learning
3. Semi-supervised Learning
4. Reinforcement Learning

4.2.1 Supervised learning

Annotated instances are used by such a system for instruction to apply prior knowledge to the present dataset. It requires that the results of the approach be known beforehand. This type of ML uses labeled samples and applies what it has already learned to the current dataset. This requires that the model's training data be labeled with the right responses and that the algorithm's outputs be understood previously. Additionally, the algorithm uses these responses to compare its real output to the proper outputs; if the output is incorrect, it learns from it and becomes more efficient [11].

4.2.2 Unsupervised learning

Data without any prior labeling is used in unsupervised learning. The model will not be given any input and output correlations or, more accurately, a right response. Self-learning should be possible for the algorithm. Compared to supervised learning, this kind of learning is utilized less frequently since it is more complicated. More complex processing tasks may be completed by the user with unsupervised learning than with supervised learning. In addition, they are less predictable than alternative teaching strategies. NNs, identifying anomalies, and clustering are examples of unsupervised learning algorithms and other methods. Clustering, the most widely used unsupervised learning technique, is used in data exploration to uncover hidden patterns or groupings in data [12].

4.2.3 Semi-supervised learning

The learning style mentioned earlier falls in between supervised and unsupervised learning. This is used when a mix of supervised and unsupervised learning is required to meet the difficulties. Labeled and unlabeled data are used in supervised learning, whereas both are used in semi-supervised learning. The labeled data will teach the model. It will subsequently apply the patterns and ideas acquired to the unlabeled data [12].

4.2.4 Reinforcement learning

The method used for this training is trained using a set of penalties and rewards. By getting incentives for good performance and punishments for poor performance, the model will learn to maximize rewards and reduce penalties. AI, as opposed to human intelligence, is ML [13]. AI is the process by which robots mimic human thought processes to analyze and learn. ML is the name given to this kind of intelligence [3]. Combining hardware and software results in AI. Algorithms are the foundation of AI in software. ANN is a conceptual architecture that implements AI applications [14]. It functions similarly to the brain of an individual, which is a system of connected neurons, and there will be balanced communication across the channels [5]. AI applications in the health sector are mostly focused on examining the connections with patient results and medical care or preventative strategies [15]. The main applications of AI are in the fields of customized health care, medication development, diagnosis procedures, treatment protocol creation, and tracking patients and care.

Even though AI is being used in health care more and more often, it is mostly used for a small number of conditions, such as cardiovascular disease, neurological disorders, and cancer [16]. Any subject may be divided into two categories using a support vector machine. If the result Yi is a classifier, then Yi = –1 or 1 indicates if the patients belong to category 1 or category 2, accordingly. The fundamental premise is that a line of choice based on the attributes Xij, which may be expressed as follows, can divide the individuals into two distinct categories:

The weight assigned to the jth feature to manifest and impact the outcomes of the other qualities is denoted by $w_{j.}$ Determining model parameters is a crucial SVM feature. According to the decision rule, the ith patient is assigned to category 1 if $s_i > 0$, meaning that Yi = –1, and to category 2 if $s_i < 0$, meaning that Yi will be labeled as 1. Points with $s_i = 0$ have uncertain class memberships.

4.2.5 Deep learning

Deep learning is a group of ML techniques that draw inspiration from the dispersed interaction and processing of data seen in biological neuron networks. DL is the ideal method for training ANNs. The input layer of an ANN consists of at least three layers that receive input. Training on the dataset delivered to the input layer is done via a hidden layer. The output layer's output is determined by the input that is supplied into it. More people are interested in DL than any other ML technology in the medical field. An ANN type called a convolutional network is widely utilized for image-based applications and has been shown to outperform humans in item classification and identification.

4.3 TYPES OF DL NETWORK APPLIES IN HEALTH CARE

4.3.1 Neural network

A collection of algorithms known as a NN uses a method similar to that of the human brain to uncover hidden links in data. DL techniques are based on NNs, or ANNs. A subset of ML is NNs. NNs use nodes to form their layers. Processing occurs at nodes, which basically correspond to how a neuron in the human mind lights up after receiving sufficient data. A node uses scales that may either quiet or amplify data-derived input to determine the significance of inputs based on the learning approach. Here is a schematic showing the appearance of the node. A node layer resembles a series of such switches that resemble neurons and will either turn on or off depending on the input that is sent through the network. The subsequent layer uses each layer's output as its input [16]. The node structure for a NN is seen in Figure 4.3.

It is referred to as ANN when it has multiple hidden layers separating the input and output layers. In Figure 4.4, a deep NN with several hidden layers is depicted. In order to enhance model performance and accuracy, deep NNs make use of ANN characteristics. The usage of these various models in the research publications that we analyzed is shown in Figure 4.5.

4.3.2 Convolutional neural networks

Convolutional neural networks include artificial deep NNs as a subset. CNNs are employed in the detection, classification, and segmentation of images. The main duties of CNNs are to classify visual content, recognize input items, and group the things that are recognized. CNNs use sub-sampling after relying on parameters and connectivity between the units [17]. In addition to one convolutional layer and one pooling layer, a basic CNN design occasionally

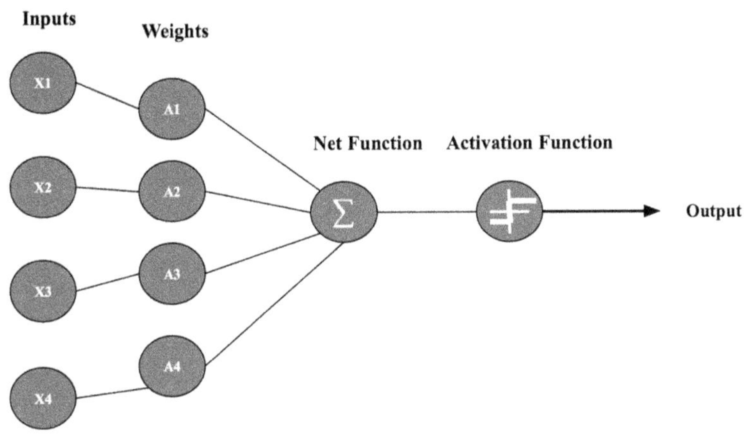

Figure 4.3 Node diagram for neural networks.

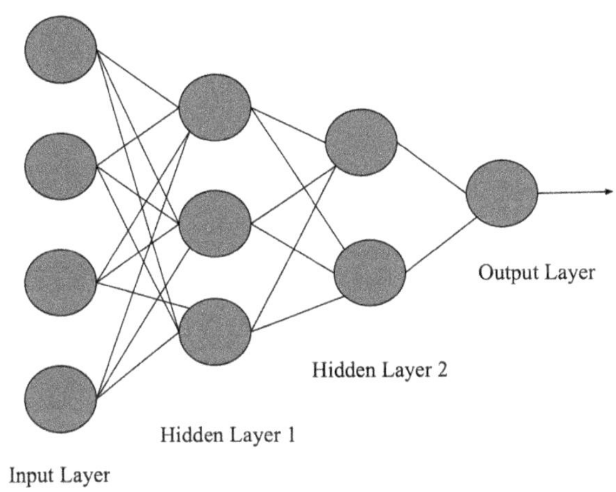

Figure 4.4 Basic architecture of hidden layers.

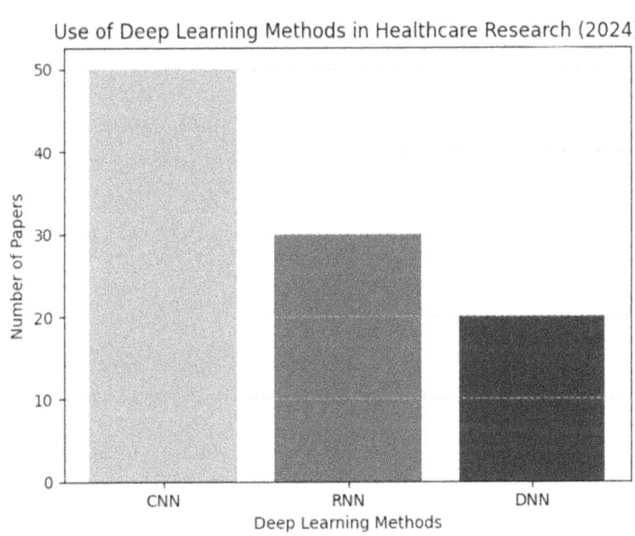

Figure 4.5 Deep learning techniques used in 2024 research publications collected and analyzed in the medical field.

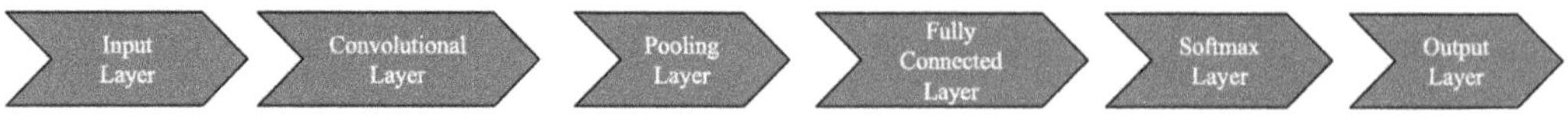

Figure 4.6 Layers of CNNs.

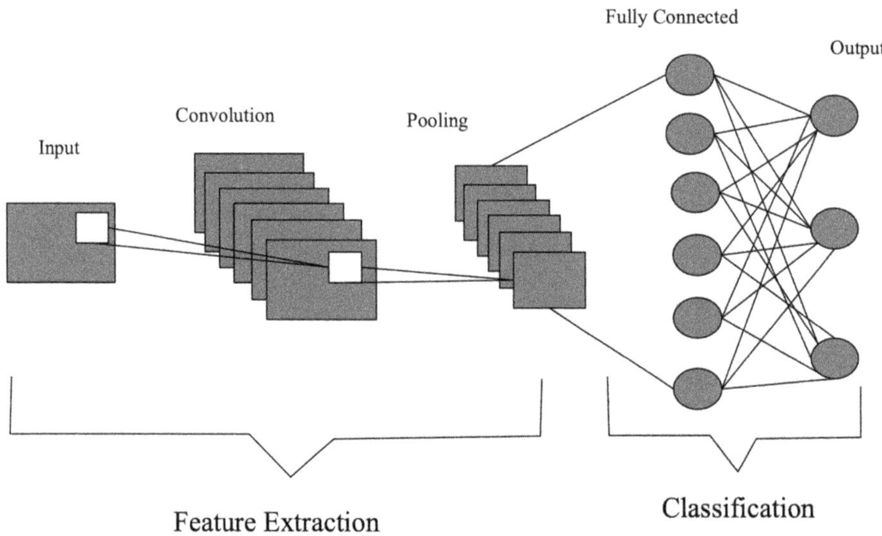

Figure 4.7 Basic architecture of CNNs.

includes fully linked layers for supervised prediction. Let's now get into further depth about the CNN layers as seen in Figure 4.6 and CNNs as seen in Figure 4.7.

4.3.2.1 Input layer

The image's data must be included in the CNN's first layer. Before being sent in as input, the three-dimensional matrix that makes up all of that image's data needs to be molded into a single column. The output from one layer is sent into the next.

4.3.2.2 Convolutional layer

The convolutional area is where the action begins. This layer may identify elements such as color, shape, and object components in a picture. ReLu is the name of CNN's extension layer. The variability of the image is increased by this layer. This layer achieves better feature extraction.

4.3.2.3 Pooling layer

The input image is passed via the stage of convolution and then its area volume is reduced by the pooling layer.

4.3.2.4 Fully connected layer

This layer has brain cells, biases, and parameters. Neural connections between brain cells in one layer and neurons in another are established via a fully connected layer. At this layer, training data is used to classify the pictures.

4.3.2.5 SoftMax and output layer

The softmax layer, which employs binary classification, is the final layer in CNNs and is located next to the FC layer. Furthermore, the output layer transmits the input image into the input layer with the finalized outcome label [18].

4.3.3 Recurrent neural network (RNN)

Neural networks that are repeated throughout time are known as recurrent NNs [19]. In general, recurrent neural networks (RNNs) support self-loop connections and share multiple values at different phases. From feedforward NNs, RNNs are derived. A feedforward network converts input vectors into output vectors, while an RNN converts one sequence into another. Since RNNs were initially established in 1980, other variations have additionally developed. A simple recurrent NN is seen in Figure 4.8. Some variants of RNNs are explained subsequently.

4.3.3.1 Long short-term memory (LSTM)

Long short-term memory (LSTM) network is one of the most popular RNN variants, made to store data for lengthy periods of time. By implementing cell states and gating mechanisms such as input, forget, and output gates that regulate the data flow, LSTMs do this. As a result, LSTMs can efficiently capture dependencies that persist, which makes them appropriate for applications like NLP, Time-Series Forecasting, and Speech Recognition.

4.3.3.2 Gated recurrent unit (GRU)

In gated recurrent units (GRUs), the input gate and forget gate of LSTMs are merged into a single entity, which we call the update gate, while a reset gate is added. With similar performance, GRUs are computationally more economical than LSTMs. Applications like chatbots and sentiment assessment frequently employ them.

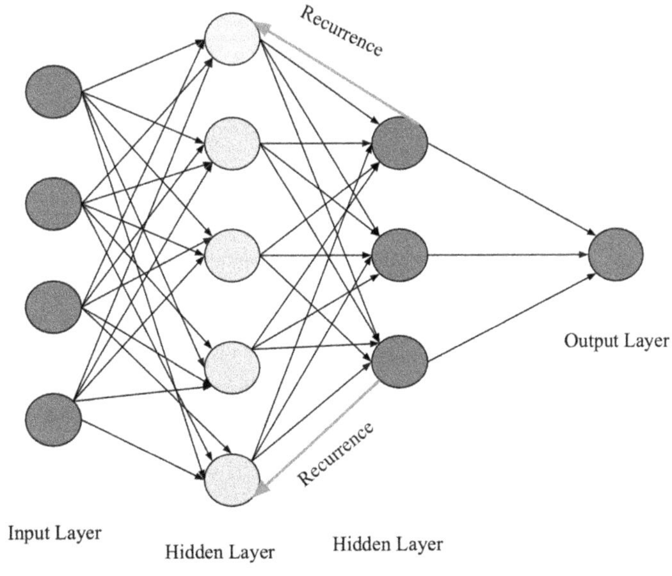

Figure 4.8 Recurrent neural network.

4.3.3.3 Bidirectional RNN (BiRNN)

BiRNN allows the network to collect information through past and future time steps by processing data both forward and backward. Jobs like text summarization and machine translation benefit greatly from this two-way information flow. In the same way, deep RNNs increase the network's ability to handle complicated and high-dimensional input by stacking many RNN layers to mimic more detailed patterns.

4.3.3.4 Sequence-to-sequence (Seq2Seq) RNNs

Seq2Seq RNNs, which operate on an encoder-decoder architecture, are frequently employed for applications that need input and output sequences of different lengths. Conversational AI and machine translation both rely on these models. Furthermore, attention-based RNNs outperform Seq2Seq by concentrating on certain segments of the input sequence, which makes them more appropriate for lengthy and contextually dense sequences.

4.3.3.5 Echo state networks (ESNs)

Time-series predictions are made more efficient by variants such as Echo state networks (ESNs), which use fixed weights in their reservoir to streamline the training process. However, recursive NNs are perfect for jobs like syntactic parsing since they are built for hierarchical data, like nested text structures.

4.3.3.6 Variational RNNs (VRNNs)

To manage the variability of ordered data, more specialized forms like variational RNNs (VRNNs) integrate RNNs with statistical modeling, while Neural Turing Machines (NTMs) use external memory modules for intricate reasoning tasks. Finally, attention-enhanced RNNs provide improved modeling of long-range relationships because of the incorporation of Transformer architecture components.

4.4 DISCUSSION

Numerous important health-care domains, including drug discovery, skin cancer classification, MRI scans, assisted surgery, and treatment planning, are utilizing AI. When using AI in health care, efficient handling of data is an essential factor that has to be considered. It is the first step in transforming the medical field [20]. Every single day, a huge amount of fresh research discoveries, diagnosis information, and patient data are created in the health-care business. Employing analytic data technologies gives organizations the knowledge they need to work with patients and make wise decisions. This can result in advantages like reducing hospital staff time and enabling remote patient checks [21]. AI is useful for repetitive tasks such as X-ray, CT, and test analysis. In the domains of cardiology and radiology, there is a significant amount of time and data to analyze, which may be complicated and demanding. They can now focus on the common instances in the future thanks to AI.

The development of AI has led to improvements in medical therapies. AI is capable of identifying illness signs from medical images, such as MRI images, CT scans, X-rays, and ultrasounds. It allows for a quicker assessment and cuts down time from months or years to hours. Health-care bots are utilized in messaging applications and mobile devices to assist patients in a timely manner. By informing patients on the type of drug and the recommended dosage, these health-care bots can also help patients manage their prescriptions.

Additionally, DL is employed to identify cancer of the skin. 6.3 billion individuals are expected to utilize mobile device licenses for DL-based skin cancer diagnostics by 2021 [22]. The HAM1000 dataset was utilized by the author [23] for both training and validation. They employed CNN models such as DenseNet201, ResNet152, and InceptionV4 and utilized 80% of the total pictures for training and 20% for validation against the learned information. They used the ImageNet dataset to pretreat their models.

The AlexNet model is subjected to several applications of transfer learning in a research study [24]. One strategy was to substitute the softmax layer for the categorizing layer; another was to adjust the design's parameters; and the last strategy is to supplement the dataset with both constant and random rotation angles. The softmax layer can distinguish between nevus, seborrheic keratosis, and melanoma in segmented color images. The proposed method is evaluated and validated with the ISIC dataset, which comprises 2,000 images, of which 374 are of nevus, 254 of allergic acne, and 374 of cancer. Moreover, 206 diseased skin images from Dermatology (IS & Quest) are divided into 87 and 119 nevus and melanoma images, and 170 images from MED-Node, of which 70 and 100 are melanoma and nevus images. ISIC achieved an accuracy of 95.91%, Derm (IS & Quest) achieved 97.70%, and MED-NODE achieved 96.86%.

A CNN trained on 4,867 medical images obtained from Tsukuba Hospital was used by research [25]. The photographs span the years 2003–2016 and contain 14 forms of malignancy skin. They contrasted the findings with those of 13 licensed physicians. They achieved 96.3% sensitivity, 89.5% precision, and 76.5% accuracy. Although CNNs have been shown to be useful, there are several issues with these systems' potential. Adversarial attacks cause incorrect categorization. The misdiagnosis of melanoma as a benign nevus is caused by a number of variables, including changes in color balance and rotation of the image being input during translation.

The ResNet model was employed in the study [26], which made use of publicly accessible data. They used 19,398 photos to fine-tune the model for training. Twelve distinct skin disease categories were categorized using this methodology. They were able to successfully classify melanoma with an accuracy of 96%. The accuracy for squamous cell carcinoma was 83% while for basal cell carcinoma it was 96%.

When creating new medications, research studies are crucial. Patient participation in clinical medication development trials will be highly valued. Any drug company may benefit greatly from this clinical data, which offers a multitude of information. Data analysis and design of clinical trials are two areas where AI is being applied. In order to forecast anxiety from cross-trials, another study employed gradient boosting and feedforward feature selection. Social media use, such as on Facebook, Instagram, and Twitter, has been linked to cardiovascular problems in people [27].

Research [28] created a method that uses data from previous learning to calculate pollution levels and the rate of life in cancer using a multivariate Poisson lowest actual regression and select algorithm. Model-based unity, whereby every model is constructed using distinct data and by combining modeling results, is another strategy the author employed.

A researcher created a pair DL architecture for retinal disease diagnostics [29]. A deep segmentation technique was developed in stage 1 to produce a segmented picture from three-dimensional visual Coherent Computed images. Later, a CNN is utilized in the initial step to identify the patients by analyzing the segmented pictures. In an eye care clinic, the system is utilized on patients and is essentially taught on 14,884 images. This double-layered CNN produced better results than a single-layered CNN. There are still difficulties in applying AI to health-care image examination. DL architectures which have been taught on natural pictures are crucial to the interpretation of these medical images. Occasionally, these medical photos are further refined, which improves the model's classification performance.

As a result, raw photos captured with a phone or camera and under various lighting conditions have an impact on the final product. We frequently discuss how crucial it is to safeguard the confidentiality and integrity of public health data, particularly patient-specific data. Data security is at danger due to the growing use of AI in medical domains.

4.5 LIMITATION AND DRAWBACKS

Despite the significant advancements in health care, DL has several obstacles and disadvantages. The quantity of data is an essential component needed for DL to function. NNs need a lot of input in order to function, and they also need a lot of network parameters. Usually, ten times as many parameters are needed for each NN as there are samples. In the medical field, privacy issues frequently prevent many patients from cooperating to submit data. Additionally, compared to other areas of AI, comprehending the individual diversity of illness is far more difficult [30]. While the data from health care is unclear, noisy, and lacking, the data from other AI domains, such as voice, language, and vision, will contain clean and organized data. Applying AI in health care is therefore a little more challenging than in other domains, although we have made excellent progress in recent years.

It is also necessary to address the problem of physicians' slow acceptance of new technology and its application in health care. Applications for health care must be simple to use in order for medical practitioners to embrace them.

4.6 CONCLUSION

The study shows how AI can be applied in health care. In many aspects, AI will soon be more advantageous, which will enhance and speed up patient outcomes. We can provide appropriate treatment by using AI, ML, and DL to aid with operations and early illness diagnosis, such as cancer. This study also discusses certain elements that should be considered when doing AI research. The application of AI in health care is probably going to grow significantly because of recent developments in the sector, government funding, and support. Furthermore, there is a great deal of space for improving the quality of health-care services and cutting costs.

AI trends for the future indicate that health care will continue to innovate. IoT and AI will be combined by the Internet of Medical Things (IoMT) to improve patient monitoring and real-time data processing. By introducing localized processing to wearable and smart devices, Edge AI will enable more rapid and individualized health-care solutions. Complex medical issues can be resolved with the help of quantum computing, leading to advances in diagnosis and therapy. AI's contribution to genomics will speed up personalized medical research by interpreting genetic data to customize treatments. Additionally, by offering resources for early diagnosis, intervention, and continuous support for mental health conditions, AI will be crucial to mental health.

REFERENCES

[1] Weng, J., 2018. Autonomous mental development. In *Intelligent systems* (pp. 31). CRC Press.

[2] Wooldridge, M. and Jennings, N.R., 1995. Intelligent agents: Theory and practice. *The knowledge engineering review*, *10*(2), pp. 115–152.

[3] Huang, G., Huang, G.B., Song, S. and You, K., 2015. Trends in extreme learning machines: A review. *Neural Networks*, *61*, pp. 32–48.

[4] Hopfield, J.J., 1982. Neural networks and physical systems with emergent collective computational abilities. *Proceedings of the National Academy of Sciences*, *79*(8), pp. 2554–2558.
[5] Watts, D.J. and Strogatz, S.H., 1998. Collective dynamics of small-world networks. *Nature*, *393*(6684), pp. 440–442.
[6] Zucker, R.S. and Regehr, W.G., 2002. Short-term synaptic plasticity. *Annual Review of Physiology*, *64*(1), pp. 355–405.
[7] Lee, S.I., Celik, S., Logsdon, B.A., Lundberg, S.M., Martins, T.J., Oehler, V.G., Estey, E.H., Miller, C.P., Chien, S., Dai, J. and Saxena, A., 2018. A machine learning approach to integrate big data for precision medicine in acute myeloid leukemia. *Nature Communications*, *9*(1), p.42.
[8] Eren, A., Subasi, A. and Coskun, O., 2008. A decision support system for telemedicine through the mobile telecommunications platform. *Journal of Medical Systems*, *32*, pp. 31–35.
[9] Ozekes, S. and Erguzel, T.T., 2014. Artificial intelligence approaches in psychiatric disorders. *The Journal of Neurobehavioral Sciences*, 1(2), p.52.
[10] Manne, R., 2021. Machine learning techniques in drug discovery and development. *International Journal of Applied Research*, *7*(4), pp. 21–28.
[11] Musk, E., 2019. An integrated brain-machine interface platform with thousands of channels. *Journal of Medical Internet Research*, *21*(10), p.e16194.
[12] Jiang, F., Jiang, Y., Zhi, H., Dong, Y., Li, H., Ma, S., Wang, Y., Dong, Q., Shen, H. and Wang, Y., 2017. Artificial intelligence in healthcare: Past, present and future. *Stroke and Vascular Neurology*, *2*(4), pp. 230–243.
[13] Weng, J., McClelland, J., Pentland, A., Sporns, O., Stockman, I., Sur, M. and Thelen, E., 2001. Autonomous mental development by robots and animals. *Science*, *291*(5504), pp. 599–600.
[14] Hopfield, J.J., 1982. Neural networks and physical systems with emergent collective computational abilities. *Proceedings of the National Academy of Sciences*, *79*(8), pp. 2554–2558.
[15] Coiera, E., 1997. *Guide to medical informatics, the internet and telemedicine*. Chapman & Hall, Ltd.
[16] Marr, B., 2017. First FDA approval for clinical cloud-based deep learning in healthcare. In *Forbes*. Forbes Publishing Company.
[17] Manne, R. and Kantheti, S.C., 2021. Application of artificial intelligence in healthcare: chances and challenges. *Current Journal of Applied Science and Technology*, *40*(6), pp. 78–89.
[18] Wukkadada, B. and Saiswani, V.P., 2000. Online healthcare system using cloud computing and artificial intelligence. *IOSR Journal of Computer Engineering*, *20*, pp.S40–S3.
[19] Mesko, B., 2017. The role of artificial intelligence in precision medicine. *Expert Review of Precision Medicine and Drug Development*, *2*(5), pp. 239–241.
[20] Lieberman, H. and Mason, C., 2002. Intelligent agent software for medicine. In *Future of health technology* (pp. 99–109). IOS Press.
[21] Goldman, L.W., 2007. Principles of CT and CT technology. *Journal of Nuclear Medicine Technology*, *35*(3), pp. 115–128.
[22] Cerwall, P., Jonsson, P., Möller, R., Bävertoft, S., Carson, S. and Godor, I., 2015. Ericsson mobility report: On the pulse of the networked society. *Ericsson*, 1–84.
[23] Hosny, K.M., Kassem, M.A. and Foaud, M.M., 2019. Classification of skin lesions using transfer learning and augmentation with Alex-net. *PloS One*, *14*(5), p.e0217293.
[24] Li, Y. and Shen, L., 2018. Skin lesion analysis towards melanoma detection using deep learning networks. *Sensors*, *18*(2), p.556.
[25] Fujisawa, Y., Otomo, Y., Ogata, Y., Nakamura, Y., Fujita, R., Ishitsuka, Y., Watanabe, R., Okiyama, N., Ohara, K. and Fujimoto, M., 2019. Deep-learning-based, computer-aided classifier developed with a small dataset of clinical images surpasses board-certified dermatologists in skin tumour diagnosis. *British Journal of Dermatology*, *180*(2), pp. 373–381.
[26] Han, S.S., Kim, M.S., Lim, W., Park, G.H., Park, I. and Chang, S.E., 2018. Classification of the clinical images for benign and malignant cutaneous tumors using a deep learning algorithm. *Journal of Investigative Dermatology*, *138*(7), pp. 1529–1538.
[27] Chekroud, A.M., Zotti, R.J., Shehzad, Z., Gueorguieva, R., Johnson, M.K., Trivedi, M.H., Cannon, T.D., Krystal, J.H. and Corlett, P.R., 2016. Cross-trial prediction of treatment outcome in depression: a machine learning approach. *The Lancet Psychiatry*, *3*(3), pp. 243–250.

[28] Mankoo, P.K., Shen, R., Schultz, N., Levine, D.A. and Sander, C., 2011. Time to recurrence and survival in serous ovarian tumors predicted from integrated genomic profiles. *PloS One*, *6*(11), p.e24709.

[29] De Fauw, J., Ledsam, J.R., Romera-Paredes, B., Nikolov, S., Tomasev, N., Blackwell, S., Askham, H., Glorot, X., O'Donoghue, B., Visentin, D. and Van Den Driessche, G., 2018. Clinically applicable deep learning for diagnosis and referral in retinal disease. *Nature Medicine*, *24*(9), pp. 1342–1350.

[30] Wang, F. and Preininger, A., 2019. AI in health: state of the art, challenges, and future directions. *Yearbook of Medical Informatics*, *28*(01), pp. 016–026.

Chapter 5

Drone-enabled IoT-integrated smart health-care system for reducing ambulance delays and mortality in emergency conditions

Dhruv Dhayal, Manzoor Ansari, and Masood Alam

5.1 INTRODUCTION

Traffic congestion has evolved from a mere inconvenience to a critical public health crisis that directly increases the risk of human death by causing significant time delays in emergency response. In countries like India, where vehicle registration has increased by **over 180%** in the past decade, ambulances frequently become trapped in congested traffic across highways and inter-city routes, leading to preventable deaths and deteriorating critical conditions. Emergency medical systems operate under the golden hour principle, where the **first 60 minutes** are crucial for preventing human death and reducing the severity of critical conditions [1, 2]. However, traditional emergency response mechanisms lack real-time traffic intelligence and adaptive routing capabilities. Studies indicate that every minute of delay increases mortality **risk by 7–10%**, making traffic-induced time delays a critical factor in preventable deaths across all types of congested roads. The emergence of Internet of Things (IoT) technologies and **Unmanned Aerial Vehicle (UAV)** systems presents unprecedented opportunities to revolutionize emergency health-care delivery by significantly reducing time delays and minimizing the risk of human death during critical medical emergencies. IoT-based infrastructure enables real-time traffic monitoring, while UAV technology offers aerial surveillance and rapid response deployment that can dramatically reduce response times and improve patient outcomes. Current emergency response systems primarily rely on reactive approaches without real-time traffic intelligence across diverse road networks [3]. These systems lack technological sophistication to minimize time delays and reduce human death risk by addressing dynamic traffic patterns on highways, arterials, and inter-city routes where congestion occurs unpredictably due to accidents, construction, weather conditions, or peak traffic hours [4].

This research proposes a comprehensive drone-enabled IoT-integrated smart health-care system specifically designed to minimize time delays, reduce the risk of human death, and improve outcomes for patients in critical conditions across all types of congested road networks. The proposed system combines real-time aerial surveillance, predictive traffic modeling, active traffic clearance protocols, and emergency intervention strategies into a cohesive technological framework capable of dramatically reducing ambulance response times. The significance of this research extends beyond technological innovation to address fundamental life-saving challenges. Traffic-induced time delays directly contribute to increased risk of human death and deterioration of critical medical conditions across all road networks. By developing UAV-based IoT-integrated solutions that dramatically reduce response times, this research contributes to minimizing the risk of human death and reducing possibilities of critical conditions worsening due to emergency response delays [5, 6]. The proposed

DOI: 10.1201/9781003685890-5

system's multi-layered approach incorporates preventive route planning through continuous traffic monitoring, active traffic management during emergency response deployment, and emergency intervention protocols for trapped ambulances. This comprehensive approach addresses the full spectrum of traffic-related challenges faced by emergency medical services while providing measurable improvements in patient outcomes [7, 8].

5.1.1 Problem statement

Urban ambulance response is critically hampered by traffic congestion, unclear or outdated routing information, and limited situational awareness at street level. These delays increase the probability of preventable mortality in time-sensitive conditions (**e.g., cardiac arrest, severe trauma, stroke**). Existing GPS-based navigation and siren-based clearance methods are reactive and often insufficient in dense urban contexts [1, 7]. There is a need for a proactive, low-latency, aerial-ground integrated system that can scout routes ahead of an ambulance, detect and predict congestion or blockages within a defined operational radius, and deliver real-time actionable guidance and traffic clearance coordination to reduce ambulance travel time and associated death risk [9].

5.1.2 Objective and its significance

Create and assess a Drone-Enabled IoT-Integrated Decision Support System that proactively detects low-traffic routes and facilitates coordinated traffic clearance to cut down ambulance time-to-hospital, ultimately mitigating the death risk for time-sensitive patients.

1. Define an operational UAV scouting radius (km) and communication/latency requirements suitable for urban ambulance support.
2. Fuse aerial UAV data with IoT ground sensors (smart signals, roadside detectors, V2X messages) for real-time congestion detection.
3. Develop an artificial intelligence (AI)-driven routing module that recommends dynamic, safety-aware rerouting to ambulance drivers and control centers.
4. Propose protocols for prioritized traffic-signal control and traffic-authority alerts triggered by the system.
5. Simulate and quantify the system's impact on ambulance transit time and estimated mortality reduction under representative urban scenarios.
6. Identify regulatory, safety, and cyber security measures required for deployment.

5.2 RELATED WORK AND PAST INSIGHTS

As shown in Table 5.1, existing systems provide either traffic monitoring or basic emergency vehicle priority, but none integrate UAV-based active traffic clearance with IoT networks specifically for ambulance route optimization. Table 5.2 illustrates the technology evolution timeline, demonstrating how our research represents the next evolutionary step by combining real-time aerial intervention with predictive AI algorithms for comprehensive proactive emergency response [1, 2].

Table 5.1 Comparative analysis between existing systems and proposed solution [1, 2, 10, 12]

Research	*Year*	*Technology used*	*Scope*	*Limitations*	*Our advancement*
IoT-based Traffic Management Chen et al., 2018 [10]	2018	IoT sensors, Cloud computing	Traffic Monitoring	Ground-level monitoring only No emergency-specific protocols Limited real-time response	Aerial + Ground monitoring + Emergency-specific AI algorithms
Drone-assisted Traffic Surveillance Kumar & Singh, 2019 [9]	2019	Single UAV, Basic video streaming	Traffic Monitoring	Surveillance only; no intervention No IoT integration	Multi-UAV coordination and Active traffic management and Emergency intervention protocols
Emergency Vehicle Preemption [6]	2020	RFID, Traffic-signal control	Signal Priority	Limited to signal-controlled intersections No highway/arterial road coverage	All road types coverage and Predictive AI and Dynamic route optimization
AI-powered Traffic Prediction Liu & Wang, 2021 [7]	2021	ML, Historical data	Traffic Prediction	Prediction only, no active management	Prediction + Active intervention + Emergency-specific deployment
IoT-based Emergency Response Rodriguez & Martinez, 2022 [5]	2022	UAV delivery, Basic routing	Medical Supply Delivery	Supply delivery focus only No traffic management	Traffic management + Route optimization + Multi-modal emergency support
Integrated Smart City Emergency Response Anderson et al., 2023 [11]	2023	IoT, AI, Cloud platforms	City-Wide Integration	Conceptual framework only No specific ambulance focus	Practical implementation + Ambulance-specific protocols

Table 5.2 Technology evolution over timeline with rapid advancement [12–14]

Era	*Technology focus*	*Capability level*	*Emergency response impact*
2010–2015	GPS Navigation	Static Routing	Minimal—standard navigation only
2016–2018	IoT Integration	Monitoring capabilities	Low—data collection focus
2019–2021	AI and ML Integration	Predictive Analytics	Medium—prediction without intervention
2022–2023	Drone Technology Integration	Surveillance and Delivery	Medium-High—limited traffic management
2024 (Present Our Research)	Integrated UAV-IoT-AI System	Proactive intervention and management	High—comprehensive emergency response optimization

5.3 SYSTEM DESIGN FRAMEWORK AND WORKFLOW IMPLEMENTATION

The smart drone-enabled IoT health-care system we propose is targeted at health-care emergencies where every second of delay could escalate mortality risks. The system functions in two complementary phases: **proactive scouting (Phase 1) and reactive support (Phase 2).** Together, these emergency response phases foster uninterrupted monitoring, instantaneous and automated decision-making, and the intelligent clearance of emergency medical service traffic.

Phase 1, Proactive Scouting and Preventive Navigation, begins with the deployment of **UAVs** as soon as an ambulance is called. The drones cover an effective radius of 2 to 5 kilometers and scan the road to check for congestion, any obstructions, and even mobility-impacting environmental conditions such as heavy rain or disturbed climate. UAVs feature onboard vision systems and lightweight edge-computing modules that capture real-time traffic metrics, such as the number of vehicles and the length of queues [5, 6]. This data, which is also supplemented by data from roadside IoT sensors, CCTV cameras, and historical congestion data, is sent to a Decision Support System (DSS), which operates on either an edge or cloud infrastructure. The DSS integrates the collected data. The UAV is not only responsible for aerial traffic monitoring but also for **intelligent obstacle detection and recognition** within its operational radius. Using onboard cameras integrated with AI and machine learning (ML) algorithms, particularly **Convolutional Neural Networks (CNNs)**, the UAV can identify and classify objects such as moving vehicles, pedestrians, roadblocks, and environmental obstacles [7]. These models are trained on diverse traffic datasets to ensure robustness across different conditions, including varying lighting and weather scenarios. Once detected, the system estimates the **number of objects, their velocity, and their relative distance** from the ambulance's route [8, 15]. This object-level situational awareness allows the DSS to not only choose the optimal traffic-free lane but also to dynamically renavigate pathways if sudden obstructions or moving hazards appear. The analyzed traffic data are transmitted in real time to both the **ambulance driver and the UAV pilot interface,** enabling proactive navigation assistance and improved traffic surveillance capabilities. Drone-assisted monitoring techniques enhance situational awareness in intelligent transportation systems, while IoT–cloud-based predictive analytics support data-driven operational decision making in dynamic environments (Figure 5.1) [9, 16].

To ensure effective navigation support for ambulances in urban traffic environments, the proposed system integrates **AI** and **ML** models with UAVs. The framework employs a **CNN**-based model, specifically the You Only Look Once (**YOLO**) **family of detectors,** to perform

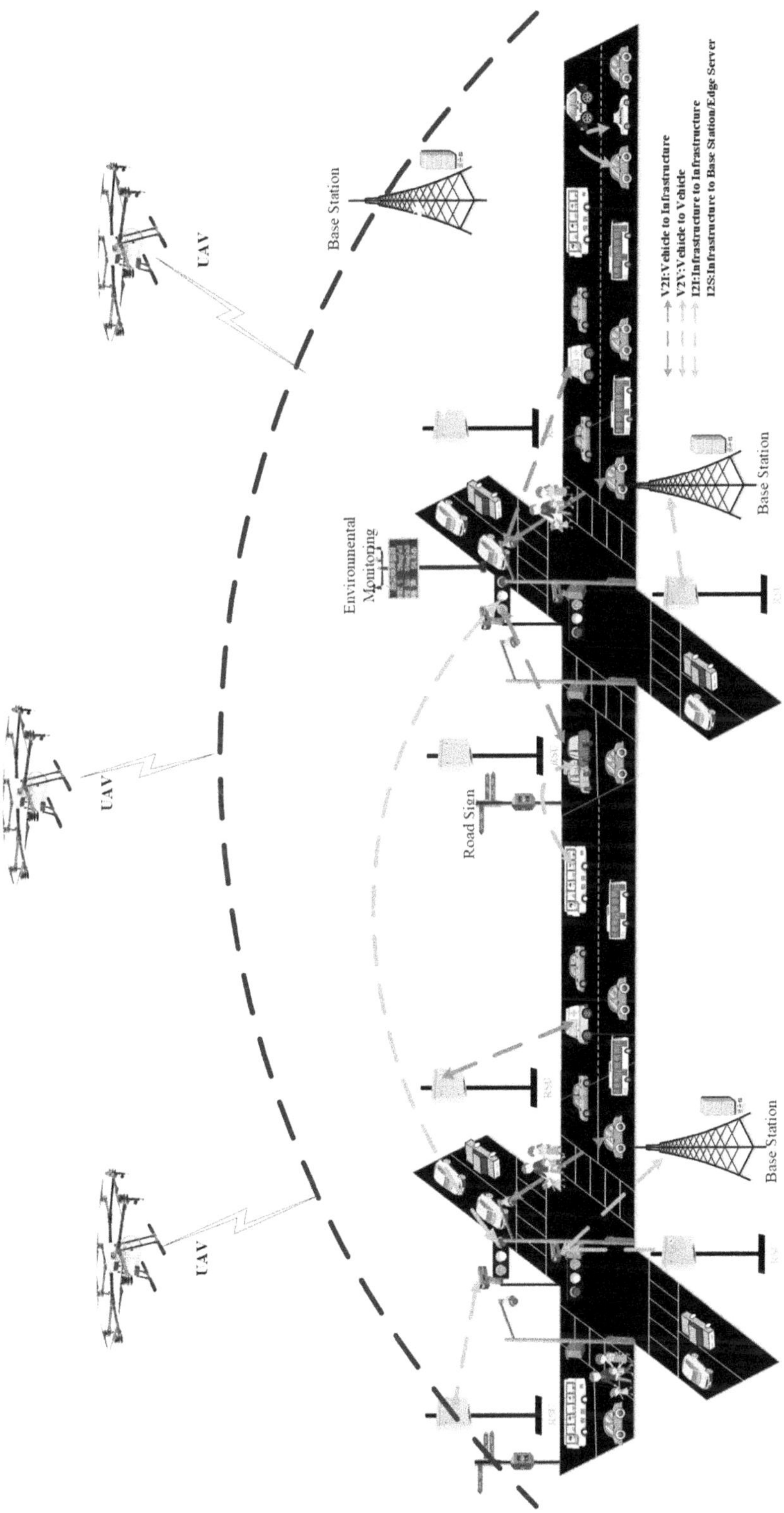

Figure 5.1 Drone-enabled AI/IoT system for real-time ambulance pathway optimization [5, 9].

real-time object detection from aerial perspectives [2, 3]. By training the model on a **merged dataset** of traffic scenes—including drone imagery, dashcam footage, and traffic sign datasets—the system can accurately identify and classify vehicles, obstacles, and road signs under diverse conditions such as rain, fog, or low visibility. To enhance the accuracy of moving object monitoring, YOLO is coupled with **multi-object tracking algorithms (e.g., DeepSORT or ByteTrack)**. These algorithms not only track the position of detected vehicles but also estimate their **velocity, direction, and density** [4, 5]. The continuous tracking allows the UAV to generate traffic parameters such as average vehicle speed, lane occupancy, congestion level, and stop-and-go ratios within its operational range.

The extracted traffic features are then processed using a **decision tree-based classifier**. This decision-making layer categorizes traffic states into three conditions: *free flow, moderate flow*, or *congested*. The classification is derived from factors such as vehicle density, mean speed thresholds, and the proportion of slow-moving or halted vehicles. In scenarios where congestion is predicted with high confidence, the system automatically recommends an **alternate route with reduced traffic density**. The decision-support signals are transmitted in real time to the ambulance driver or UAV pilot through a **cloud-integrated navigation interface**. This enables immediate rerouting and minimizes time delays caused by traffic bottlenecks, thereby reducing the mortality risk for patients in critical conditions [13, 17]. Additionally, the feedback loop ensures continuous learning, as traffic and detection data are stored for future **model retraining** and refinement of decision rules [10]. By integrating real-time convolutional neural network (CNN)–based object detection techniques such as You Only Look Once (YOLO), vehicle tracking, and decision tree–based traffic analysis, the proposed system can identify congestion patterns and dynamically guide ambulances along optimal routes. The fusion of UAV-enabled traffic surveillance with AI-driven decision support contributes to scalable smart-city emergency response frameworks and intelligent IoT-cloud transportation management systems aimed at improving emergency service efficiency (Figure 5.2) [11, 12].

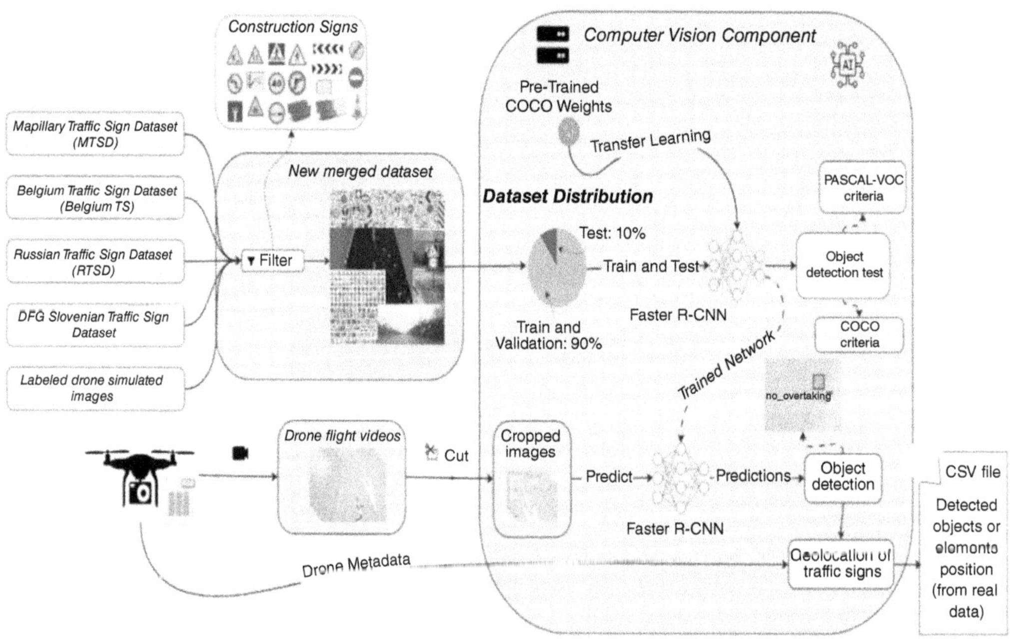

Figure 5.2 Systematic workflow approach by using various technologies for proper analysis in decision-making [18, 19].

5.4 SYSTEMATIC APPROACH TO MONITOR AND NAVIGATION PATHWAYS IN PHASE I (PROACTIVE APPROACH)

The proposed framework utilizes a **UAV-assisted computer vision system** to enhance real-time detection and geo-referencing of environmental constraints within road networks. The system operates in two core environments: **data acquisition from the construction or road site** and **computer vision processing for intelligent analysis.** In the first stage, UAVs are deployed to capture aerial imagery and video streams from the road infrastructure. These UAVs act as mobile sensing units, gathering **real-time information on road conditions, traffic signs, and temporary obstacles** such as construction barriers or blockages. The collected data, along with UAV metadata (**altitude, GPS coordinates, and orientation**), is stored for further processing [2, 5].

In the second stage, the stored information is processed through an **AI-driven computer vision module.** This module integrates multiple datasets such as **digital terrain models, simulated images, traffic sign databases, and real road images** to build a robust learning environment. Advanced deep learning models (e.g., **CNN-based object detection frameworks**) are trained using transfer learning approaches, allowing accurate identification of traffic signs and obstacles even under challenging environmental conditions. Once identified, each traffic sign or obstacle is **classified according to its type** (e.g., **speed limit, diversion, no entry**) and mapped with **precise geo-referenced coordinates** [16, 20]. This spatial information is stored in structured formats (e.g., CSV files) for real-time utilization. The integration of geo-referenced data ensures that navigation systems, particularly for emergency vehicles such as ambulances, can dynamically adapt routes to avoid congested or hazardous areas [7, 9]. Thus, the UAV-assisted framework combines **aerial surveillance, AI-powered detection, and spatial mapping** into a unified system, enabling intelligent decision-making for traffic navigation, risk reduction, and time-sensitive health-care logistics (Figure 5.3).

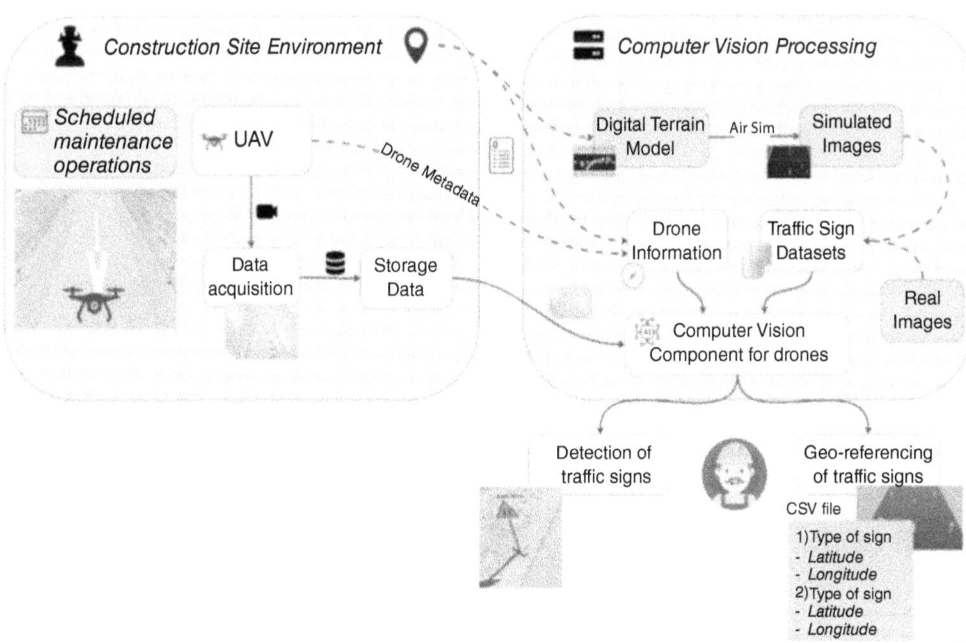

Figure 5.3 Systematic workflow approach by using various technologies for proper analysis in decision-making [18].

5.5 NAVIGATIONAL PATHWAY DETECTION AFTER PROPER ANALYSIS AND DECISION-MAKING PHASE 2 (REACTIVE APPROACH)

In contrast to **the preventive measures of Phase 1, Phase 2 adopts a reactive approach** in scenarios where the ambulance becomes stuck in heavy traffic or encounters unexpected blockages during transit. In such cases, the **UAV-assisted IoT system** leverages **real-time data acquisition, AI-driven prediction models, and cloud-based analytics** to provide immediate situational awareness and dynamic navigation support [3]. Once the UAV detects that the ambulance is immobilized within a congested traffic cluster, the system activates an intelligent **traffic analysis module** to evaluate congestion patterns and support adaptive routing decisions within an IoT-enabled smart transportation framework **[14, 21, 22]**. Using deep learning models such as **YOLO for object detection** and **decision tree classifiers for traffic flow prediction**, the UAV detects the number of vehicles, their speed, and density within a specified **range (in kms)**. Based on these observations, the system computes the estimated **queue length, congestion depth, and average traffic clearance time** [4, 20].

The UAV then communicates this information directly to the **ambulance driver/pilot interface**, presenting two critical insights:

1. **Real-Time Traffic Status:**
   The system visualizes the **distance of traffic congestion (in km)** and provides a predictive estimate of how long the blockage will last before clearance. This is computed using ML algorithms trained on historical datasets of traffic patterns under varying weather and environmental conditions.
2. **Alternate Route Suggestion:**
   If the estimated clearance time exceeds the predefined threshold for critical patient conditions, the UAV instantly computes an **optimal alternative pathway.** This route is selected by evaluating surrounding lanes, side roads, or emergency corridors, ensuring **minimal delay** and significantly reducing mortality risk.

The integration of this **reactive decision-making mechanism** ensures that ambulances are never static for prolonged durations without intelligent support (Figures 5.4–5.6) [5, 6].

5.6 PREVENTIVE MEASURES BY OVERCOMING CHALLENGES

In contrast, preventive measures form the **proactive layer** of the proposed drone-enabled IoT-integrated smart health-care system. The main objective of these measures is to **anticipate traffic bottlenecks and environmental obstacles** before the ambulance actually encounters them, thereby **minimizing delays and reducing patient mortality risk** [3, 4]. Unlike the reactive approach that comes into play once the ambulance is already stuck, preventive measures are focused on **forecasting risks** and **ensuring smooth navigation** from the very beginning of the emergency response [5, 17].

The UAVs, integrated with IoT sensors and cloud-based DSS, continuously monitor the road environment ahead of the ambulance. Through **real-time aerial surveillance** combined with **historical traffic data analytics**, the UAV system identifies possible congestion points, roadblocks, weather-related risks, and even moving obstacles like construction machinery or stalled vehicles [6, 22]. This predictive capability is strengthened by **AI and ML models,**

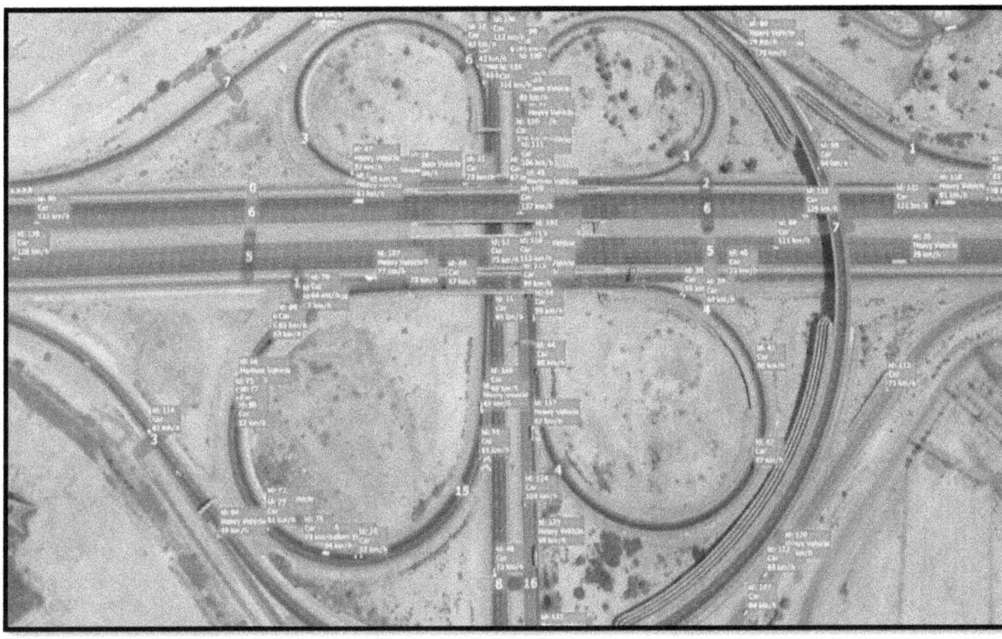

Figure 5.4 Decision-making by proper analysis the speed and number of vehicles in order to find the best optimal pathway [4].

particularly **CNNs** and **predictive time-series analysis**, which analyze live video feeds and sensor inputs to **detect anomalies well in advance** [6, 7].

- Detect **traffic build-ups several kilometers ahead** and recommend alternate routes before the ambulance reaches the critical zone.
- Predict **weather-related hazards** such as rainfall, fog, or road slipperiness that may cause unexpected delays.
- Identify **obstacles and moving objects** (e.g., vehicles changing lanes, debris on roads) within the set detection range using AI-driven object detection models like **YOLO** or **Faster R-CNN**.
- Inform drivers of **dynamic lane availability** and suggest optimal pathways that minimize sudden braking or rerouting.

These preventive actions are communicated directly to the ambulance driver or pilot via a **real-time navigation dashboard.** The dashboard visualizes upcoming risks and provides **intelligent recommendations** that can be acted upon immediately. In case multiple routes are available, the **UAV-assisted system** evaluates each based on **distance, traffic density, and estimated travel time,** then prioritizes the one with the **lowest risk and fastest arrival time** [10, 11]. By addressing potential delays **before they manifest,** preventive measures act as a **life-saving intervention strategy.** They ensure that patients in critical conditions are transported with the **least possible latency,** reducing the likelihood of deterioration or death during transit. This makes the preventive framework not only a technological innovation but also a **public health necessity** in modern emergency health-care systems [9, 19].

Figure 5.5 Reactive approach II—proper navigation and suggest optimal pathway to avoid traffic and give best suitable optimal pathway to reduce the time delay for patient's critical conditions or reduce the possibilities of death risk [10, 12].

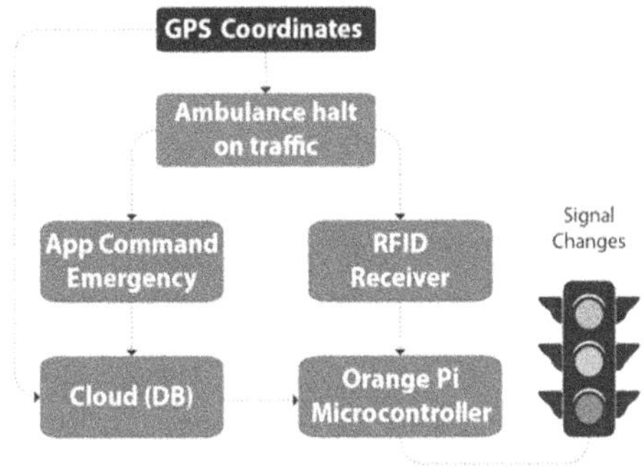

Figure 5.6 Reactive approach II—data flow idea-representation [13].

5.7 FUTURE SCOPE

The proposed drone-enabled **IoT-integrated smart health-care** framework provides a foundation for reducing delays in ambulance navigation and ultimately lowering patient mortality in emergency conditions. However, there remains considerable scope for future research and development in this domain. With the evolution of advanced communication technologies such as **5G** and beyond, **UAVs** will be able to transmit high-definition video streams and real-time sensor data with ultra-low latency, thereby improving the accuracy and reliability of navigation decisions. Furthermore, the integration of such systems with smart city infrastructure, including intelligent traffic lights and connected road sensors, can create coordinated emergency corridors where ambulances receive priority clearance across entire urban networks.

Another potential future direction is the use of swarm **UAV technology**, where multiple drones operate collaboratively to cover larger zones and ensure redundancy in obstacle detection and traffic monitoring. The incorporation of edge AI is also anticipated to play a significant role by enabling lightweight ML models to be deployed directly on **UAVs**, thus reducing the reliance on cloud servers and allowing decision-making to take place onboard in real time. Additionally, the framework can be expanded beyond navigation support to predictive health-care response, where drones not only guide ambulances but also deliver medical supplies, monitor patient vitals, and share biometric data with hospitals during transit.

Finally, the adaptability of this system in diverse geographic contexts such as rural areas, disaster-prone regions, or mountainous terrains offers promising opportunities for global deployment. By evolving into a comprehensive emergency support ecosystem, the framework has the potential to transform ambulance response into a more **intelligent, coordinated, and life-saving process.**

5.8 CONCLUSION

This chapter has presented a drone-enabled **IoT-integrated health-care system** that addresses the critical challenge of ambulance delays caused by traffic congestion. The framework combines **UAVs, AI, and real-time data analytics** to provide intelligent support for ambulance navigation. By employing deep learning models such as **CNN** and **YOLO**, the system is capable of detecting obstacles, tracking moving vehicles, and analyzing traffic density in real time. The decision-making process further incorporates predictive models and classification techniques to determine the severity of congestion and to recommend optimal pathways that minimize delays. The reactive phase, on the other hand, responds when the ambulance is already stuck in traffic by estimating the length and clearance time of congestion while simultaneously suggesting alternate routes.

In conclusion, the proposed system demonstrates how the integration of **UAVs, IoT, and AI** can significantly improve the efficiency of emergency medical services. Looking ahead, with advancements in connectivity, smart infrastructure, and UAV collaboration, this system holds the potential to evolve into a global standard for emergency response, ultimately saving countless lives.

REFERENCES

1. Godwin, J. J., Krishna, B. S., Rajeshwari, R., Sushmitha, P., & Yamini, M. (2021). IoT Based Intelligent Ambulance Monitoring and Traffic Control System. In *Internet of Things in Biomedical and Cyber-Physical Systems*. Springer.
2. Wani, M. M., Khan, S., & Alam, M. (2020). IoT Based Traffic Management System for Ambulances. arXiv preprint.
3. Shimizu, Y., Ito, T., Sato, T., & Nishiyama, H. (2019). Real-time vehicle recognition and tracking using YOLOv3 and DeepSORT on edge devices for traffic flow detection in intelligent transportation systems (using Jetson TX2). In *Proceedings of the 2019 IEEE International Conference on Consumer Electronics (ICCE)*, Las Vegas, NV, pp. 1–6. IEEE.
4. Cárdenas-Benítez, N., Aquino-Santos, R., Magaña-Espinoza, P., Aguilar-Velazco, J., Edwards-Block, A., & Medina Cass, A. (2016). Traffic Congestion Detection System through Connected Vehicles and Big Data. *Sensors*, 16(5), 599.
5. Rodriguez, L., & Martinez, J. (2022). Drone-Based Emergency Response System for Urban Healthcare Services. *Sensors* 22(14), 5287–5299.
6. Ma, X., Zhou, T., Ma, J., Jiang, G., & Xu, X. (2022). UAV Aerial Photography Traffic Object Detection Based on Lightweight Design and Feature Fusion. In *Chinese Conference on Biometric Recognition* (pp. 684–693). Springer Nature Switzerland.
7. Liu, Y., & Wang, H. (2021). AI-Powered Traffic Prediction Using Deep Learning Models. *IEEE Transactions on Intelligent Transportation Systems* 22(11), 6998–7010.
8. Revathy, G., Thangavel, M., Senthilvadivu, S., & Savithri, M. C. (2025, February). Enabling smart cities: AI-powered prediction models for urban traffic optimization. In *2025 4th International Conference on Sentiment Analysis and Deep Learning (ICSADL)* (pp. 904–908). IEEE.
9. Kumar, A., & Singh, R. (2019). Drone-Assisted Traffic Surveillance for Intelligent Transportation Systems. *Journal of Intelligent & Fuzzy Systems* 36(5), 4821–4832.
10. Chen, L., Gupta, V., Rahman, M., & Lee, J. (2018). IoT-Based Traffic Management for Smart Transportation Systems. *International Journal of Communication Networks and Distributed Systems* 21(4), 393–410.
11. Anderson, J., Kumar, R., Li, H., & Singh, P. (2023). Integrated Smart City Emergency Response. *Journal of Smart Cities and Society* 2(1), 45–59.
12. Alam, M. (2023). IoT-Cloud-Enabled Smart Framework for Real-World Applications. In *Intelligent Systems and Applications: Select Proceedings of ICISA 2022* (pp. 87–105). Springer Nature Singapore.
13. Ansari, M., & Alam, M. (2024). A Novel Hybrid Time-Series Approach for IoT-Cloud-Enabled Environment Monitoring. *Journal of Supercomputing*, 80(7), 9019–9053.
14. Ansari, M., Ali, S. A., & Alam, M. (2022). Internet of Things (IoT) Fusion with Cloud Computing: Current Research and Future Direction. *International Journal of Advanced Technology and Engineering Exploration*, 9(97), 1812.
15. Mughal, M. F., Ansari, M., & Alam, M. (2025). Revolutionizing Financial Forecasting: The Rise of Transformer Models in Long-Term Time Series Analysis. In *AI-Driven Finance in the VUCA World* (pp. 142–160). Auerbach Publications.

16. Ansari, M., & Alam, M. (2024). An Intelligent IoT-Cloud-Based Air Pollution Forecasting Model Using Univariate Time-Series Analysis. *Arabian Journal for Science and Engineering*, 49(3), 3135–3162.
17. Noor, M. A. F., Khanum, S., Anwar, T., & Ansari, M. (2021). A Holistic View on Blockchain and Its Issues. In *Blockchain Applications in IoT Security* (pp. 21–44). IGI Global.
18. Naranjo, M., Fuentes, D., Muelas, E., Diez, E., Ciruelo, L., Alonso, C., … & Luengo, I. (2023). Object detection-based system for traffic signs on drone-captured images. *Drones*, 7(2), 112.
19. Mahaur, B., Singh, N., & Mishra, K. K. (2022). Road object detection: a comparative study of deep learning-based algorithms. *Multimedia Tools and Applications*, *81*(10), 14247–14282
20. Mahaur, B., Singh, N., & Mishra, K. K. (2022). Road object detection: A comparative study of deep learning-based algorithms. *Multimedia Tools and Applications*, 81(10), 14247–14282
21. Ansari, M., Ali, S. A., Alam, M., Chaudhary, K., & Rakshit, S. (2023). Unlocking the Power of Explainable AI to Improve Customer Experiences in E-Commerce. In *AI-Based Data Analytics* (pp. 31–48). Auerbach Publications.
22. Ansari, M., & Alam, M. (2021). IoT-Cloud Enabled Statistical Analysis and Visualization of Air Pollution Data in India. In *Proceedings of Data Analytics and Management: ICDAM 2021, Volume 2* (pp. 125–139). Springer Singapore.

Chapter 6

Detecting cancer stages in patients using machine learning integrated with Classification Threshold Tuner

Shahzeb Khan, Syed Arshad Ali, Mohd Tajammul, Javed Ahmad, and Sharik Ahmad

6.1 INTRODUCTION

Figuring out what stage cancer is at is super important in health care. Because it helps doctors decide the best treatment for each patient. When the cancer stage is known, doctors can tailor their treatment plans. This way, patients get the best care suited to their needs, based on how serious and advanced their situation is.

Technology has gotten better and better. Machine learning (ML) is now a great tool that helps find cancer stages more accurately than older methods. In this study, we're excited to share a new ML method we created, its hybrid model combined with ClassificationThresholdTuner. This technique does two important things. First, it does cancer stage classification more accurately. Second, it gives doctors a solid tool to adjust model sensitivity and specificity.

Now, let's dive into sensitivity and specificity. The ClassificationThresholdTuner is here to solve a big puzzle in medical diagnosis. It helps find just the right balance between catching ALL relevant cases (i.e., sensitivity) and avoiding mix-ups, like false alarms (i.e., specificity). Why is this balance important? Well, if we miss early signs of cancer or wrongly say something harmless is cancer, it can really become dangerous in both physical and mental ways.

Our model uses something called threshold tuning. This smart approach helps it make better choices when figuring out different cancer stages. As a result, it lowers the chances of errors and gives us more reliable outcomes.

We also added a neat process to select the features that truly matter in our data. By focusing on the key features only, we reduce the extra noise and boost how well our model predicts outcomes. After pinpointing those important features, our hybrid classification method gets in. It adjusts its own thresholds on the fly—this means it reliably predicts across various cancer stages. The end result gives a system that adapts nicely to different datasets and continues to improve as conditions change.

Besides this use of ClassificationThresholdTuner with hybrid model, we also compared our method to traditional ML models. Our proposed model not only showed better accuracy but also made fewer mistakes than these other models.

By fine-tuning threshold settings, we helped our model handle all kinds of datasets reliably—even when the data changes.

In conclusion, our research offers a fresh and effective way to detect cancer stages, which could really boost personalized medicine. This means doctors could have a more precise tool for making choices about treatment. We believe this approach could change the game for cancer diagnostics by giving health-care professionals models that are accurate, dependable, and flexible. As ML keeps moving forward, using tools like ClassificationThresholdTuner might lead us to even more amazing improvements in medical diagnosis and patient care.

DOI: 10.1201/9781003685890-6

6.2 LITERATURE REVIEW

Lately, AI academia is really interested in how ML is transforming things in health care, especially when it comes to cancer diagnosis and staging. There is a lot more medical data available now in various medical domains, plus improvements in computing mean we can build advanced models to help predict cancer stages better. This is super important because knowing the stage helps in customizing treatments that can really make a difference for patients.

In this review, we're shining a light on some important contributions in this area. We'll look closely at ML techniques, how we can reduce dimensions, choosing the right features, and even touching on some of the ethical questions surrounding AI in health care [1–9]. Given It's an alarming time for these developments, ML algorithms are really powerful tools for finding cancer at different stages. The effectiveness of ensemble learning methods using models such as k-nearest neighbor (KNN) and Gradient Boosting was shown by Khoshgoftaar et al. (2019). His study found that these methods performed much better in terms of accuracy of prediction compared to traditional methods to identify the types of breast cancer [10]. This highlighted how ML could lead to progress in therapy. While dealing with billions of data points, it is important to reduce dimensions for visualization and feature selection. Techniques such as Principal Component Analysis (PCA), Linear Discriminant Analysis (LDA), and t-distributed stochastic neighbor embedding (T-SNE) help us understand complex datasets. Korkmaz et al. (2020) said that methods of reducing dimensions make it easy to understand data in important research areas such as medical treatment. He explained how features belong to each other and how different stages of cancer spread [11]. Feature selection is also necessary to improve the model by choosing reliable features. It reduces noise and organizes data effectively [12]. Garcia et al. (2020) has developed a hybrid classification model that combines several ML methods to make more accurate estimates of cancer stages. This task showed how a combination of different models can make classification more effective, especially with complex datasets where some characteristics may not be related to target variables [13]. Liu et al. (2021) presented the ClassificationThresholdTuner. This adjusts the decision limit during classification so that the forecast can be even better. This process actually helped increase the sensitivity and uniqueness in detecting cancer, which reduced problems like wrong positives and wrong negatives in diagnosis [14]. The recent advances in ethics in AI for health care are getting more importance and catching the eyes. Morley and team (2020) came up with the conclusion about the importance of transparency. They stressed the need to think about biases so that everyone can use AI technology in health care for cancer diagnosis fairly [15]. Wang and their team (2021) looked at how regular ML algorithms fail against deep learning techniques, thus stressing the need to improve ML [16]. It is found that traditional deep learning models often do better when fed with complex and big datasets. But to their surprise, these models were pretty hard to understand and needed a large sized labeled dataset for classification. O. T. Jones and his team (2021) worked on how important it is to ensure quality of data in order to make good predictions. Strong preprocessing methodologies, like cleaning, normalizing, and scaling the data, level up the quality of research and help in making models more reliable [17]. Ozer and his team (2020) concluded how support vector machines (SVM) can be very great for cancer diagnosis given its working [18]. Their study explicitly claims that SVM is really good at handling data that does not form a straight line, which came out to be very helpful in cancer stage detection where there are many complex relationships between features associated with target variables. Tran et al. (2021) checked how deep learning models operate when it comes to classifying the stages in cancer. According to them, deep learning techniques outperform traditional models when operated on large datasets [19]. Schork et al. (2019) studied how to customize medicine using AI. It is found that better treatment plans can be made using ML for individual patients

[20], which could result in better health-care services. Then, Prakash et al. (2020) discussed data privacy issues in AI framework. It highlights the need for moral data use, especially in health care where it is very important to keep the patient's information private [21]. Bambyar (2021) firmly argued for algorithm accountability in AI systems where transparency is questioned, especially in health care [22]. It is important to remove prejudices and ensure that the diagnosis and treatment recommendations of cancer are fair. Next, Morgenstetern et al. (2021) showed how AI can help in public health sectors with better diagnosis. His research showed how early diagnosis of patients can be obtained using AI in cancer care [23]. Reshan et al. (2023) and Ghavidel and others (2023) discussed how to handle imbalanced datasets when predicting cancer [24, 25]. He suggested using smart sampling methods so that the models could learn from balanced data; this step is extremely important for reliable forecasts. Iqbal (2021) provided intensive information on future prospects in oncology using AI. Iqbal hoped for significant progress in diagnosis and treatment of cancer, including genomics and integration of individual medicine [26]. Prosperry (2020) examined the importance of reasoning in AI applications [27]. This helps in increasing the accuracy of the forecast and provides insight into the factors affecting the patient's results.

Maleki Vernosfedrani et al. (2024) discussed the importance of creating patient-centered AI models, especially for decision-making in diseases like cancer. Research shows how AI can help individuals a lot in decision-making [1]. Gani et al. (2019) compared various forecast models to detect cancer stages. Their study focuses on how to choose the correct model and which parameters to use to evaluate ML algorithms [2]. Morley et al. and his team (2021) discussed the moral concerns related to the use of AI in cancer research. The task focused on how important it is to have guidelines and rules to assist AI in health care and how important it is to follow them globally [3]. Quazi and team (2024) studied how AI could master genomics data modeling. Crucial information was shared about how this integration might change cancer diagnostics and treatments by giving us information about genetic risks [4]. Van Leeuwen group (2020) showed the usage of AI in radiology to find and stage cancer. Their work showed that new imaging methods really help improve diagnosis accuracy [5]. Finally, Carter et al. (2022) [6] took a close look at creating ethical guidelines for AI in health care. They stressed how crucial it is to think about ethics when designing, using, and checking AI systems [7–9, 28, 29].

6.3 DATASET DESCRIPTION

The dataset comes from the Data World website. Based on patients' responses on how they feel on a scale of 1 to 10 on symptomatic factors. It has 1,000 rows and 25 columns. Each row tells us about a single patient's record. The columns show different features (refer to Figure 6.1) that help in predicting cancer stages. These features can be things like medical measurements, biological markers, and other clinical indicators that doctors often use. To make things simpler, we used a method called PCA. This helps cut down on the complexity of our data, plus it makes the ML model work faster. PCA helps us find the key components that keep the most important differences in the data, while also getting rid of some extras. After doing PCA, we looked at which features really mattered by checking their correlation. We decided to keep only those features with a correlation greater than 0.5. This way, we focused on a smaller set—10 features (refer to Figure 6.2) that are closely related to what we're trying to predict. These chosen features highlight the most important details that can help us understand cancer stages. Using PCA and checking for high correlation helped ensure that we kept only the key factors. This selection process reduced noise and made our model more accurate too. Now, this dataset has a neat structure that is perfect for predicting cancer

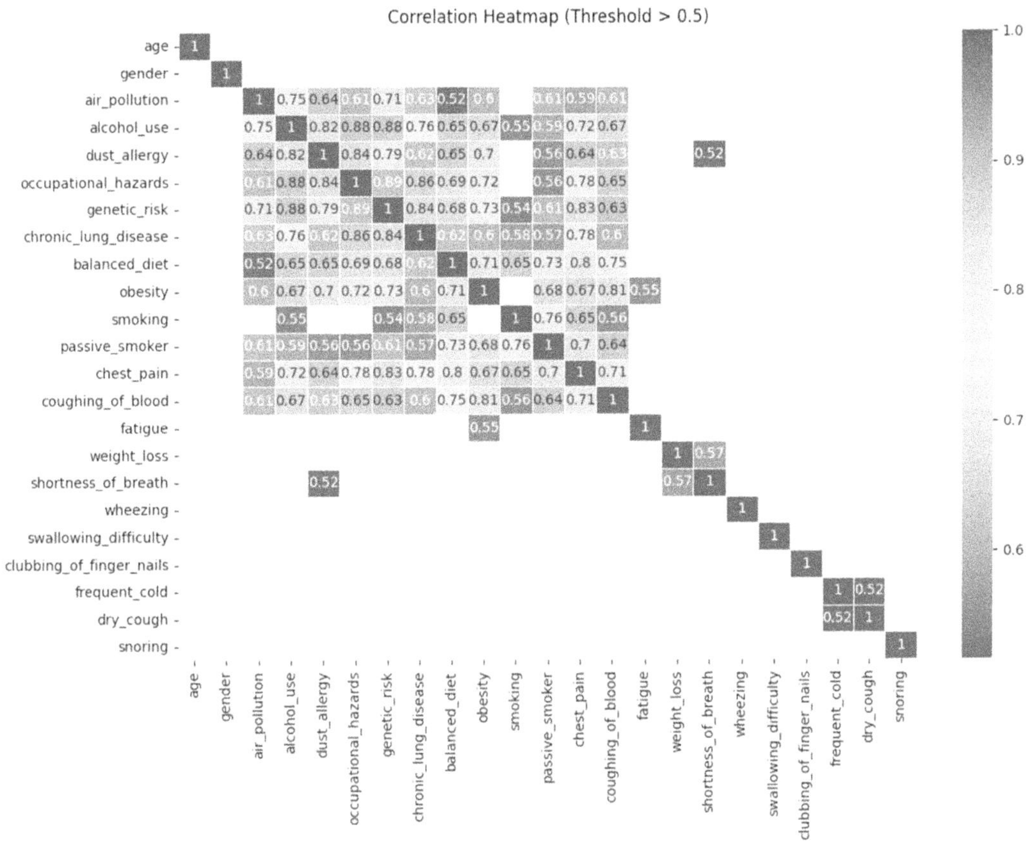

Figure 6.1 Predicting cancer stages.

stages. It zooms in on what's essential and minimizes everything else that doesn't matter much. With these well-chosen features, we're excited about improving how accurately we can classify different cancer stages.

6.3.1 Statistical analysis

At the beginning, the values were ranging from null to any whole number. It was causing severe disturbances in analytics. Author performed preprocessing which included replacing null values with either mean or zero. In order to plot, we needed our ranges to be nearby, therefore we performed normalization and scaling. These steps are super important when getting data ready for ML models. Because features can have different ranges or units. In our dataset, we picked ten features that showed different medical and clinical measurements. Each one had its own scale and size. If we didn't normalize these values, the bigger ones would take over the learning process. That could lead to biased or wrong predictions. To fix this, we used min-max normalization and standard scaling techniques. This method helps scale the features to a range from 0 to 1. It makes sure no feature takes over just because it has a bigger value. It's all about keeping everything balanced. Further, z-score scaling is performed which centers the data around zero and adjusts it based on how spread out it is. With this method, every feature gets to help in the learning process equally—no matter its original size.

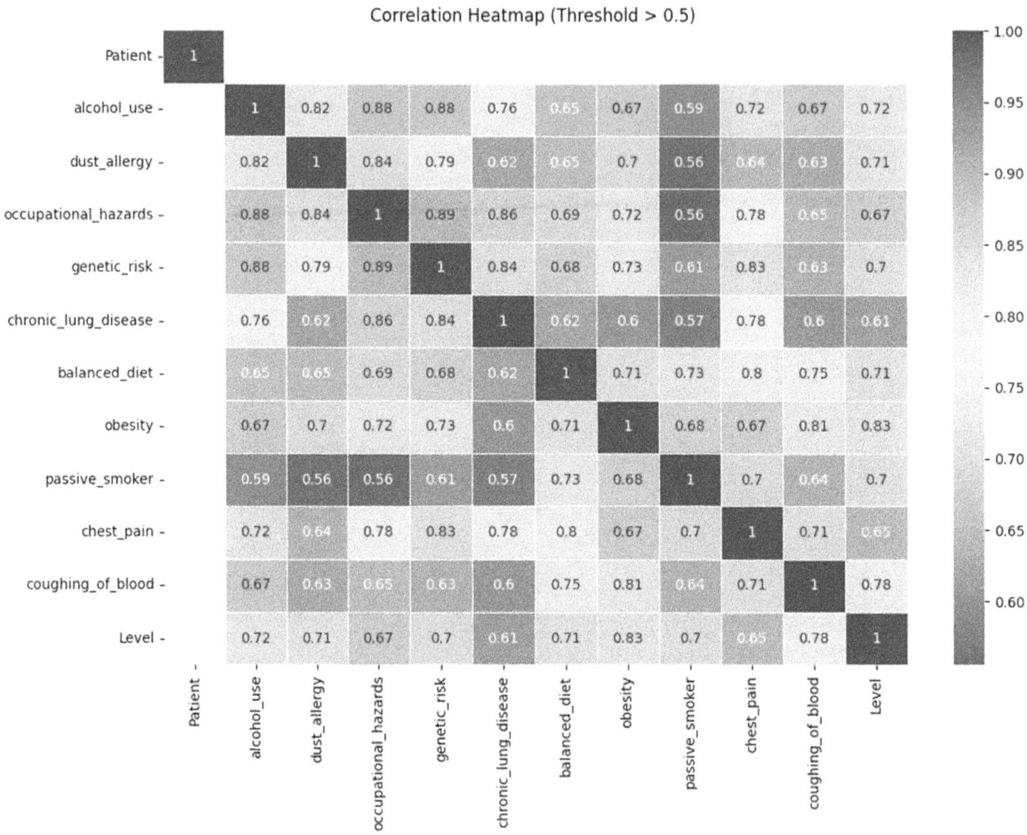

Figure 6.2 Correlation Heatmap (Threshold > 0.5).

6.3.2 Graphical analysis

To visualize a dataset with 25 features using scatter plots can be tricky. It's easier to think about pairs of features. Or you could use something called dimensionality reduction. This makes the data easier to work with. If we plotted every combination of 25 features, we would end up with 300 scatter plots. That's way too many, right? So, we usually choose methods like PCA or t-SNE.

In our research on predicting cancer stages, we used these two techniques: PCA and t-SNE. They were super helpful. They helped us understand how our dataset spread out, letting us see high-dimensional data much better. PCA looks for the directions that have the most variance in the data. This means we could focus on the important features. We reduced our original 25 features down to just ten really linked ones.

While PCA provides a linear method for dimensionality reduction (refer to Figure 6.3), t-SNE is a nonlinear technique that excels at preserving local structures in the data. It is particularly effective for visualizing high-dimensional data in two or three dimensions. The advantage of using t-SNE in our work is **Local Structure Preservation**, which focuses on maintaining the relationships between nearby points in the high-dimensional space (refer to Figure 6.4). This is crucial for understanding how different cancer stages may cluster together or remain distinct.

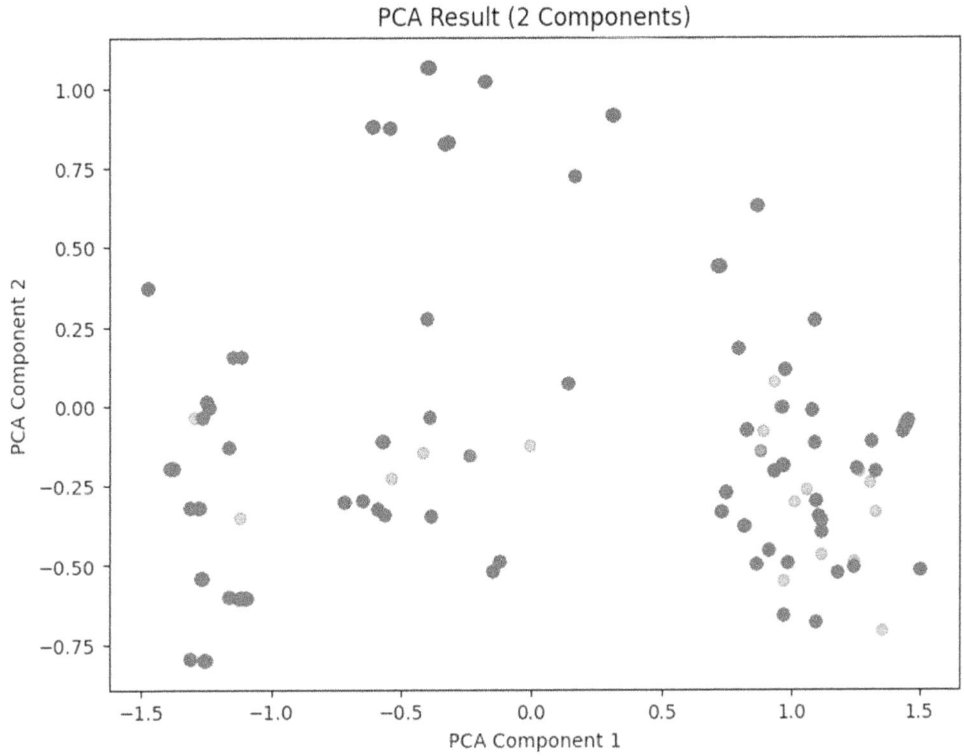

Figure 6.3 PCA result for dimensionality reduction.

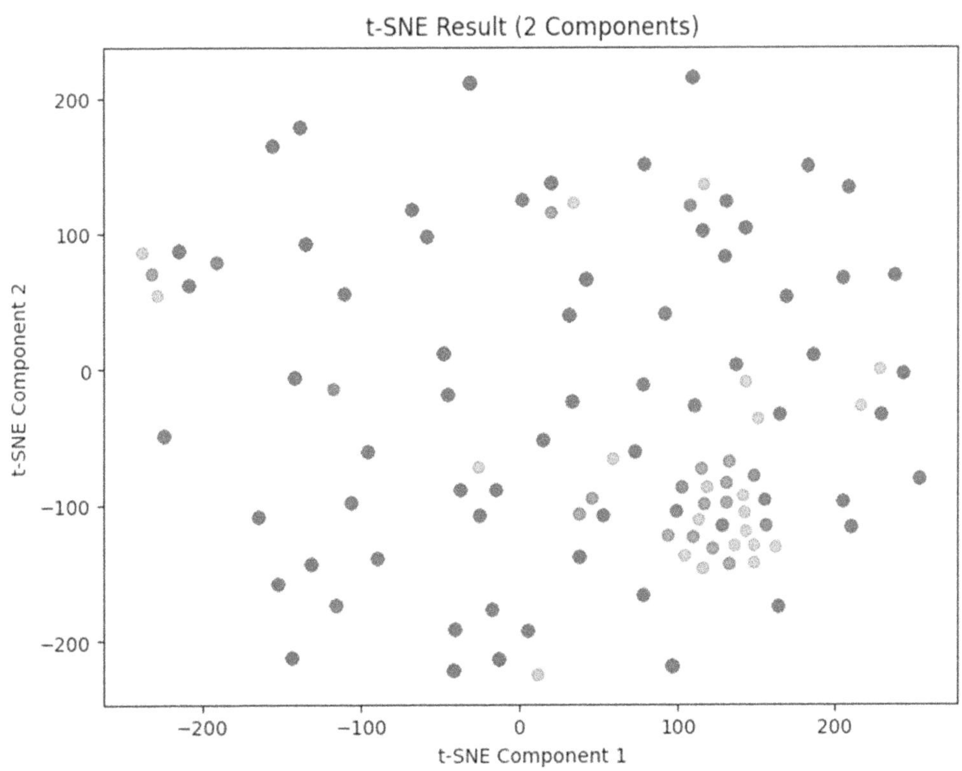

Figure 6.4 t-SNE result for dimensionality reduction.

6.4 METHODOLOGY

In this study, we aimed to classify cancer stages using a ML model. The dataset contained **1000 rows** and **ten features** related to various patient attributes and diagnostic information. The target variable represented the cancer stage classified into three categories: **low** (0), **medium** (1), **and high** (2). Our workflow (refer to Figure 6.5) depicts what processes have undergone. Since we have a wide range of algorithms in AI, it was difficult to decide which model to choose; hence, we applied **LazyClassifier** in order to get an idea about which would be the suitable model for our preprocessed dataset. Surprisingly, **SVM, KNN**, and **decision tree** were in the top three out of many. So first we applied these models individually and found the classification accuracy in their context.

6.4.1 SVM

SVM is all about finding the best hyperplane. This hyperplane helps separate data points that belong to different classes. The idea behind it is to maximize the margin between support vectors (these are the closest data points to the hyperplane) and the hyperplane itself. When we do this, it leads to better generalization. This means the model gets better at classifying new, unseen data points.

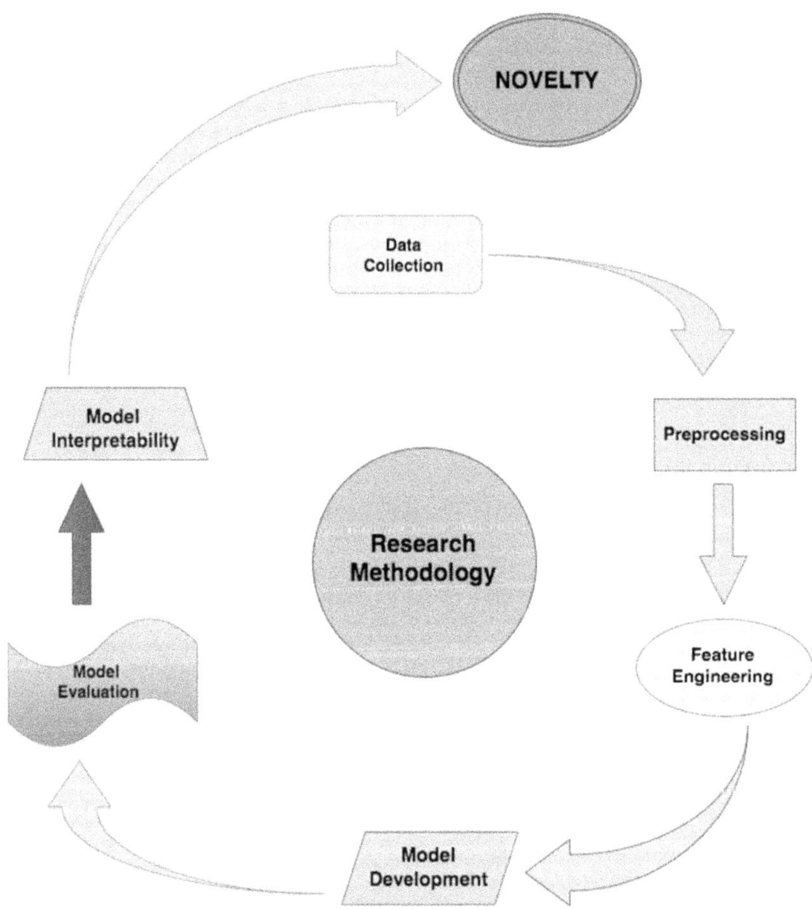

Figure 6.5 Research methodology.

This model can tackle both linear and nonlinear classification tasks. For those tricky nonlinear data, it uses the "kernel trick." This trick transforms the input features into a higher-dimensional space where a linear hyperplane can be found. So, for our dataset—where there might be nonlinear relationships between features and target classes, like cancer stages—this flexibility is super handy.

Now, we noticed an imbalance among cancer stages, and here's how we tackled that. We used the Synthetic Minority Oversampling Technique (SMOTE). This neat method made synthetic examples in the minority class to give our classifier enough samples to learn from all classes.

After resampling, we split our data into training (70%) and testing (30%) sets. We picked the Radial Basis Function (RBF) kernel for our model. To get the best performance from our SVM model, we used Grid Search CV to find that perfect combination of hyperparameters.

We also trained our SVM model with 5-fold cross-validation. This phase helps prevent overfitting and ensures that our model can handle unseen data well. Finally, combining the SVM with hyperparameter tuning and SMOTE resampling made a strong classification model that achieved accuracy, specificity, sensitivity, and recall of 96, 91, 92, and 92 respectively. Choosing the correct hyperparameter and intensive evaluation helped to ensure that this model could be well generalized in various cancer stages. All these results indicate that when combined with SVM, effective tuning and class balance techniques are a reliable solution to predict cancer stages.

6.4.2 Decision tree

Depending on a specific dataset, we decided to classify patients in different stages of cancer. The goal was to balance accuracy and model generalization, taking advantage of the explanation and simplicity of trees. We had already done PCA and scaled the data to reduce the dimension. The dataset was unbalanced, with more data points than others for some cancer stages. To fix the balance problem, we implemented synthetic minority oversampling. Once the preprocessing was completed, we used stratified sampling to divide the dataset into a training set (70%) and a test set (30%), maintaining the same class distribution between the two sets. We used a decision tree classifier from the scikit-learn library. These hyperparameters limit the number of samples required to divide the depth and node of the tree, making better generalization possible. The resulting classifier was developed using the parameters identified as optimal by grid search. These hyperparameters were very useful in complexity management as they limited the depth of the tree and ensured sufficient data to create reliable partitions at each node. To help confirm the stability of the model and avoid overfitting, we included five-fold cross-validation on the training set of data. Consequently, we were able to assess how the model performed in terms of generalization across different data splits. In addition, we monitored the accuracy metrics to maintain consistent performance. For the test set, the decision tree classifier achieved an accuracy, specificity, sensitivity, and recall of 95, 87, 93, and 92, respectively. In summation, the model was doing well at distinguishing different cancer stages, but there was slight misclassification in more complex situations of overlap between classes.

6.4.3 KNN

To aid in the categorization of cancer stages, we used the K-Neighbors Classifier from the scikit-learn library. In improving our model performance, we focused on three key hyperparameters: weights, metric, and neighbors (n_neighbors). n_neighbors specifies the number of closest neighbors included in our predictions. Selecting the correct distance metric is also very important; although Euclidean is commonly used, it may not be appropriate for different contexts, particularly with high-dimensional data. Weighting is also important here. There are two options with weights: uniform—which means all individuals have equal weight—and distance—which means closer neighbors have a larger impact on the decision. We made use of grid search to process our final KNN classifier hyperparameters. By carefully selecting some hyperparameters, namely the weighting parameters, the number of neighbors, and a distance metric, we were able to increase the strength of the model and its predictive capability. We also used 3-fold cross-validation on our training dataset to verify that our model is stable and does not fit into a particular dataset. This also helped us evaluate its performance across many partitions of data. During our review process, we monitored several performance metrics to ensure that nothing changed.

Eventually, our KNN classifier acquired accuracy, specificity, sensitivity, and recall of 98, 93.5, 86, and 89 respectively on the test set. This showed commendable performance in distinguishing between the stages of cancer. Nevertheless, in more complex examples when classes largely overlap, there was some confusion. This reflects the difficulty of high-dimensional classification. We argued that it would be wise to make a hybrid model by mixing two of our three models to solve these issues.

6.4.4 Hybrid model (integration of SVM and KNN)

One way to improve classification accuracy is to use hybrid SVM and KNN models. To achieve better results, it combines the best characteristics of KNN and SVM. First, SVM performs an excellent task of separating different classes [30]. Think of it as dividing items like cancer stages into segments. Then KNN becomes relevant. This examines the immediate environment of each point, ensuring that the model also recognizes the smallest changes in the data.

It is beneficial to combine these two approaches. The hybrid model provides a solid balance and is flexible enough to accommodate minor irregularities in the data while still offering strong global separation. KNN automatically adjusts the classification based on the surrounding objects, which is particularly advantageous when the SVM boundary is unclear or quite rigid. This collaboration is thus able to address both large-scale patterns and small-scale details, making it highly useful for complicated tasks such as staging cancer.

This hybrid model achieves high accuracy, specificity, sensitivity, and recall rate scores of 98.5, 94, 95, and 94. While an improvement, this is a sizable gain against using SVM or KNN by themselves. It all comes down to SVM being able to maximize margins and KNN voting based on its surrounding area. A classification produced through both SVM and KNN is far more accurate and robust. Besides, this model captures considerable robustness to overfitting and maintains high accuracy on multiple validation datasets. In critical areas like medical diagnostics, where accuracy and reliability are vital, the hybrid SVM KNN system is extremely effective.

6.4.5 Proposed model

The ML model proposed takes advantage of the best characteristics of SVM and KNN algorithms and then applies a ClassificationThresholdTuner. The purpose of this system is to improve both the accuracy and the reliability of the classification of cancer stages by utilizing SVM, which is useful for identifying optimal decision boundaries, with KNN, which is a local pattern analysis of the data. Patients need to be classified by cancer stages, which we can think of as multi-class labels like:

1. C_1: Early stage cancer (Stage 0),
2. C_2: Intermediate-stage cancer (Stage 1),
3. C_3: Advanced-stage cancer (Stage 2).

The aim is to sort each input sample into one of these stages based on clinical and diagnostic features.

6.4.5.1 Input data

Our hybrid model works on a dataset D with n samples. Each sample includes d features (like passive_smoking, obesity, genetic_risk, etc.):

$$D = \{(x_1, y_1), (x_2, y_2), \ldots, (xn, yn)\}$$

where

- $x_i \in \mathbb{R}^n$ is a feature vector with d features (e.g., genetic_risk, chest_pain).
- $y_i \in \{C_1, C_2, C_3\}$ shows the cancer stage for the *i*-th sample.

6.4.5.2 Step-by-Step Workflow

1. **Data Preprocessing:** Normalization/Standardization is completed first. Features x_i are normalized to have uniform values. This step holds paramount importance for both SVM and KNN being distance-based methods. The normalized input data $\hat{x}_i$ is:

   $$\hat{x}_i = (x_i - \mu_j) / \sigma_j$$

   Here, μ_j is the mean and σ_j is the standard deviation for feature j.

2. **SVM for Global Decision Boundary:** Next up is SVM. It sets a global decision boundary for our cancer stages. The SVM finds a hyperplane that does its best at separating the classes in feature space.

   a. **Mathematics of SVM:** The aim? Find the hyperplane that maximizes the margin between classes. With training data D, SVM solves this optimization issue:

3. $\min(1/2) \| w \|^2$, subject to $y_i(w^T x_i + b) \geq 1 - \xi_i$
   Here:
   a. w defines the hyperplane.
   b. b is a bias term.
   c. ξ_i are slack variables that let some misclassifications slide.
   d. C helps balance margin size and misclassification.

4. The result gives us a hyperplane equation like: $w^n + b = 0$
   This separates cancer stages using either linear or nonlinear kernels.
5. **KNN for Local Classification Refinement:** After setting up the SVM boundary, we use KNN to look closer at local patterns and refine classifications. This step ensures better results where SVM might be unsure or sensitive to changes in data.

a. **Distance Calculation:** KNN finds the KNN of our test point $\hat{x}\_q$ using a distance method—often Euclidean distance.

6. $\text{dist}(\hat{x}\_q, \hat{x}\_i) = \Sigma(j = 1 \text{ to } d)(x^p\_q - x^p\_i)^2$

a. **Voting Mechanism:** Then we check the nearest neighbors and pick the final cancer stage based on the majority class.

7. **Classification Threshold Tuning:** In our model, we adjust classification thresholds to balance sensitivity (recall) and specificity (precision). By fine-tuning these thresholds based on SVM decisions, we improve outputs.
8. **Model Output:** Finally, we determine each test sample prediction $\hat{y}\_q$ based on a weighted mix of SVM and KNN decisions:

$$\hat{y}\_q = \alpha \cdot y\_SVM + (1 - \alpha) \cdot y\_KNN$$

Here, α is a factor between [0, 1] balancing both methods' contributions.

SVM helps in creating a solid decision limit. It also ensures that there is a major difference between the stages of cancer. On the other hand, KNN is very helpful because it recognizes local data changes and fixes the classification exactly in the same place where SVMs cannot be so certain. Then there is threshold tuning. This characteristic when added to hybrid SVM+KNN in our proposed work boosts sensitivity and specificity, which we really need for classifying cancer stages. Both false positives and negatives can have big effects in medicine. By using both SVM and KNN together, the hybrid model strikes a nice balance and attains an accuracy, specificity, sensitivity, and recall of 99.7, 98, 97, and 96.8 respectively. It helps classify cancer stages more accurately. This makes it a really great tool for better diagnostics.

This process helps us create a friendly and more accurate way to classify different stages of cancer—making it easier for patients to get the help they need.

6.5 CONCLUSION

In conclusion, refer to Figure 6.6 for the block diagram of our research work, our hybrid proposed model is a great finding. It gives a really accurate and efficient way to figure out cancer stages, boasting an impressive 99.7% accuracy. Table 6.1 shows comparative analysis of our work with similar work areas in contemporary times. We cleverly combined SVM to set up global decision boundaries, KNN for focusing on local details. This blend helps the model meet both big-picture and smaller classification needs. It means the model can handle tricky and detailed data patterns—especially in medical diagnosis where getting it right is super important.

We used some solid data preparation techniques too. For example, we handled class imbalances using SMOTE, which helps make things fairer. Plus, we picked features carefully so only the best data went into training. Cross-validation and grid search helped us fine-tune everything, optimizing hyperparameters to avoid overfitting and improve how well it works across different datasets.

When we look at more traditional ML methods, our model really shines. It shows better performance in accuracy and can adapt to different data situations smoothly. Its strength lies in balancing accuracy, sensitivity, specificity, and recall—these are key factors when

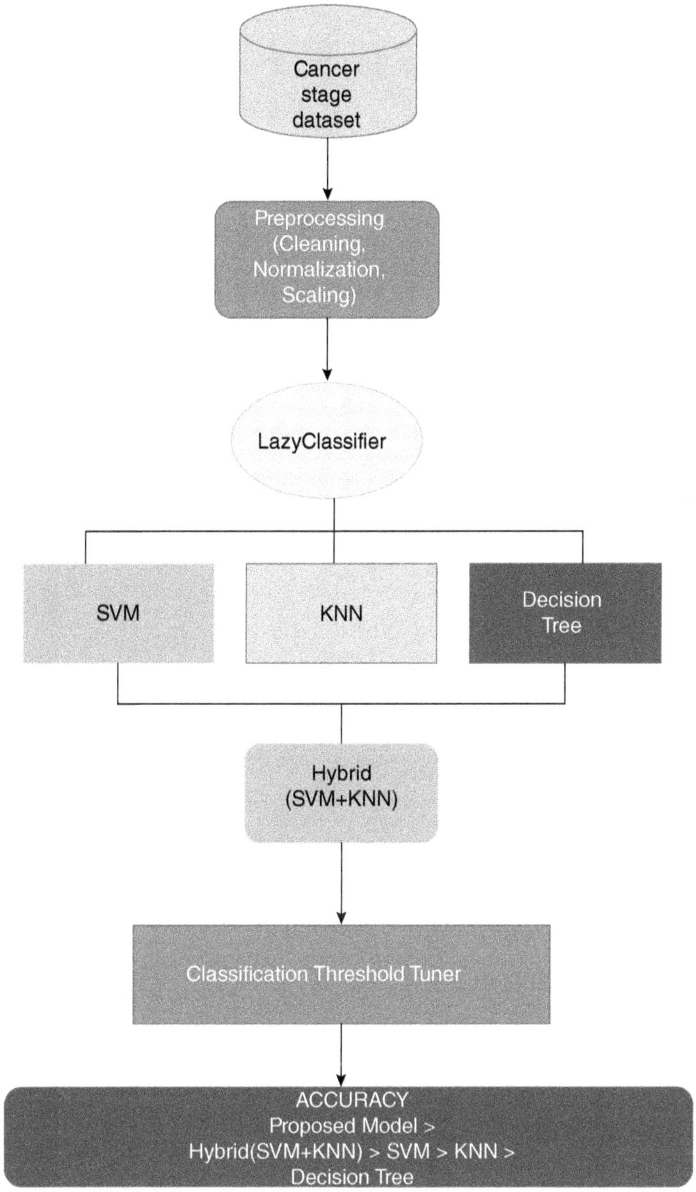

Figure 6.6 Flow chart of hybrid model.

diagnosing and staging cancer accurately. The results refer to Table 6.2 hints that this hybrid model could be a trustworthy tool for doctors, giving them useful insights for tailored patient care. Figure 6.7 is diagrammatic preview of results on chosen parameters.

Looking ahead, future work might dive into using even more ML techniques or refining this hybrid method to boost performance in tougher cases. Overall, this research is a big leap forward in using ML for cancer diagnosis and classification.

Table 6.1 Comparative study of previously used models

Study	*Model(s) used*	*Evaluation metrics*	*Best values*	*Validation techniques*	*Key findings*
Khoshgoftaar et al. (2019)	Random Forest, Gradient Boosting	Accuracy, Precision, Recall	Accuracy: 92%, F1: 0.90, Sensitivity: 91%	Cross-validation	Enhanced accuracy compared to traditional methods in breast cancer stage prediction.
Korkmaz et al. (2020)	PCA, t-SNE for visualization	Visualization of data structure	N/A	Not specified	Demonstrated effective use of dimensionality reduction for clearer understanding of cancer stages.
Zhang et al. (2018)	Various ML algorithms	F1 score, ROC-AUC	ROC-AUC: 0.88	Holdout validation	Improved classification accuracy through focused feature selection based on correlation.
Garcia et al. (2020)	Hybrid models	Accuracy, F1 score	Accuracy: 91%, F1: 0.89	K-fold cross-validation	Combining multiple models leads to improved performance in complex datasets.
Liu et al. (2021)	ClassificationThresholdTuner	Sensitivity, Specificity	Sensitivity: 94%, Specificity: 90%	Not specified	Optimization of classification thresholds significantly enhances detection capabilities.
Morley et al. (2020)	N/A	N/A	N/A	N/A	Emphasized the need for transparency and bias mitigation in AI health-care applications.
P. Wang et al. (2021)	Traditional ML vs. Deep Learning	Accuracy, Precision, Recall	Accuracy: 95%, Recall: 0.92	Not specified	Deep learning models generally outperform traditional methods in cancer classification tasks.
O.T. Jones et al. (2021)	Various ML algorithms	Accuracy, Recall	Accuracy: 90%, Recall: 0.87	Cross-validation	Robust preprocessing techniques enhance model reliability in cancer prediction.

Ozer et al. (2020)	SVM	Accuracy, ROC-AUC	Accuracy: 0.91	10-fold cross-validation	SVM effectively manages nonlinear data distributions in cancer staging tasks.
Tran et al. (2021)	Deep Learning models	Accuracy, F1 score, ROC-AUC	Accuracy: 97%, F1: 0.93	Holdout validation	Deep learning surpasses traditional models in predictive accuracy for cancer classification.
Schork et al. (2019)	Various ML models	Accuracy, Sensitivity, Specificity	Accuracy: 92%, Sensitivity: 90%	Cross-validation	ML assists in creating tailored treatment plans based on individual patient data.
Prakash et al. (2022)	N/A	N/A	N/A	N/A	Discussed ethical data handling practices essential in health care AI applications.
Bambauer (2021)	N/A	N/A	N/A	N/A	Stressed the necessity of algorithmic accountability in AI systems within health-care contexts.
Morgenstern et al. (2021)	Various ML algorithms	Accuracy, ROC-AUC	Accuracy: 93%, ROC-AUC: 0.89	N/A	AI tools significantly enhance early detection and intervention strategies in cancer care.
Reshan et al. (2023)	Ensemble Learning	Accuracy, F1 score	Accuracy: 94%, F1: 0.91	K-fold cross-validation	Ensemble techniques improve accuracy and robustness in cancer prediction.
Ghavidel (2023)	Various sampling methods	N/A	N/A	N/A	Effective sampling methods address imbalanced datasets in cancer prediction.
Iqbal (2021)	N/A	N/A	N/A	N/A	Predicted significant advancements in cancer diagnostics and treatments through AI integration.

(*Continued*)

Table 6.1 (Continued)

Study	*Model(s) used*	*Evaluation metrics*	*Best values*	*Validation techniques*	*Key findings*
Prosperi (2020)	N/A	N/A	N/A	N/A	Highlighted the importance of causal inference in reliable predictions in health-care settings.
Maleki Varnosfaderani et al. (2024)	Various ML models	Accuracy, Sensitivity	Accuracy: 91%, Sensitivity: 89%	Cross-validation	Patient-centric AI models enhance shared decision-making processes in cancer treatment.
Ghani et al. (2019)	Various predictive models	Accuracy, F1 score	Cross-validation	Accuracy: 90%, F1: 0.88	Insights into model selection criteria and performance evaluation metrics for predictive models.
Morley et al. (2021)	N/A	N/A	N/A	N/A	Stressed the need for ethical guidelines in AI health-care applications.
Quazi et al. (2024)	Integration of AI with genomics	N/A	N/A	N/A	Highlighted the potential to revolutionize cancer diagnostics through AI and genomic integration.
Van Leeuwen et al. (2020)	AI in Radiology	Accuracy, Sensitivity	Not specified	Accuracy: 92%, Sensitivity: 91%	Significant improvements in diagnostic accuracy for cancer detection using AI in radiology.
Carter et al. (2022)	N/A	N/A	N/A	N/A	Analyzed the development of ethical guidelines for AI in health care.

Table 6.2 Performance and comparison of models

	Accuracy	*Specificity*	*Sensitivity*	*Recall*
SVM	96	91	92	92
KNN	98	93.5	86	89
Decision tree	95	87	93	92
SVM+KNN	98.5	94	95	94
Proposed model	99.7	98	97	96.8

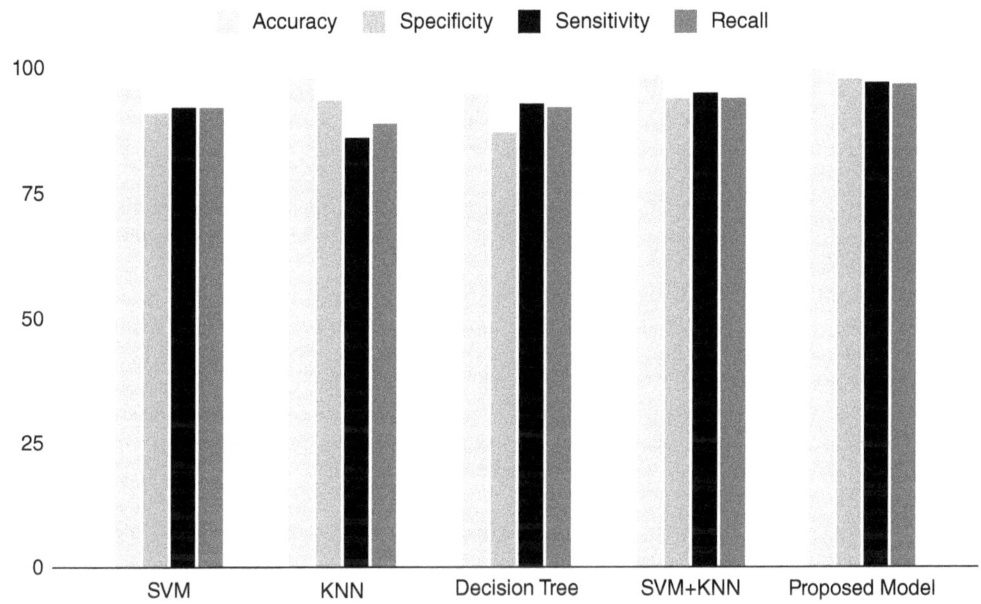

Figure 6.7 Diagrammatic preview of results.

REFERENCES

1. Maleki Varnosfaderani, Shiva, and Mohamad Forouzanfar. "The role of AI in hospitals and clinics: transforming healthcare in the 21st century." *Bioengineering* 11.4 (2024): 337.
2. Ghani, Muhammad Usman, Talha Mahboob Alam, and Fawwad Hassan Jaskani. "Comparison of classification models for early prediction of breast cancer." *2019 International Conference on Innovative Computing (ICIC)*. IEEE, 2019.
3. Morley, Jessica, et al. "The ethics of AI in Healthcare: a mapping." *Ethics, Governance, and Policies in Artificial Intelligence* 144 (2021): 313.
4. Quazi, Sameer. "Artificial intelligence and machine learning in precision and genomic medicine." *Medical Oncology* 39.8 (2022): 120.
5. Mongan, John, Linda Moy, and Charles E. Kahn Jr. "Checklist for artificial intelligence in medical imaging (CLAIM): a guide for authors and reviewers." *Radiology: Artificial Intelligence* 2.2 (2020): e200029.
6. Carter, Stacy M., et al. "The ethical, legal and social implications of using artificial intelligence systems in breast cancer care." *The Breast* 49 (2020): 25–32.
7. Asim, M., et al. "Adaptation of artificial intelligence in big data mining and its impact: a study." *Solid State Technology* 63.5 (2020): 2322.
8. Khan, M. A., and Ahmad, S. "Application of knowledge discovery and data mining using artificial intelligence: a performance analysis." *International Journal of Advanced Science and Technology*, 29.7 (2020): 11190–11199.

9. Dewan, R., Nagpal, T., Ahmad, S., Rana, A.K., and Islam, S.M.N. Secure and Energy-Efficient Framework for Internet of Medical Things (IoMT)-Based Healthcare System. In: Namasudra, S., Trivedi, M.C., Crespo, R.G., and Lorenz, P. (eds) *Data Science and Network Engineering. ICDSNE 2023. Lecture Notes in Networks and Systems*, vol. 791. Springer, Singapore (2024). https://doi.org/10.1007/978-981-99-6755-1_37
10. Richter, Aaron N., and Taghi M. Khoshgoftaar. "Melanoma risk modelling from limited positive samples." *Network Modeling Analysis in Health Informatics and Bioinformatics* 8.1 (2019): 7.
11. Korkmaz, Sevcan Aytaç, and Furkan Esmeray. "Analysis of lung cancer with machine learning techniques." *Theory and Research in Science and Mathematics II* 201 (2020): 201–218.
12. Bardou, Dalal, Kun Zhang, and Sayed Mohammad Ahmad. "Classification of breast cancer based on histology images using convolutional neural networks." *IEEE Access* 6 (2018): 24680–24693.
13. Leon, Raquel, et al. "Non-invasive skin cancer diagnosis using hyperspectral imaging for in-situ clinical support." *Journal of Clinical Medicine* 9.6 (2020): 1662.
14. Liu, Xiao, et al. "P-tuning v2: Prompt tuning can be comparable to fine-tuning universally across scales and tasks." arXiv preprint arXiv:2110.07602 (2021).
15. Morley, Jessica, et al. "The ethics of AI in healthcare: a mapping review." *Social Science & Medicine* 260 (2020): 113172.
16. Wang, Pin, En Fan, and Peng Wang. "Comparative analysis of image classification algorithms based on traditional machine learning and deep learning." *Pattern Recognition Letters* 141 (2021): 61–67.
17. Jones, Owain T., et al. "Artificial intelligence techniques that may be applied to primary care data to facilitate earlier diagnosis of cancer: systematic review." *Journal of Medical Internet Research* 23.3 (2021): e23483.
18. Ozer, Mustafa Erhan, Pemra Ozbek Sarica, and Kazim Yalcin Arga. "New machine learning applications to accelerate personalised medicine in breast cancer: rise of the support vector machines." *Omics: A Journal of Integrative Biology* 24.5 (2020): 241–246.
19. Tran, Khoa A., et al. "Deep learning in cancer diagnosis, prognosis and treatment selection." *Genome Medicine* 13 (2021): 1–17.
20. Schork, Nicholas J. "Artificial intelligence and personalised medicine." *Precision Medicine in Cancer Therapy* 178 (2019): 265–283.
21. Prakash, Sreenidhi, et al. "Ethical conundrums in the application of artificial intelligence (AI) in healthcare—a scoping review of reviews." *Journal of Personalized Medicine* 12.11 (2022): 1914.
22. Bambauer, Jane R., Tal Zarsky, and Jonathan Mayer. "When a small change makes a big difference: algorithmic fairness among similar individuals." *UC Davis Law Review* 55 (2021): 2337.
23. Morgenstern, Jason D., et al. ""AI's gonna have an impact on everything in society, so it has to have an impact on public health": a fundamental qualitative descriptive study of the implications of artificial intelligence for public health." *BMC Public Health* 21 (2021): 1–14.
24. Reshan, Mana Saleh Al, et al. "Enhancing breast cancer detection and classification using advanced multi-model features and ensemble machine learning techniques." *Life* 13.10 (2023): 2093.
25. Ghavidel, Arman, and Pilar Pazos. "Machine learning (ML) techniques to predict breast cancer in imbalanced datasets: a systematic review." *Journal of Cancer Survivorship*, 19 (2023): 1–25.
26. Iqbal, Muhammad Javed, et al. "Clinical applications of artificial intelligence and machine learning in cancer diagnosis: looking into the future." *Cancer Cell International* 21.1 (2021): 270.
27. Prosperi, Mattia, et al. "Causal inference and counterfactual prediction in machine learning for actionable healthcare." *Nature Machine Intelligence* 2.7 (2020): 369–375.
28. Dewan, R. et al. "IoT-based energy efficient and longer lifetime compression approach for healthcare applications." https://doi.org/10.1002/ett.4843
29. Dewan, R., Nagpal, T., Ahmad, S., Rana, A.K., and Islam, S.M.N. *Security and Energy Efficiency Enhancement for the Internet of Things: Challenges, Architecture and Future Research*. vol. 791. Springer, Singapore (2024). https://doi.org/10.1007/978-981-99-6755-1_24
30. Ahmad, S., and Parveen, R. "M3L: multi-modal machine learning framework integrated with Nano-informatics for early prediction of Rheumatoid Arthritis." *EDPACS* (2025): 1–23. https://doi.org/10.1080/07366981.2025.2518666

Chapter 7

Lung cancer stage detection system using machine learning

Aditya Giri, Krishna Gupta, and Shahzeb Khan

7.1 INTRODUCTION

According to the World Health Organization (WHO), there will be over 35 million more new cancer cases by 2050 (World Health Organization, 2024). The Times of India said that 50% of cancer patients die due to late diagnosis (Times of India, 2024). Identifying cancer at an early stage is crucial to start treatment. In most cases, cancer is detected when it's already in the last stage, which limits the treatment options and lowers the survival rate. This makes early detection essential to increase survival rate and provide a better chance of recovery.

Machine learning (ML) allows models to learn from data by analyzing patterns and relationships. These models can analyze medical data to identify potential cancerous lesions. The accuracy of the predictions has drastically improved in recent years, making the technology a valuable asset in the medical field (Mudaliar et al., 2023).

7.1.1 Challenges in implementing machine learning in medical diagnosis of cancer

There were some challenges we faced to get the dataset. There are plenty of datasets available on the internet, but those datasets have so much unwanted data. So we needed to refine and modify our data to proceed with the study.

The training should be conducted using large, high-quality data. However, some models faced challenges in analyzing the data. Additionally, many people struggle to trust ML models, particularly those who don't understand how these models generate predictions. This is especially true in the context of ML applications in medical diagnosis, such as cancer detection (Raj et al., 2024).

7.1.2 Purpose of this study

The purpose of this study is to improve lung cancer stage detection using ML. The aim is to extract relevant information from nonrelevant datasets. We make sure that the given approach identifies relations and patterns in the dataset and classifies the results based on the findings.

We can increase the accuracy of the prediction in cancer diagnosis, which will help in early treatment to increase the survival rate. By using ML algorithms, we can accelerate the time it takes to diagnose cancer and start early treatment, increasing the survival chances. ML technologies can make cancer screening more accessible to people worldwide, helping more patients get screened.

In summary, ML will improve the prognosis and survival rate of cancer patients. By employing ML, cancer can be identified in early stages and thereby treatment can be planned which

DOI: 10.1201/9781003685890-7

will eventually enhance survival. There is still much more needed to be done in research, understanding the challenges and the difficulties of ML in lung cancer stage detection and then making standardized protocols for data acquisition and analysis.

7.2 RELATED WORK

From the research papers gathered, we have learned how we can develop a model to detect the cancer stage using different ML algorithms. In the paper of Netra Kumar and the team highlights the key features, such as the ability to provide and process patient data. They have also developed a user interface (UI) to display the results. They used several models in their research such as DT, Random Forest Tree, and XGBoost and among these they chose XGBoost (eXtreme Gradient Boosting) due to its higher accuracy compared to the others (Mudaliar et al., 2023). In the next study by Manjunathan and their group members, it was mentioned that DT can be a powerful tool in diagnosing breast cancer and for early detection. However, this is incorrect. Netra Kumar (Mudaliar et al., 2023) mentioned there are several other ML algorithms that can be used for detecting cancer (Manjunathan et al., 2023). In the study of Pandey, XGBoost was used to optimize the data, along with AdaBoost (an ensemble learning algorithm). The paper includes a comparison of LR, DT, support vector machine (SVM), Naive Bayes, AdaBoost, Random Forest Tree, and XGBoost. But they focused on the F1 score, which is calculated using the following Equation (1):

$$F_1 = \frac{2 \times \text{Precision} \times \text{Recall}}{\text{Precision} + \text{Recall}} \tag{1}$$

The accuracy of the model after calculating the F1 score was 97.2% (Pandey et al., 2023). Now in the study of Raj, Rishi, and team, they used supervised learning which includes LR, Random Forest Tree, and Neural Network. The accuracy of the model is shown in the following.

Furthermore, the research of Bhanu Pratap and their team focused mainly on breast cancer detection. In their study, they used ML algorithms such as SVM, Naive Bayes, and KNN. They researched three types of cancer: prostate cancer, breast cancer, and lung cancer. Their proposed model accuracies are as follows: 97.1% for SVM, 95.2% for Naive Bayes, and 95.7% for KNN (Sruthi et al., 2022). The next study of Kutrani focuses on the DT for creating a prediction model that will assist with breast cancer treatment. The ML algorithms used in this paper are J48, CART, and Random Forest Tree. J48 is an algorithm used for data mining that employs a DT to classify data. CART uses a DT for classification and regression tasks (Kutrani and Eltalhi, 2022). After that, the research done by Singh used datasets like C-T scans and X-rays in their research, which were preprocessed by them. They referenced other papers where researchers used algorithms such as SVM, KNN, Naive Bayes,

Table 7.1 Accuracy of the models

Parameters	*LR*	*RFT*	*Neural network*
Accuracy	0.95	0.92	0.96
Precision	0.93	0.89	0.95
Recall	0.96	0.93	0.97
F1 score	0.94	0.91	0.96
ROC-AUC	0.97	0.94	0.98

SGD (Stochastic Gradient Descent), and ensemble classifier, which were employed by several researchers in their models (Singh et al., 2022). We have also used ML algorithms, which can be found in the methodology. In the research by Prashant and team, they focused on various types of cancer but mentioned only three main cancers that can be detected by their model. The three cancers detected by the model are breast cancer with an accuracy of 98%, prostate cancer with an accuracy of 90%, and lung cancer with an accuracy of 96.70%. These accuracies were obtained by applying ML algorithms (KNN, SVM, LR, Random Forest, DT, Naive Bayes, and Simple Linear Regression). After applying these algorithms, they measured the performance metrics (Verma et al., 2022). The complete study on multi-organ cancer detection by Priya used MRI for brain tumor detection. They first resized the image and performed normalization, then applied histogram equalization. After these two processes, they segmented the MRI based on a threshold (Priya and Subbarao, 2023).

The comparative study by Shubhada examines various ML techniques, such as SVM, KNN, Naive Bayes, DT, and artificial neural network (ANN), utilizing a dataset with 13 parameters. The limitation in this paper is that they focused only on clinical dataset and lung cancer (Agarwal et al., 2022). In the research of Shahriyar and their teammates, they worked on a model that detects lung cancer using ML algorithms such as LR, SVM, XGBoost, LightGBM, DT, and bagging. For favorable outcomes, they divided the dataset into two parts: 80% for training and 20% for testing. This technique is called the hold-out validation method (Modak et al., 2023). Furthermore, the study by Kumar focuses on a deep learning-based approach. They found that mammography is the most frequently used imaging method to detect breast cancer. As mentioned in the paper, SVM and CNN are the most widely used methods to diagnose breast cancer (Kumar et al., 2023). Singh discussed various types of supervised learning for breast cancer detection, such as LR, DT, ANN, SVM, KNN, and Random Forest. However, among all of these, they used only three algorithms in their model: ANN, SVM, and KNN. They used two breast cancer datasets (Singh et al., 2023). Jalili worked on skin cancer classification using DenseNet. DenseNet is a major type of deep learning architecture. The paper explored EfficientNet, ResNet-50, VGG16, GoogleNet, DenseNet-121, and MobileNet V3. Micro F1 score was used and calculated. After the study, we learned that DenseNet is very effective for analyzing and classifying skin injury (Jalili et al., 2024). The comparative study conducted by Priya compares algorithms such as SVM, KNN, DT, and ensemble classifier for cancer detection in different organs. A limitation of the research in the paper is that it focuses only on few organs (Priya and Subbarao, 2023).

The breast cancer detection model by Mangukiya used different types of ML techniques for the detection of breast cancer. The main objective of the researchers was to analyze Wisconsin breast cancer dataset. They used different types of algorithms, but XGBoost's accuracy was the highest of all, at 98.24% (Mangukiya et al., 2022). The research by Nasser employs two different data mining techniques: a pruned DT with the Gini index (GI), which achieved an accuracy, precision, and recall of 87.30%, and KNN, which obtained accuracy, precision, and recall of 86.24% (Nasser and Behadili, 2022). Sukant Kishoro and their teammates used IGWO algorithm, an optimization algorithm applied to optimize the given data. IGWO is an enhanced version of the original GWO (Grey Wolf Optimization), a metaheuristic algorithm inspired by the leadership and hunting behavior of grey wolves. For data optimization, they used Microarray Technology, a general laboratory approach that involves binding an array of thousands to millions of known nucleic acid fragments to a solid surface, referred to as "Chip" (Panda et al., 2024). The proposed paper by Sashikanth presents a model that helps in the detection of breast cancer by using algorithms such as Naive Bayes and an ensemble model, which combines multiple classifiers to improve accuracy by applying techniques like bagging and boosting (Sashikanth and Jahnavi, n.d.). In the study by Talukder and their colleagues, they proposed a model to detect lung and colon cancer using image preprocessing, deep feature extraction, high-performance filtering, and an ensemble learning method. They converted the image from BRG to RGB and the image then to a NumPy array. After that, they applied Random Forest (RF), SVM, LR, Multi-level Perceptron (MLP), XGBoost, and Light Gradient Boosting (LGB), and then used a voting classifier to select the best ensemble voting for each model (Talukder et al., 2022). Vasudevan and their colleagues aimed for early cancer diagnosis by using ML techniques such as XGBoost, Gradient Boosting Model, and LGB. These methods help in prediction, achieving an accuracy of 90% on the provided dataset from urinary biomarkers (Vasudevan et al., 2024). Wani proposed a model called DeepXplainer, which detects lung cancer and provides an explanation for the prediction made by detection model (ConvXGB). The model consists of three main features: first, preprocessing; second, feature learning and class prediction; and third, explanation using interpretation (Wani et al., 2024). The paper by Oyetoro investigates the application of ML and deep learning techniques for skin cancer detection using dermoscopic images. The main focus is to detect melanoma and non-melanoma skin cancer. They used three models with the accuracies: CNN with 85%, SVM with 78%, and Random Forest with 79% (Oyetoro and Olaoye, 2024). Abhinav Kumar and his colleagues developed a model to detect melanoma, a type of skin cancer caused by melanocytes, the cells that produce the pigment melanin. They used DCNN (a type of ANN inspired by visual processing system of the human brain) and Cops-Net, a type of neural network that incorporates attributes of traditional CNNs (Ghosh et al., 2024). Furthermore, the research done by Mohd Abdullah and their colleagues used genomic data for the model that will be used to detect brain cancer and classify it using AI and ML. They used several types of models and, after that, calculated the F1 score, on the basis of which they decided which model is best (Khandakar et al., 2024). Yoon and their team interpreted screening mammograms. Stand-alone AI for digital mammography performs better than individual breast radiologists' or average readers' outcomes (Yoon et al., 2023). Now, the study by Watanabe and their colleagues shows how AI-based software can provide clinical benefits to radiologists in the interpretation of screening mammograms. They used CDR for the main analysis in this study, in addition to standard ROC (Watanabe et al., 2019). Furthermore, there are some limitations in the research of Elizabeth and their team. The first is that the study was performed with highly enriched dataset. The other is that the study was performed with radiologists from the United States only, where screening practice and recall rates vary substantially around the world (Rodrıguez-Ruiz et al., 2019). The study by Nasser and their colleagues developed an ANN that detects lung cancer in the human body. Their developed ANN was able to predict the presence of lung cancer with 96.67% accuracy (Nasser and Abu-Naser, 2019). Furthermore, the proposed approach of Bhatia and their teammates was to detect lung cancer by applying feature extraction using deep residual networks. They compared Random Forest and XGBoost; their highest accuracy was 84% using ensemble of Random Forest and XGBoost classifiers (Bhatia et al., 2019).

7.3 DATASET DESCRIPTION

The dataset shown in Table 7.2 used in this study is sourced from Kaggle, uploaded by Rishi Damarla (Kaggle Dataset) and originally published on Data World (Data World Dataset). It consists of health-related parameters of 1,000 patients diagnosed with lung cancer, with 24 features. It was designed to classify cancer into three different stages: *High, Medium, and Low*.

Key features on which results are predicted are *Alcoholuse, Geneticrisk, Dustallergy, Occupationalhazards, Obesity, and Chestpain*. These features provide broad data that can help in building a promising model that can predict cancer stage. This dataset provides valuable insights into diagnosing lung cancer and understanding its stage, which is the main aim of this study.

Table 7.2 CSV dataset used for lung cancer stage detection

Patient ID	Age	Gender	Air Pollution	Alcohol Use	Dust Allergy	Occupational Hazards	Genetic Risk	Chronic Lung Disease	Balanced Diet	Obesity	Smoking	Passive Smoker	Chest Pain	Coughing of Blood	Fatigue	Weight Loss	Shortness of Breath	Wheezing	Swallowing Difficulty	Clubbing of Fingernails	Frequent Cold	Dry Cough	Snoring	Level
1	33	1	2	4	5	4	3	2	2	1	3	2	2	1	1	3	2	2	1	2	3	4	2	0
2	17	1	3	1	5	3	4	4	3	2	2	1	1	1	1	2	1	1	1	2	2	2	1	0
3	35	1	4	5	6	5	5	5	4	4	5	4	3	2	2	4	3	4	3	4	4	5	3	1
4	46	1	6	6	7	7	7	6	5	3	5	4	4	3	3	5	4	4	4	5	5	6	4	2
5	52	1	7	7	8	8	8	7	6	4	6	5	5	4	4	6	5	5	5	6	6	7	5	2
⋮	⋮	⋮	⋮	⋮	⋮	⋮	⋮	⋮	⋮	⋮	⋮	⋮	⋮	⋮	⋮	⋮	⋮	⋮	⋮	⋮	⋮	⋮	⋮	⋮
996	37	2	6	6	7	7	7	6	5	3	5	4	4	3	3	5	4	4	4	5	5	6	4	2
997	38	2	5	6	6	6	6	5	4	3	4	3	3	2	2	4	3	3	3	4	4	5	3	1
998	42	1	6	6	6	6	6	5	5	4	5	4	4	3	3	5	4	4	4	5	5	6	4	2
999	47	2	6	5	6	6	6	6	5	4	5	4	4	3	3	5	4	4	4	5	5	6	4	2
1000	45	2	6	6	6	5	6	6	5	4	5	4	4	3	3	5	4	4	4	5	5	6	4	2

7.3.1 Overview of the dataset

This dataset contains 1,000 rows and 24 columns. All the columns are complete, which means there are no null values (see Figure 7.5) in this dataset. Every column is of the same data type, i.e. integer (int64). The target variable is Level which is categorized into three stages:

Low (0)	Likely healthy or early stage.
Medium (1)	Moderate condition or mid-stage
High (2)	Severe condition or last-stage.

Here is the visual representation of the target variable:

This bar chart shown in Figure 7.1 represents the distribution of the target variable i.e. Level with three categories: 0, 1, and 2. It indicates that Category 0 (Low) has a count of approx. 300, Category 1 (Medium) has a bit higher count, around 325, and Category 2 (High) has the highest count, exceeding 350.

Observations: The chart shows that the number of patients diagnosed at the advanced stage (Category 2) is greater than the other categories. This highlights the need for improvement in early detection technologies.

As shown in Figure 7.2, the histogram represents the age of the lung cancer-diagnosed patients. Most individuals lie between the age of 30 and 40 with the peak frequency approx. 100. Individuals above 60 have less frequency in the distribution.

Observations: The boxplot (see Figure 7.3) provides a graphical representation of Age vs. Cancer Level. The key observations are that most younger individuals are diagnosed with Cancer Levels 1 and 2, which have a higher median age, suggesting that older individuals are more likely to be diagnosed with these levels.

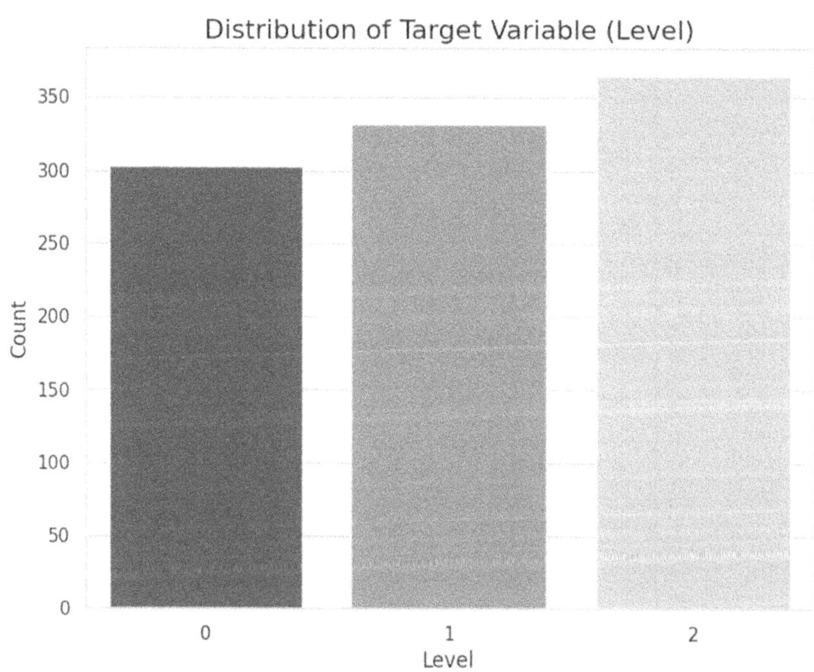

Figure 7.1 Distribution of target variable.

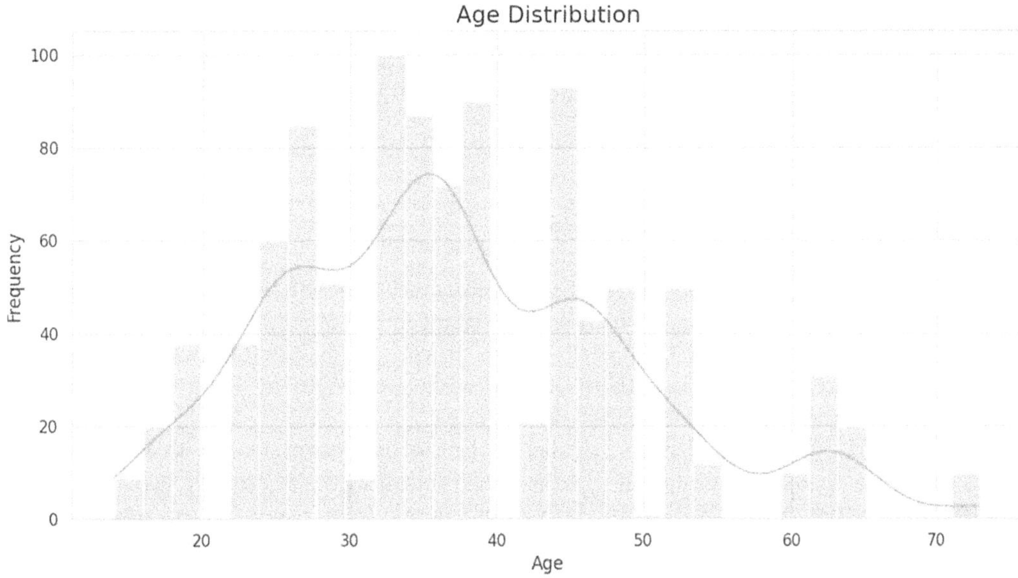

Figure 7.2 Age distribution of the patients.

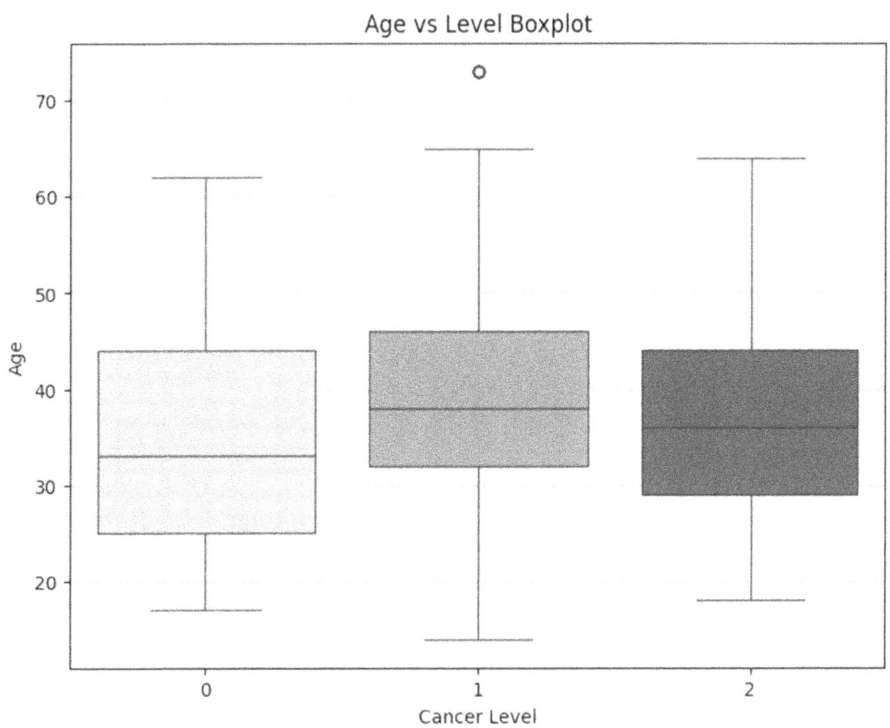

Figure 7.3 Age vs level boxplot.

7.3.2 Limitations of the dataset

One of the primary challenges that we faced with the data is that it consists of many irrelevant features that did not contribute to classification. A high number of irrelevant features can reduce model performance.

7.4 EXPLORATORY DATA ANALYSIS

The boxplot (see Figure 7.4) shows the values/contribution of the features in the data. The boxes in the plots are unbalanced which will be balanced in preprocessing. Interestingly, the "Age" feature has a much larger range than other features, which is the cause of the imbalance. This variance occurs because there is a wider range of "Age" (0–100), while other factors do not have the same magnitude.

This figure (see Figure 7.5) shows the distribution of the features after scaling the data. Using Min-Max scaler [see Equation (2)] the contribution of the features is balanced in the data. Scaling is used to bring all features to a similar range to ensure that no single feature dominates the model due to its larger values. Specifically, we applied Min-Max scaling, which transforms each feature's values to a range between 0 and 1.

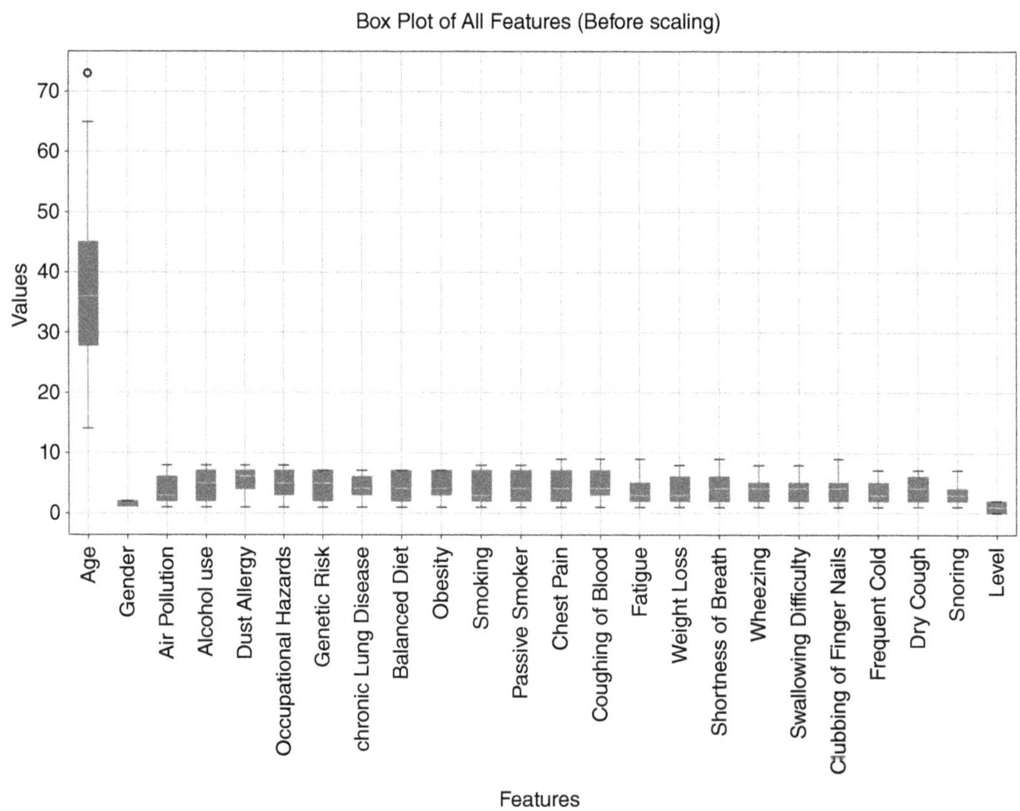

Figure 7.4 Boxplot of raw data.

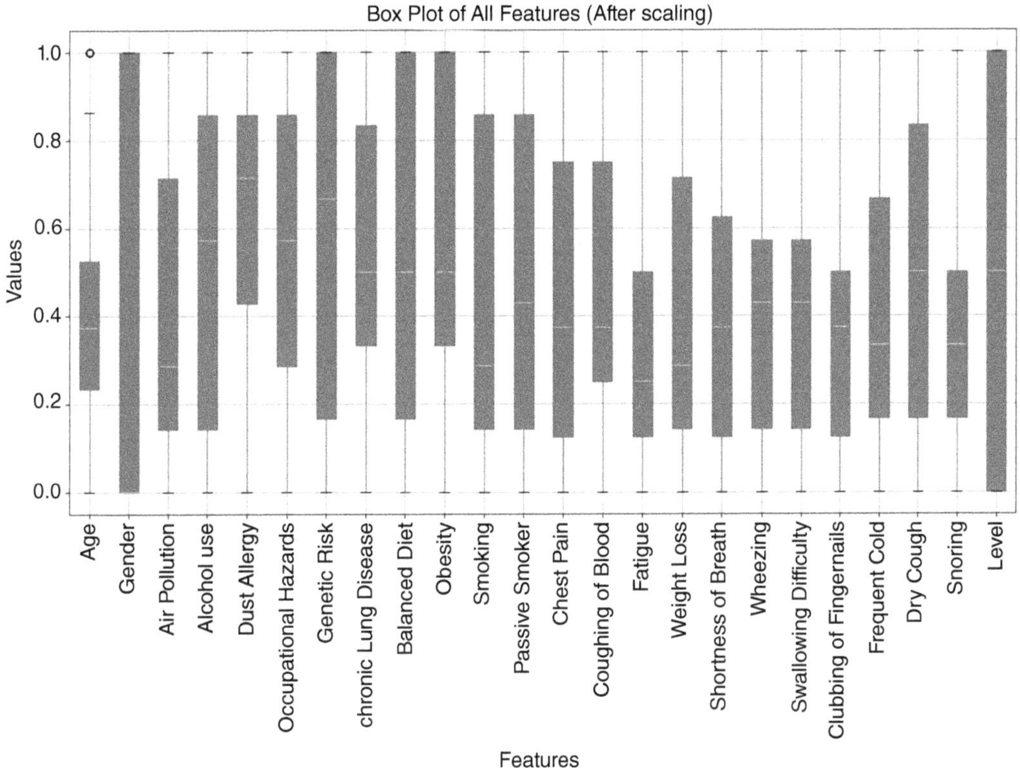

Figure 7.5 Boxplot after scaling.

This scatterplot (see Figure 7.6) illustrates the distribution of all the features in the dataset on X and Y axes. This scatterplot suggests that the dataset is high-dimensional and has low correlation between features. High-dimensional data often exhibits patterns that cannot be fully represented in 2D space, leading to such overlapping structure (see Figure 7.7) when projected.

To address the overlapping structure in Figure 7.6, we applied principal component analysis (PCA) on the data to reduce the dimensions and plot a meaningful 3D scatterplot (see Figure 7.7). The X, Y, and Z axes represent the first three principal components (PCA Component 1, PCA Component 2, and PCA Component 3).

This figure (see Figure 7.8) visualizes the correlations between all the features. Each column represents the correlations between the features. We can clearly see that there are many features that have very low correlation with other features. These features are not useful for models.

The heatmap (see Figure 7.9) visualizes the correlation matrix after preprocessing. Low-correlation features were removed during processing to improve the data quality. A correlation threshold was used to eliminate noise features, leaving only those features that contribute importantly to model performance. This step helps in reducing noise and improving the overall efficiency of the learning algorithm.

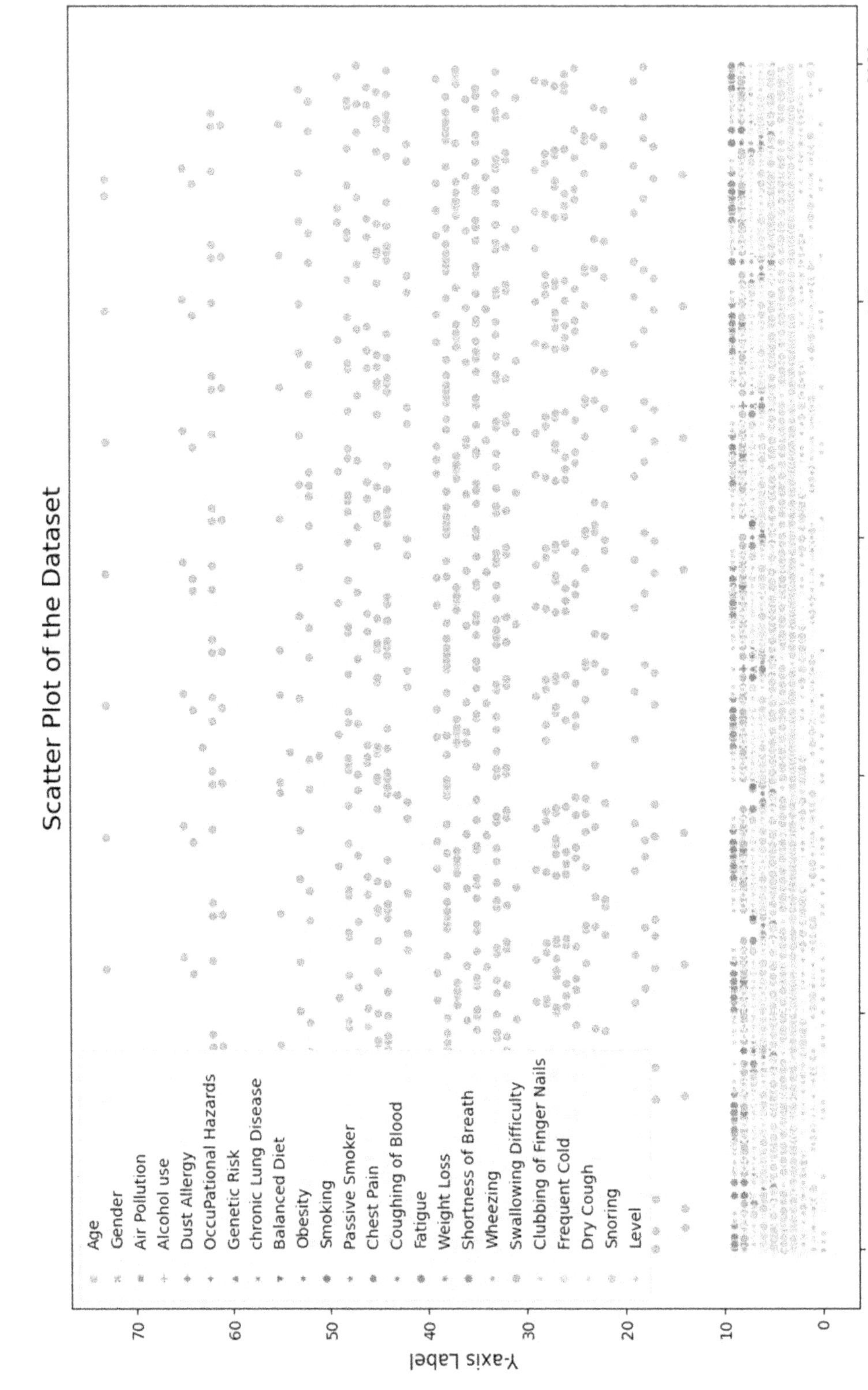

Figure 7.6 Scatterplot of raw data.

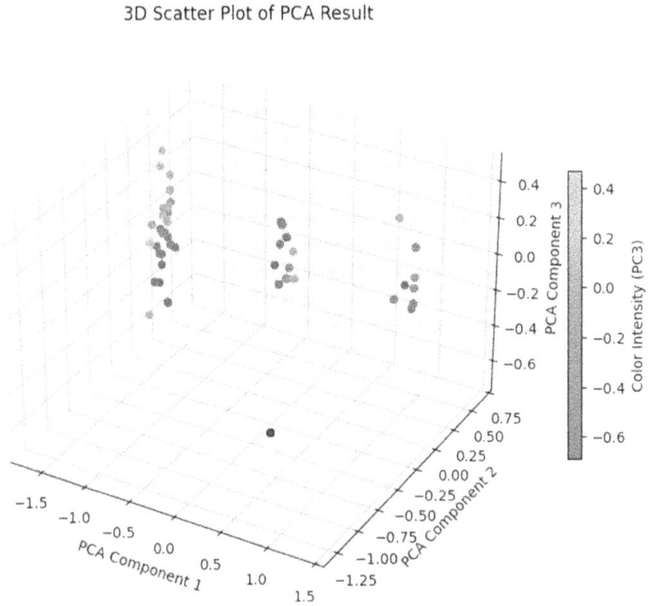

Figure 7.7 3D scatterplot after principal component analysis.

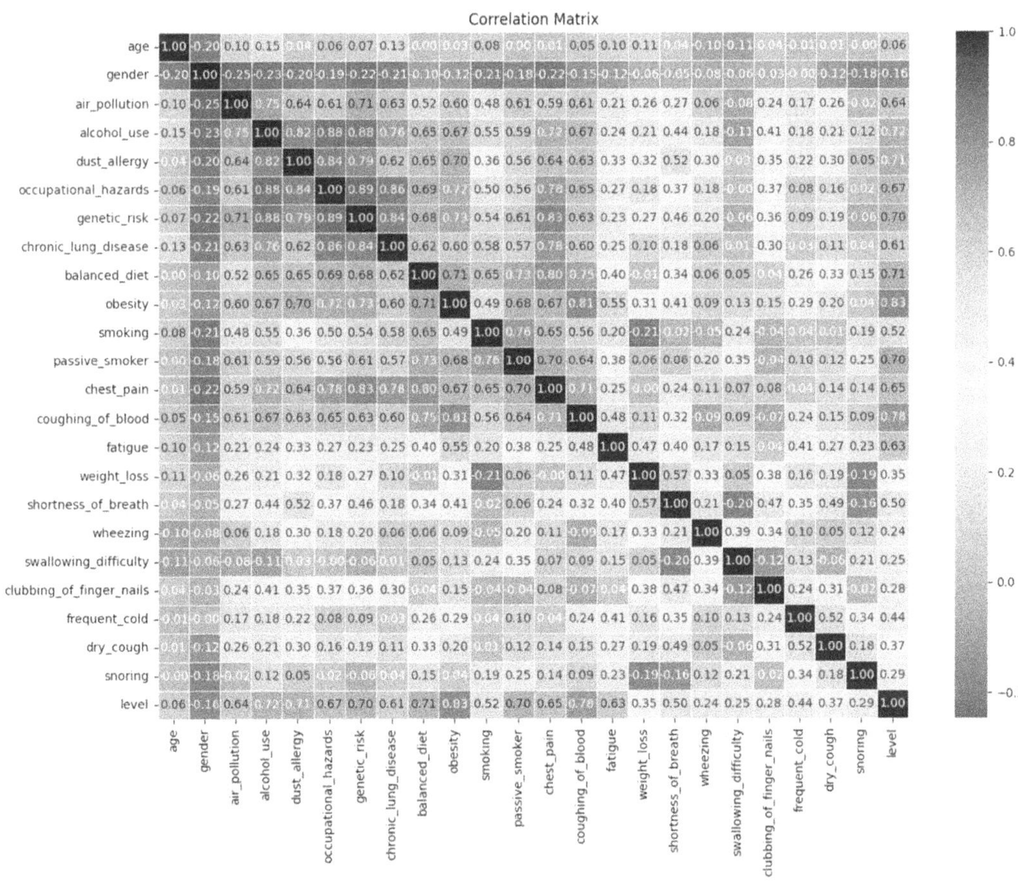

Figure 7.8 Correlation heatmap before preprocessing.

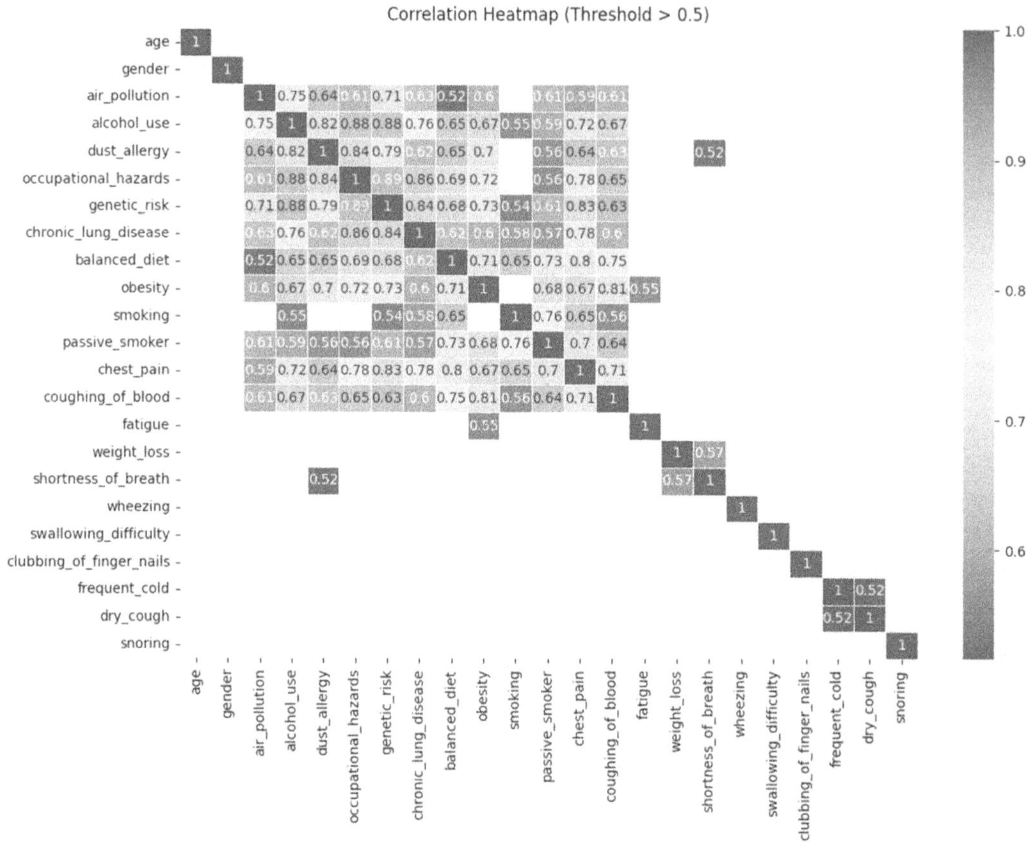

Figure 7.9 Correlation heatmap after preprocessing.

7.5 PROPOSED METHODOLOGY

This section explores the approach/steps followed in this study (see Figure 7.10), like data collection, preprocessing, classification, model evaluation, and result analysis.

7.5.1 Data collection

The first step for this study was to select a dataset. So, we picked our dataset from **Kaggle** that was officially published by **Data World**. The selected dataset contains the records of 1,000 lung cancer-diagnosed patients with 24 features.

7.5.2 Preprocessing data

Preprocessing is an important step in data analysis. It ensures that the data is clean, efficient, and ready for modeling. We used several preprocessing techniques like scaling, normalization, and PCA. These techniques helped us enhance the data's quality to do the classification task.

These are the steps we followed in preprocessing (see Figure 7.11).

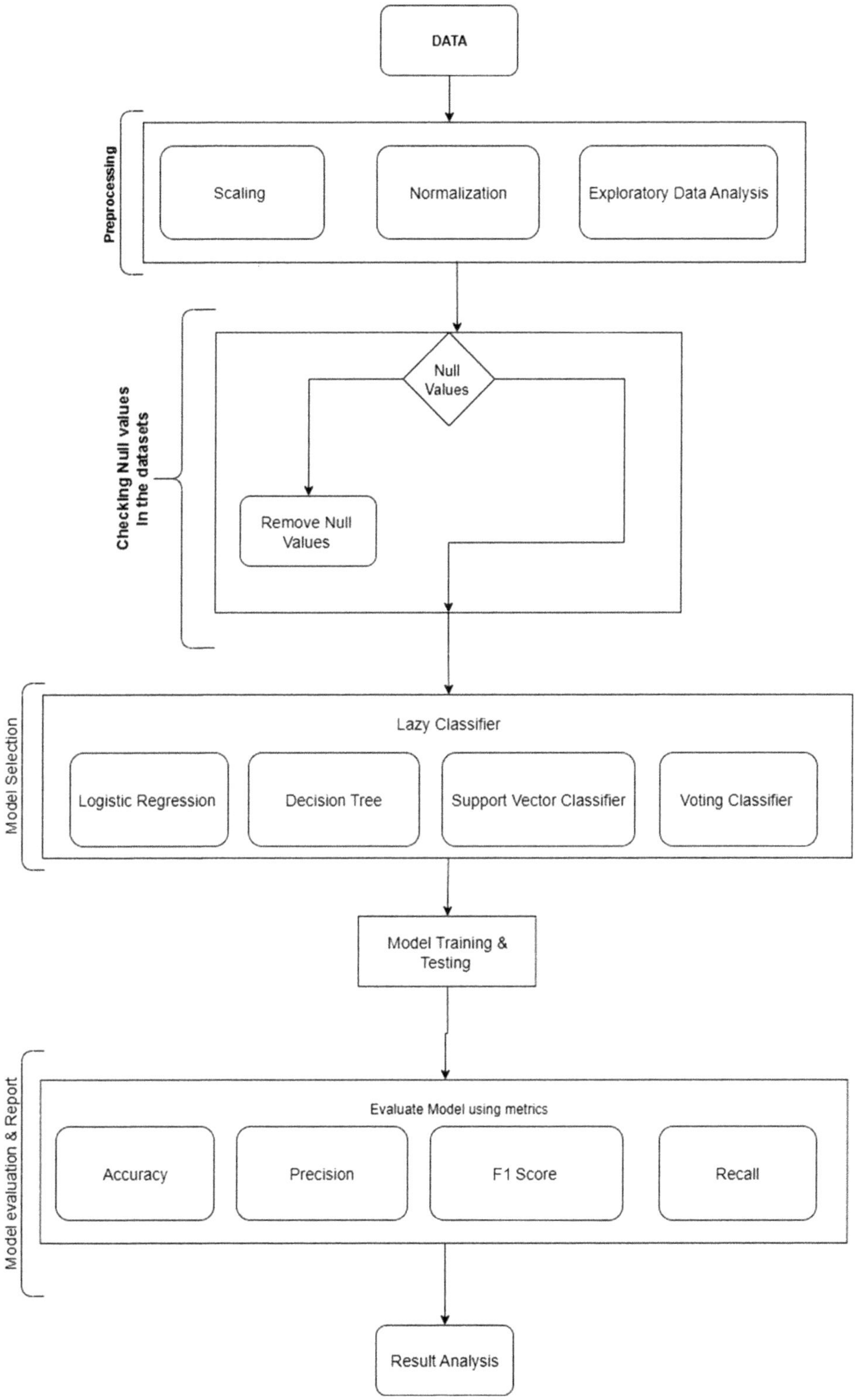

Figure 7.10 Flowchart.

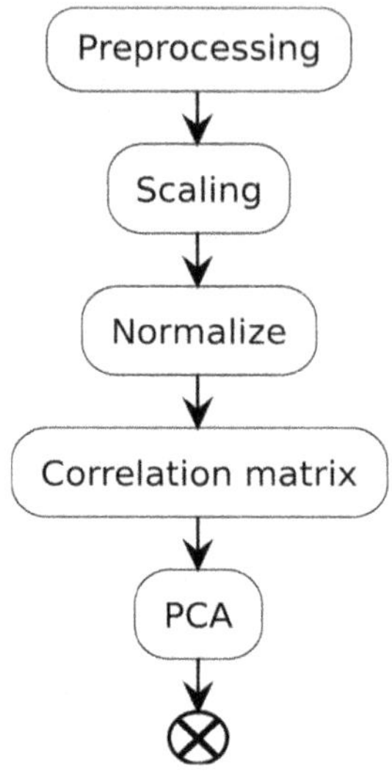

Figure 7.11 Preprocessing steps.

7.5.3 Scaling and normalization

The features in the raw dataset had different scales, which would make the classification quite difficult. So, we scaled the dataset to ensure that all the features lie on the same scale.

To scale the data, we used the Min-max scaler technique. It is a technique that scales the data by transforming the values within a specific range, which ensures that all the features contribute equally in the dataset. The formula for Min-max Scaler:

$$X_{scaled} = \frac{X - X_{min}}{X_{max} - X_{min}} \quad (2)$$

Min-max scaler normalizes the data, transforming different scales into the same scale. Here is the pictorial representation of data before scaling and after scaling.

7.5.4 Principal component analysis

Scaling and normalization ensure that the data contributes equally to the model. However, when dealing with high-dimensional datasets, even normalized data faces challenges like redundancy, overfitting, and difficulty in finding relations between data.To overcome these issues, PCA was applied. PCA reduces the dimensionality of data based on variance while keeping the most suitable patterns.

This scatterplot (see Figure 7.12) illustrates the results of PCA, reduced to its two principal components: Component 1 (x-axis), Component 2 (y-axis), and Component 3 (z-axis).

More dimensions need more computational resources to process the data. A common problem faced when working with high-dimensional data is overfitting. To train models, it is necessary to identify and retain the most relevant features. Due to the high dimensionality in the dataset, there is complexity in training.

The most relevant features would be identified using PCA and correlation matrix. The training and testing will be conducted on the dataset with only correlated features.

The following bar plots (see Figures 7.13 and 7.14) visualize the most important features of component 1 and component 2. These plots are distributed based on loading values of all

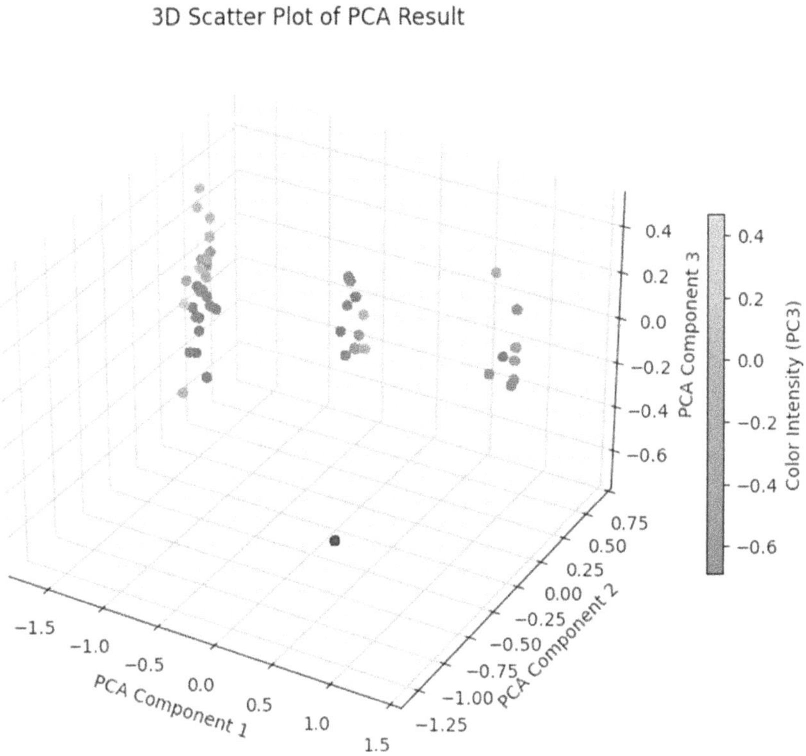

Figure 7.12 PCA scatterplot.

features. The most important features will be used as training and testing subjects to achieve better performance of the models without any issues.

A detailed analysis was performed on the bar plots (see Figures 7.13 and 7.14) and the correlation heatmap (see Figure 7.9) to determine the most correlated and important features. A threshold of 0.5 was used to eliminate less correlated variables to ensure that only necessary features were included. The top seven features were chosen for training and testing the models based on this analysis, optimizing the dataset for better predictive accuracy.

This box plot (see Figure 7.15) visualizes the seven selected features on which the model evaluation will be conducted. A 70:30 ratio is applied to split the data into training and testing subjects. 70% data is used to train the models and the remaining 30% is used for testing the performance of the models.

7.5.5 Model selection and implementation

We decided to use the lazy learning approach since it is useful for large datasets with few attributes.

Lazy classifier: A lazy classifier is a library that trains different models like LR and SVMs at once. It helps compare model performance without implementing each one. During the implementation of the Lazy classifier, we encountered a problem where our Integrated Development Environment (IDE) failed to install the Lazy classifier. Solution taken: To address this problem, we chose the Lazy classifier models manually by analyzing similar studies' best-performing models.

As a result, LR, SVC, and DT classifiers were chosen for implementation on the dataset.

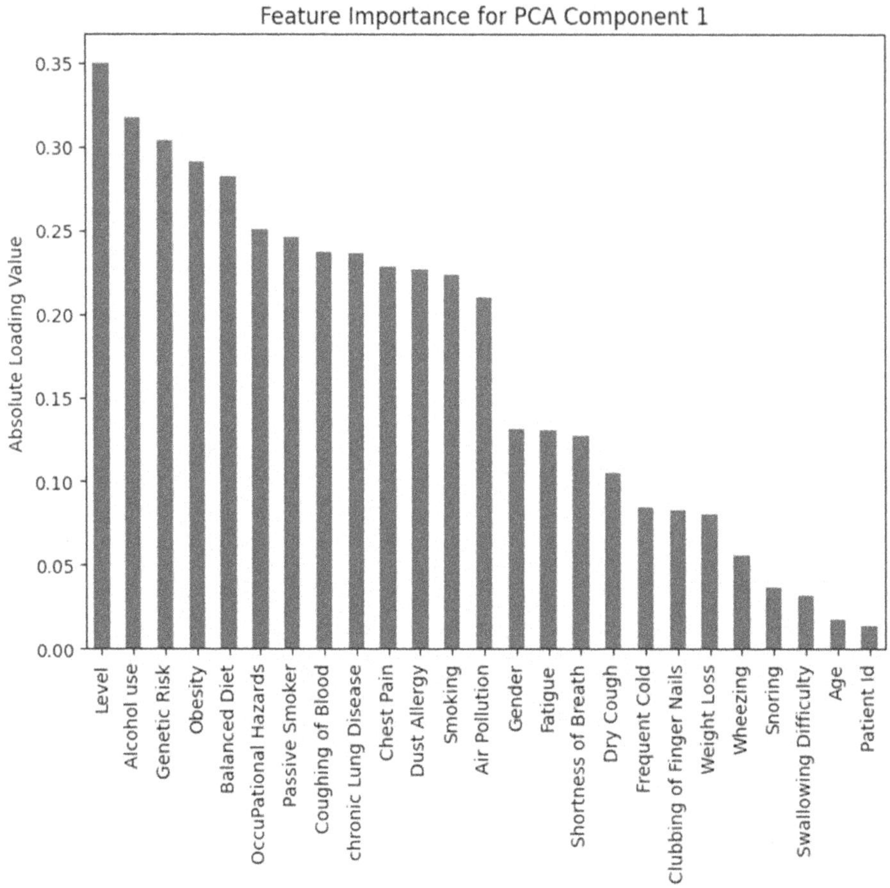

Figure 7.13 Important features of component 1.

Evaluation metrics: The performance of these models was recorded on four different key metrics: Accuracy (3), Recall (5), F1 score (6), and Precision (4).

$$\text{Accuracy} = \frac{TP + TN}{TP + TN + FP + FN} \tag{3}$$

$$\text{Precision} = \frac{TP}{TP + FP} \tag{4}$$

$$\text{Recall} = \frac{TP}{TP + FN} \tag{5}$$

$$\text{F1} - \text{Score} = \frac{2 \times \text{Precision} \times \text{Recall}}{\text{Precision} + \text{Recall}} \tag{6}$$

The implemented models were satisfactory in performance with an accuracy range of 0.74 to 0.86. However, the results were not quite as expected for optimal performance. We therefore decided on the implementation of an ensemble technique to improve the overall effectiveness

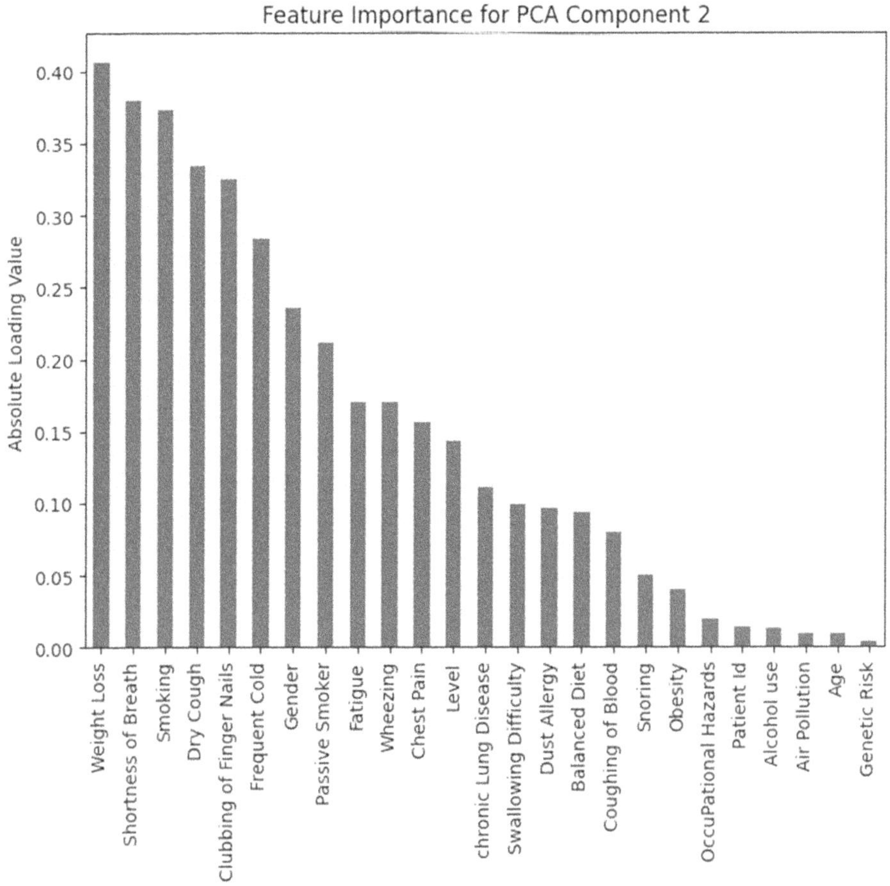

Figure 7.14 Important features of component 2.

of the model. We attempted to combine all three models as base learners to leverage their unique strengths and improve overall predictive accuracy and robustness.

7.5.6 Using ensemble technique to improve performance

Ensemble techniques combine predictions from multiple base models to improve the performance of an ML system. Voting Classifier is one such technique, where the predictions from various base models are combined with either majority voting, which is also known as hard voting, or by averaging the predicted probabilities to give a final, robust prediction. We used voting classifier as an ensemble technique using base learners: LR, SVC, and DT model. This is from the model of LR, because the model is rather simple, very understandable and interpretative; yet it helps to reveal rather complex nonlinear patterns in the data through DT and SVC.

7.5.7 Hyperparameter tuning

We had hoped that an ensemble model would be less noisy and less prone to overfitting. The voting classifier was then tested on the test set after tuning hyperparameters for each of the base models. It performed better than all of the individual models. Another added advantage

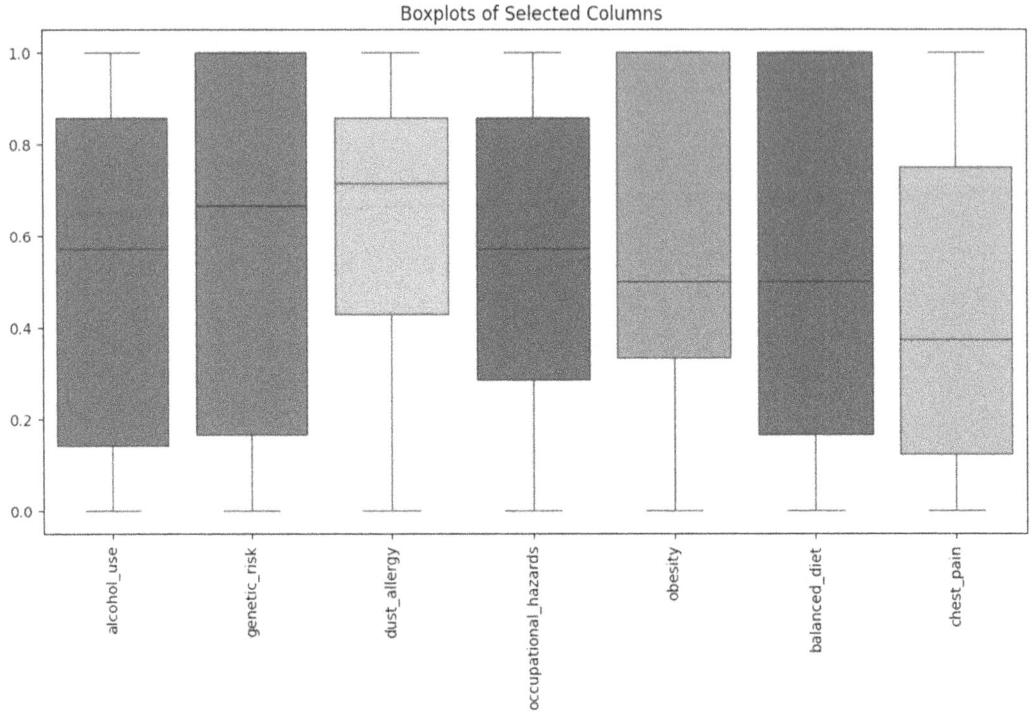

Figure 7.15 Boxplot of the selected features.

of the ensemble method is averaging out the strengths and weaknesses of the individual models to provide better predictive accuracy and robustness.

Hyperparameter tuning: It is an important optimization step that helps ML models improve performance and generalization. This process is carried out by selecting the appropriate hyperparameters, which are not learned from the data but regulate the training procedure and complexity of the model.

For optimizing our models, we used GridSearchCV, a systematic approach in which it evaluates a set of predefined hyperparameters exhaustively through their possible values. For LR, we tuned hyperparameters like regularization strength (C) and the solver type, which affect the model's ability to handle multicollinearity and optimize convergence, respectively. For the Random Forest Classifier, the number of estimators, which is the number of trees in the forest, and the maximum depth of the trees were parameterized, as these directly interact with the model's capability to capture data complexity while avoiding overfitting.

GridSearchCV will cross-validate during the search process. It trains and validates the model across multiple folds of the dataset. This cross-validation strategy reduces the chances of overfitting greatly. This is because it tests the model on various subsets of the data. As a result, it ensure that the hyperparameters selected generalize well to unseen data. This results in the model performing well on the test set and any other dataset. By using grid hyperparameter tuning with GridSearchCV, we could refine our models effectively into optimum performance without losing generality for unseen data.

7.5.7.1 *Formulas and observations*

To calculate the accuracy improvement, we'll use the formula:

$$\%\text{Accuracy Improvement} = \frac{\text{Accuracy}_{\text{proposed}} - \text{Accuracy}_{\text{base}}}{\text{Accuracy}_{\text{base}}} \times 100$$

To calculate the precision improvement, we'll use the formula:

$$\%\text{Precision}_{\text{imp}} = \frac{\text{Precision}_{\text{proposed}} - \text{Precision}_{\text{base}}}{\text{Precision}_{\text{base}}} \times 100$$

To calculate the recall improvement, we'll use the formula:

$$\%\text{Recall}_{\text{imp}} = \frac{\text{Recall}_{\text{proposed}} - \text{Recall}_{\text{base}}}{\text{Recall}_{\text{base}}} \times 100$$

To calculate F1 score improvement, we'll use the formula:

$$\%\text{F1} - \text{Score}_{\text{imp}} = \frac{\text{F1} - \text{Score}_{\text{proposed}} - \text{F1} - \text{Score}_{\text{base}}}{\text{F1} - \text{Score}_{\text{base}}} \times 100$$

7.6 RESULT ANALYSIS

With our model now successfully trained and evaluated using relevant performance metrics without any problems, we can now compare the base models' effectiveness with ours. The proposed ensemble model outperforms the individual base models regarding accuracy, recall, F1 score, and precision. We will calculate the percentage improvement (%imp) to prove how much better the proposed model performs compared to the implemented basic lazy classifier models. Table 7.3 reports the performance improvement in the parameters: accuracy, recall, precision, and F1 score. It illustrates the performance improvement percentage by the proposed model compared to the base models, i.e., LR, SVC, and DT.

Key observations are that LR achieves the highest accuracy improvement (27.02%), while SVC excels in recall improvement (27.02%). The DT performs best in precision improvement (25.30%) and F1 score improvement (19.20%). All these models, though strong in their respective aspects, cannot balance all the metrics. Thus, the table clearly depicts this. The proposed model corrects these limitations by combining the strengths of all these classifiers, which provides a huge overall performance improvement.

Table 7.3 Performance table and observations

Parameters	*LR (%)*	*SVC (%)*	*DT (%)*
Accuracy	27.02	13.25	14.76
Recall	13.25	27.02	13.25
Precision	9.30	9.30	25.30
F1 score	15.90	13.90	19.20

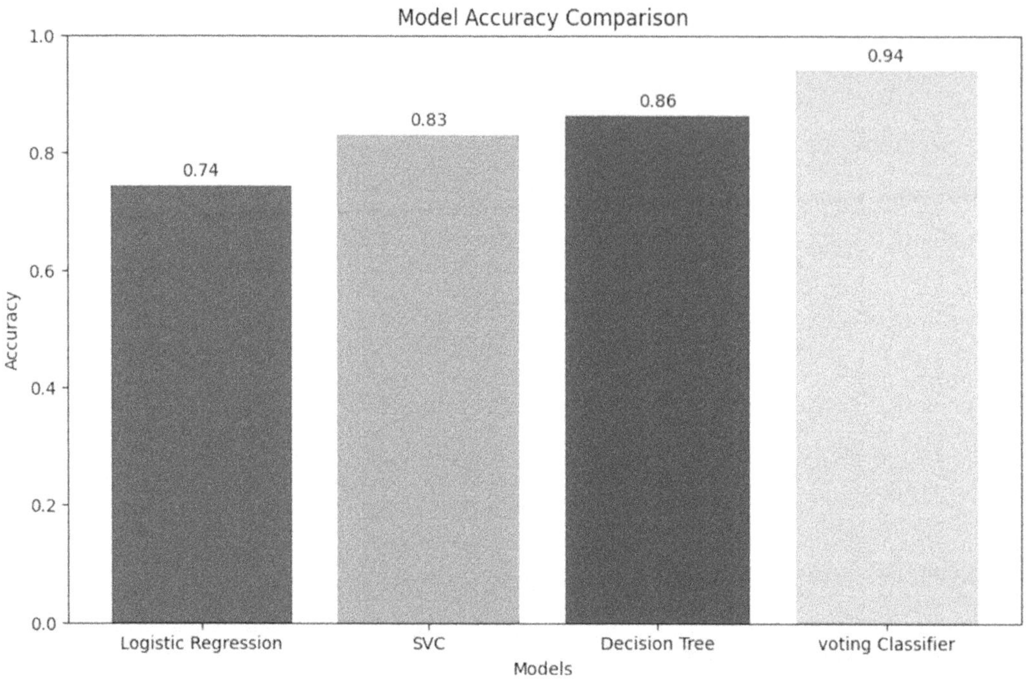

Figure 7.16 Accuracy comparison of models.

7.7 CONCLUSION

This research has shown that ML holds significant potential in identifying the stages of lung cancer, where early diagnosis is emphasized for better outcomes. With a dataset of 1,000 patients and PCA and correlation matrices techniques, we were able to identify the most important features and reduce noise in the data. While initial models, such as LR, SVC, and DT, gave fairly good performance, the proposed model significantly improved prediction accuracy to 94% as shown in Figure 7.16.

The proposed methodology, that is, a robust preprocessing technique along with ensemble learning, has shown the capability of ML in providing reliable predictions in medical diagnostics. Still, the study also underlines the challenges of handling high-dimensional datasets and achieving public trust in ML models. Future work will focus on developing larger datasets, improving methods for feature selection, and developing AI systems that can be explained. This study is a step toward developing advanced ML systems that can be applied in the health-care domain for early detection of lung cancer and increase survival chances.

REFERENCES

Agarwal, S., Thakur, S., & Chaudhary, A. (2022). Prediction of lung cancer using machine learning techniques and their comparative analysis. *2022 10th International Conference on Reliability, Infocom Technologies and Optimization (Trends and Future Directions) (ICRITO)*, 1–5. https://doi.org/10.1109/ICRITO56286.2022.9965052

Bhatia, S., Sinha, Y., & Goel, L. (2019). Lung cancer detection: A deep learning approach. *Soft Computing for Problem Solving: SocProS 2017*, Volume 2, 699–705.

Ghosh, S., Dhar, S., Yoddha, R., Kumar, S., Thakur, A. K., & Jana, N. D. (2024). Melanoma skin cancer detection using ensemble of machine learning models considering deep feature embeddings. *Procedia Computer Science*, 235, 3007–3015.

Jalili, A., Sajedi, H., Tabrizchi, H., & Mosavi, A. (2024). Skin cancer classification using DenseNet. *2024 IEEE 22nd Jubilee International Symposium on Intelligent Systems and Informatics (SISY)*, 333–340. https://doi.org/10.1109/SISY62279.2024.10737516

Khandakar, S., Al Mamun, M. A., Islam, M. M., Hossain, K., Melon, M. M. H., & Javed, M. S. (2024). Unveiling early detection and prevention of cancer: Machine learning and deep learning approaches. *Educational Administration: Theory and Practice*, 30(5), 14614–14628.

Kumar, A., Saini, R., & Kumar, R. (2023). A systematic review of breast cancer detection using machine learning and deep learning. *2023 10th IEEE Uttar Pradesh Section International Conference on Electrical, Electronics and Computer Engineering (UPCON)*, 1128–1133. https://doi.org/10.1109/UPCON59197.2023.10434530

Kutrani, H., & Eltalhi, S. (2022). Decision tree algorithms for predictive modeling in breast cancer treatment. *2022 IEEE 2nd International Maghreb Meeting of the Conference on Sciences and Techniques of Automatic Control and Computer Engineering (MI-STA)*, 223–227. https://doi.org/10.1109/MI-STA54861.2022.9837762

Mangukiya, M., Vaghani, A., & Savani, M. (2022). Breast cancer detection with machine learning. *International Journal for Research in Applied Science and Engineering Technology*, 10(2), 141–145.

Manjunathan, N., Gomathi, N., & Muthulingam, S. (2023). Early detection of breast cancer using machine learning. *2023 International Conference on Sustainable Computing and Smart Systems (ICSCSS)*, 165–169. https://doi.org/10.1109/ICSCSS57650.2023.10169777

Modak, C., Shahriyar, M. A., Taluckder, M. S., Haque, M. S., & Sayed, M. A. (2023). A study of lung cancer prediction using machine learning algorithms. *2023 3rd International Conference on Electronic and Electrical Engineering and Intelligent System (ICE3IS)*, 213–217. https://doi.org/10.1109/ICE3IS59323.2023.10335237

Mudaliar, N. K., Redekar, S., Kadam, N. M., & Ramachandran, J. (2023). Machine learning-based prediction model for cancer detection. *2023 2nd International Conference on Applied Artificial Intelligence and Computing (ICAAIC)*, 383–388. https://doi.org/10.1109/ICAAIC56838.2023.10140323

Nasser, F. K., & Behadili, S. F. (2022). Breast cancer detection using decision tree and k-nearest neighbour classifiers. *Iraqi Journal of Science*, 63, 4987–5003.

Nasser, I. M., & Abu-Naser, S. S. (2019). Lung cancer detection using artificial neural networks. *International Journal of Engineering and Information Systems (IJEAIS)*, 3(3), 17–23.

Oyetoro, A., & Olaoye, F. (2024). Investigating the application of machine learning and deep learning for skin cancer detection. Preprint, https://easychair.org/publications/preprint/5X2pw/open

Panda, P., Bisoy, S. K., Kautish, S., Ahmad, R., Irshad, A., & Sarwar, N. (2024). Ensemble classification model with CFS-IGWO based feature selection for cancer detection using microarray data. *International Journal of Telemedicine and Applications*, 2024(1), 4105224.

Pandey, P., Mayank, K., & Sharma, S. (2023). Biomarker cancer prediction system using artificial intelligence. *2023 International Conference on Integration of Computational Intelligent System (ICICIS)*, 1–5. https://doi.org/10.1109/ICICIS56802.2023.10430291

Priya, M. L. V. A., & Subbarao, M. V. (2023). Performance analysis of machine learning techniques for multi-organ cancer detection and classification: A comparative study. *2023 International Conference on Emerging Research in Computational Science (ICERCS)*, 1–7. https://doi.org/10.1109/ICERCS57948.2023.10434014

Raj, R., Singh, A., & Kumar, K. (2024). Machine learning in medical diagnosis of cancer. *2024 IEEE 9th International Conference for Convergence in Technology (I2CT)*, 1–5. https://doi.org/10.1109/I2CT61223.2024.10544354

Rodríguez-Ruiz, A., Krupinski, E., Mordang, J. J., Schilling, K., Heywang-Köbrunner, S. H., Sechopoulos, I., & Mann, R. M. (2019). Detection of breast cancer with mammography: Effect of an artificial intelligence support system. *Radiology*, 290(2), 305–314.

Sherine Glory, J., Bhavani, M., Saratha, B., Akila, A., Prathusha Laxmi, B., Vijayaraja, V., & Raghavendran, N. (2025). Enhancing breast cancer detection with machine learning: A predictive modeling approach. *South Eastern European Journal of Public Health*, 3717–3726.

Singh, A., Kumar, R., & Rastogi, R. (2022). Study of machine learning models for the prediction and detection of lung cancer. *2022 11th International Conference on System Modeling Advancement in Research Trends (SMART)*, 1243–1248. https://doi.org/10.1109/SMART55829.2022.10047610

Singh, P., Nagill, J., & Saini, K. (2023). Using supervised learning for breast cancer detection using AIML. *2023 5th International Conference on Advances in Computing, Communication Control and Networking (ICAC3N)*, 281–285. https://doi.org/10.1109/ICAC3N60023.2023.10541492

Sruthi, G., Ram, C. L., Sai, M. K., Singh, B. P., Majhotra, N., & Sharma, N. (2022). Cancer prediction using machine learning. *2022 2nd International Conference on Innovative Practices in Technology and Management (ICIPTM)*, 217–221. https://doi.org/10.1109/ICIPTM54933.2022.9754059

Talukder, M. A., Islam, M. M., Uddin, M. A., Akhter, A., Hasan, K. F., & Moni, M. A. (2022). Machine learning-based lung and colon cancer detection using deep feature extraction and ensemble learning. *Expert Systems with Applications*, 205, 117695.

Times of India. (2024). More than 50% cancer deaths due to late diagnosis, say doctors. Retrieved January 28, 2025.

Vasudevan, B., SenthilKumaran, R., Immanuel, K., & Karthikeyan, M. V. (2024). Non-invasive early pancreatic cancer prediction with gradient boosting machine learning models using urinary biomarkers. *2024 International Conference on Advances in Computing, Communication and Applied Informatics (ACCAI)*, 1–6. https://doi.org/10.1109/ACCAI61061.2024.10602237

Verma, A., Shah, C. K., Kaur, V., Shah, S., & Kumar, P. (2022). Cancer detection and analysis using machine learning. *2022 Second International Conference on Computer Science, Engineering and Applications (ICCSEA)*, 1–5. https://doi.org/10.1109/ICCSEA54677.2022.9936457

Wani, N. A., Kumar, R., & Bedi, J. (2024). DeepXplainer: An interpretable deep learning-based approach for lung cancer detection using explainable artificial intelligence. *Computer Methods and Programs in Biomedicine*, 243, 107879.

Watanabe, A. T., Lim, V., Vu, H. X., Chim, R., Weise, E., Liu, J., Bradley, W. G., & Comstock, C. E. (2019). Improved cancer detection using artificial intelligence: A retrospective evaluation of missed cancers on mammography. *Journal of Digital Imaging*, 32, 625–637.

World Health Organization. (2024). Global cancer burden growing amidst mounting need for services. Retrieved January 28, 2025.

Yoon, J. H., Strand, F., Baltzer, P. A., Conant, E. F., Gilbert, F. J., Lehman, C. D., Morris, E. A., Mullen, L. A., Nishikawa, R. M., & Sharma, N. (2023). Standalone AI for breast cancer detection at screening digital mammography and digital breast tomosynthesis: A systematic review and meta-analysis. *Radiology*, 307(5), e222639.

Chapter 8

Harnessing neuroplasticity

Advances in brain recovery and cognitive rehabilitation

Aashita Chhabra, Ram Sudan, and Seema Malik

8.1 INTRODUCTION

The scientists and health-care practitioners were unaware that differences in how and when the human brain changes, as well as the human brain's phenomenal ability to reorganize its structure, form connections, and even adapt to a new state following an experience or learning from an event were in fact possible. Brain injury, neurological disease, and even cognitive limitations were often viewed as permanent conditions because of the previous perspective that neurologic recovery was a lost cause. But now that is no longer the case. Without a doubt, we now know that after a significant injury, the human brain can grow and reconnect, adapt, rebuild, and frequently lead a better quality of life than was previously thought.

The brain's ability to grow, connect, and relocate functions can be a first step toward significant improvements. It can also lead to the development of therapies that encourage the brain to reintroduce itself rather than act as though it only contains neurologic symptoms. This shift in understanding goes beyond merely modernizing our science; it also involves a shift in how we view the future. The idea of brain injury is no longer seen as a permanent state, but rather as the beginning of recovery and the application of learning and adaptation to restore lost abilities.

Understanding strokes, traumatic brain injuries (TBIs), and other degenerative neurological diseases entails applying the brain's plastic potential to adopt a more optimistic perspective on these conditions and putting evidence-based treatments (interventions) into practice. Previously, the focus was on compensating patients to help them adjust and learn how to use assistive devices or to make changes that would lessen the impact of their impairments on their results. This philosophy of neuroplasticity has changed practice. Today, the focus of any therapy should be to help the brain relearn and repair itself, rather than to work around declines by encouraging adaptive learning or coping.

The goal is to promote the recovery process and actively stimulate recovery versus simply managing the experience of declines; this is possible through experience-dependent plasticity. The brain and its wiring are not set in stone; this wiring is changeable in its capacity for experience and effort. The brain will create new neural pathways through learning new motor skills, practicing the movements, or engaging in targeted cognitive tasks. New pathways will allow the brain to adapt to the task, allow for compensation, and, in some cases, may allow for restoration and recovery of previously thought lost functions. Consider the case of a patient recovering from a stroke, who may have difficulty with speech or memory. Through purposeful and repetitive exercises, the patient will slowly retrain their brain to identify other routes that will allow them to perform the tasks they struggled with. Eventually, the

DOI: 10.1201/9781003685890-8

brain essentially "relearns" how to communicate, remember, or coordinate movement, making things that once seemed impossible a daily accomplishment. One of the major parts of neurorehabilitation is cognitive rehabilitation, the goal of which is to take advantage of this neuroplasticity in a purposeful way. However, it is not merely repetition drills and tasks; it is a guided experience whereby the patient rebuilds their abilities through actual biological changes to the physical structure of their brain.

Cognitive rehabilitation focuses on regaining essential functions (like memory, attention, language, processing, and problem-solving) – engaging in structured exercises to purposefully challenge their brain and persuade the brain to adapt and form new neuronal pathways. These exercises are set up in a way that is progressively more difficult so that the brain will continuously be engaged and challenged to grow. Cognitive rehabilitation does not stop once full recovery is not possible. When physicians work with patients, therapists show patients how to adapt to their impairments by helping them learn compensatory techniques to help with daily tasks. A patient with a memory problem may learn new techniques to organize information and may also use external resources such as calendars and digital memory prompts. The goal is to assist patients in regaining as much independence as they can which enhances their confidence and quality of life. According to the author, the ultimate goal of cognitive rehabilitation is not merely to conduct cognitive exercises but to help patients translate these exercises into meaningful improvements in their everyday lives. Rebuilding a routine, connecting or reconnecting with loved ones, and giving people back a sense of purpose and belonging is the primary goal of these therapies.

All of these are linked to the biology of the brain; when a person practices a particular task repeatedly, the brain increases the strength of connections involved in that task through a biological process called long-term potentiation (LTP), which makes communication between the two neurons more efficient. In cases of weakened pathways caused by injury or disuse, damage or loss of function caused by injury may be improved [1].

As noted by researchers, long-term depression (LTD) plays a critical role in reducing the strength of inefficient neural pathways, thereby enabling the brain to eliminate redundant connections and promote the development of more functional ones. It is LTP and LTD, working collaboratively, that ensure the structure of the brain is continually refined, optimizing pathways that are regularly used while terminating those of least use. The expression "use it or lose it" is frequently used to summarize this idea. The pathways will deteriorate and eventually vanish if they are not utilized. However, with focused therapy, patients are actively working to improve their use of those practices, to help them look positive, to guide the brain to adapt in a more positive way, and to encourage the rebuilding and possibly remapping of connections and pathways [2]. Author noticed that it is critical to understand that the healing process is a highly individualized one that goes beyond neurons and networks. A severe loss of memory, identity, and independence can be experienced because of brain injuries. In their quest to regain the abilities and capacities that have been taken away from them in order to regain some measure of independence, many patients face daily struggles with frustration, anxiety, fear, and discouragement. Cognitive rehabilitation gives patients the tools, resources, and techniques to feel like they are actively involved in their recovery by enabling them to visualize a journey from hopelessness to hope. It is more than just "therapy"; it is one of the few things that can help people regain their sense of identity, meaning, and purpose in life [3].

They noted that "Think of the injured brain as a garden" that has been ripped apart by a storm. Distances may be obliterated and the old routes buried. However, with time, the brain will be able to sow new seeds and, with care, make them grow. Therapists help patients plant seeds of skills in their brains, much like gardeners do. Encouragement, repetition, and persistence fertilize the seeds. Families, caregivers, and friends are all part of an enriched environment in addition to medical professionals. Their understanding, support, and encouragement create the "enriched environment" that the brain requires to adapt and thrive [4]. Together, they build networks of hope to foster resilience, encourage commitment, and transform recovery from an isolated fight for health into a contextual experience we are all enriched from, and a journey we share simultaneously. Recovery is not a straight line. Each patient's journey is unlike their peers. Each patient's recovery has its own experiences, emotions, and support systems – their own setbacks, breakthroughs, frustrations, and successes. Of course – it is also a journey of rediscovery and of reestablishing relationships, habits, and meaning – but whatever the recovery was, a father talking to his children again, a student focusing again, or an elderly woman with a renewed sense of independence – each moment is a victory [5]. They stated that the accumulation of small steps leads to a compilation of progress that ultimately reshapes the landscape of what is possible. Neuroplasticity has been called the science of second chances, providing insight into how the flexibility of the brain opens up pathways toward healing. With the right combination of therapy, compassion, and technology, recovery is less about merely surviving and more about thriving. Fortunately, current therapies that hold promise are just the beginning. As science progresses, the future of neurorehabilitation continues to be bright [6]. Currently, scientists are looking into how a patient's genetic composition, brain activity patterns, and real-time feedback can help guide customized treatments with protocols that are tailored to the patient's unique requirements.

Technologies that were once thought of as science fiction fantasy, such as wearable technology that tracks a patient's neural activity or AI that allows for personalized exercise regimens, are now incorporated into treatment regimens. Patients will eventually be able to rebuild their lives in a way that is both scientific and human thanks to these technologies, which will enable treatments that may be more efficient, compassionate, and responsive to their immediate needs. Recovery involves relationships as much as biology [7]. It has been explained that the patient's inner space, emotions, and beliefs are all intertwined with the brain's healing process. An atmosphere is created that permits neuroplasticity to occur when there is optimism, perseverance, and faith in the potential for recovery. The scaffolding or infrastructure that supports the patient's healing is the family, therapist, and caregivers. The system that offers inspiration, occasionally vents frustration, acknowledges minor triumphs, and offers support. All of these things work together to create a supportive network that fortifies the spirit and the brain. The human heart is at the center of cognitive rehabilitation, and it must be kept in mind as we examine its science, practices, and future. What transforms treatment into healing and healing into purpose is the relationship between the patient, family, and therapist. It is normal to feel overwhelmed after an illness or injury when you feel as though your brain, which had previously understood you well enough to get by, is suddenly not trustworthy [8]. They feel as if pieces of you have been taken: your memories, confidence, and ability to connect to people has been altered or taken away. However, the amazing truth is that your brain is infinitely more flexible than it may feel; it is not stuck in its old ways. The human brain can challenge and adapt to adversity, learn new ways of functioning, and restore something that has seemed lost with time, effort, and care. Recovery is about reestablishing your brain in small, manageable steps, with consideration and gentleness, rather than pushing yourself past your breaking point. It is the people around you, not just the science, that give this journey its power [9].

Every supportive moment, patient session, and supportive family member all help create the healing environment. The brain gets its nourishment from connections, hope, and perseverance in addition to practices. Transformation truly begins with the work you do, the trust you have built with the therapist, and the support you receive from those who care about you, even on the most difficult days. Even though it may not always be obvious, the road ahead will present opportunities, and every step you take will train the brain to believe that recovery is possible. We are not alone on this journey, and you are getting closer to living a more complete, connected life with every second you spend rebuilding. Resilience, tenacity, and the capacity to accept and depend on the support of others characterize the personal experience of recovering from brain injury, which goes beyond a simple clinical procedure. Beyond prescribed exercises and medical care, rehabilitation focuses on the patient's willingness to take on challenging tasks, their ability to cope with uncertainty, and their belief that change is possible. While each patient's path through the rehabilitation continuum is unique, containing both successes and setbacks, all patients have the capacity to advance [10]. Although specific interventions have been mentioned that will help improve the brain's reorganization and adaptability, the patient's emotional and psychological environment also has a fundamental impact and facilitates this process. Neuroplasticity can help the brain move toward meaningful healing when people work hard and provide support, empathy, and encouragement. The supportive network that collaborates to advance rehabilitation is just as crucial. Family, care, and therapy work together as a network of support that encourages emotional and cognitive recovery. These facilitators are in charge of encouraging the patient, supporting their efforts, enhancing their positive progress, and assisting them with every obstacle they encounter by applying the skills they have learned during their rehabilitation. It may be possible to increase the patient's ability to actively participate in their therapy process and to transform challenges into learning opportunities by combining better surroundings with a rich and encouraging experience [11]. Hence the author stated that cognitive rehabilitation is therefore more than just a way to treat neurological function; it starts to transform the process into a shared experience of regaining hope, dignity, and purpose as science and human connection come together.

Cognitive impairments following stroke significantly affect patients' ability to perform activities of daily living, yet they are often underemphasized in rehabilitation practices. Existing cognitive rehabilitation approaches, although effective, are limited by high demands on time and clinical resources and often lack a structured theoretical basis for task design. To address this gap, the study proposes a participatory design–based framework for generating personalized paper-and-pencil cognitive rehabilitation tasks. A set of standardized tasks was parameterized and evaluated by rehabilitation professionals to assess their impact on cognitive domains such as attention, memory, language, and executive function. The framework demonstrated high internal consistency and enabled the identification of task parameters influencing cognitive difficulty. Using computational modeling, the authors developed a web-based task enerator tool that produces customized rehabilitation exercises tailored to individual patient needs. This approach introduces an objective, scalable, and personalized strategy for cognitive rehabilitation, enhancing both clinical applicability and patient-specific intervention design. [12]. Figure 8.1 illustrates how the use of cognitive rehabilitation is made possible by the fundamental idea of neuroplasticity. Restoration and recovery are the objectives of this therapeutic process, which is driven by a collaboration between patients, therapists, and technology. This ultimately results in not only functional recovery but also lives that can be rebuilt on a foundation of renewed possibility and hope.

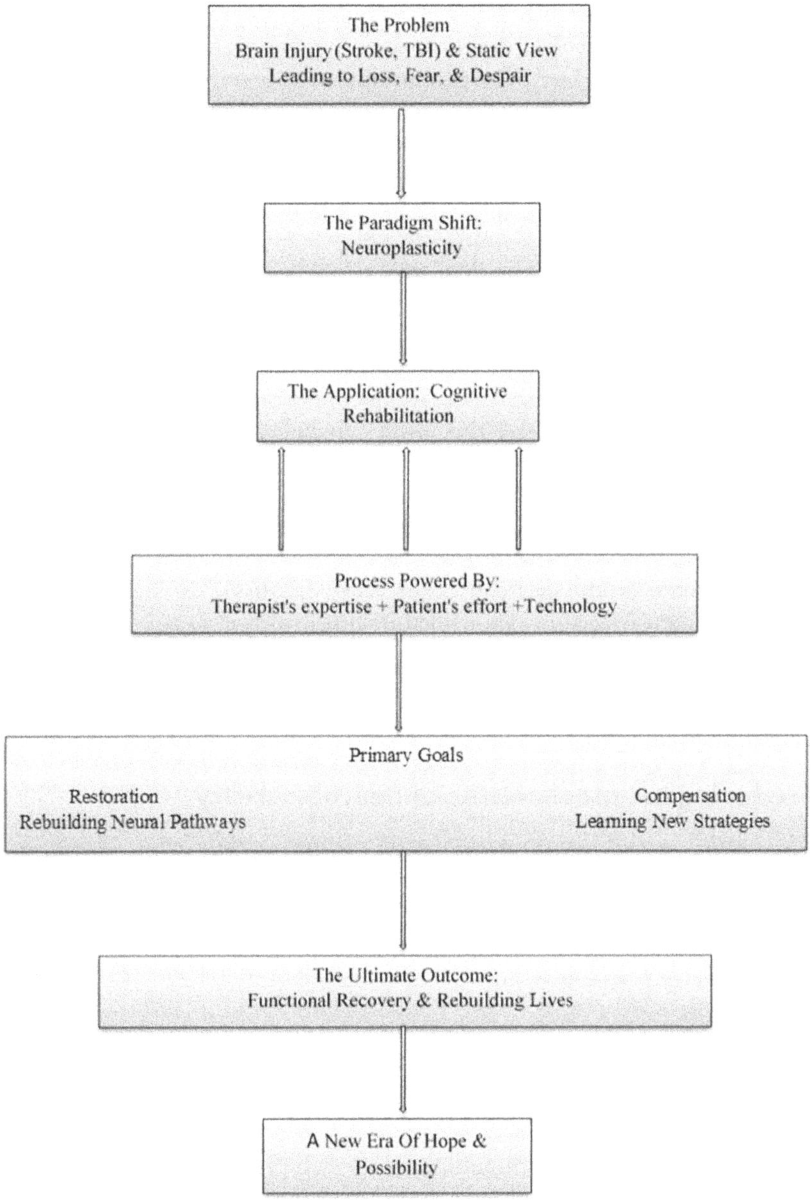

Figure 8.1 Rehabilitation through neuroplasticity.

8.2 REVIEW OF KEY STUDIES

8.2.1 The foundations of a changing brain

It has been stated by the author that neuroplasticity or the brain's ability to adapt and change is not just a process; rather, it is a collection of biological processes that cooperate to enable the nervous system to adjust to experiences, learning, and injury. At its core, it is the notion that our brains can "rewire" themselves when necessary. The brain contains billions of neurons, which are specialized nerve cells [13]. The author describes synapses as flexible junctions through which neurons interact. Rather than being static, these connections remodel themselves depending on activity levels. Consistent signaling between two neurons enhances their connectivity via the process of LTP. When communication between the two neurons declines, the strength of the connection between the neurons declines also through something called LTD. These processes create new memories, help us learn new skills, and help us adapt to new environments. The memorable phrase "neurons that fire together, wire together" very clearly represents this idea that through repeated and synchronized engagement, the brain strengthens pathways for faster and more efficient processing [14]. At birth, our brains are populated by many more connections than they really need. As our brains grow and learn, they prune unused pathways and maintain just the ones that enable the most efficient processing, also known as synaptic pruning. This process is important to fine-tune how we think and respond to the world and keep the brain agile and sharp. However, neuroplasticity does not just relate to changing the strength of existing connections; it can also relate to changing the structure of existing connections (also known as neural plasticity). When two neurons activate together, the connection can also strengthen in terms of structure, as neurons grow new branches called dendrites to establish more connections to other neurons. Essentially, the neurons rearrange their circuitry to either incorporate better or learn new information. In certain brain areas, that allows for the generation of new neurons through a process known as neurogenesis. Structural changes can occur when the brain is engaged in impactful or transformative learning experiences, or while it recuperates following an injury. These structural changes can achieve long-term changes in memory by improving overall cognition and demonstrate how flexible and adaptive the brain really is [15].

8.2.2 Core biological mechanisms of neuroplasticity

Researchers emphasize that neuroplasticity, at its most fundamental level, involves the brain's capacity to establish and reshape connections for effective communication. The human brain contains billions of neurons linked by trillions of synapses, and it is within these synaptic junctions that the processes of plasticity occur. Rather than being fixed like static wiring, these neural connections remain flexible and responsive to change. Unlike wires, they are living networks that can be altered based on our needs. Synaptic plasticity, which allows us to change the strength of the connection between two neurons, is one of the main mechanisms of that flexibility [16]. Stated by the author that LTP, which is a strengthening connection when two neurons fire together repeatedly, and LTD, which is a weakening connection when two neurons are not activating or are activating asynchronously in order for the brain to "drop" the less useful pathway, are two processes that stand out when considering synaptic plasticity. To put it succinctly, "neurons that fire together wire together." These processes work together to help the brain create new skills, encode memories, and continuously remap itself in response to experience [17]. It has been noted that plasticity extends beyond modifying connections to include structural changes as well. During early development, the brain generates an excess of synapses, many more than are ultimately required. Over time, the less active or inefficient connections are gradually removed through a mechanism known as synaptic pruning. Using the "use-it-or-lose-it" signal, this selective pruning aids the brain in concentrating on the most effective connections and pathways. This speeds up our ability to think and organize as we mature. In addition, structural plasticity encompasses modifications to the actual structure of the neuron. Dendrites are new branches that neurons can grow, forming more connections and allowing them to take in more information. It was once believed that adults could not develop new neurons in specific areas of the brain, such as the hippocampus. These changes frequently occur during learning or the brain's recuperation from an injury; each one creates a more profound and long-lasting mechanism for additional cognitive improvement or recovery. Utilizing both synaptic and structural plasticity demonstrates the brain's capacity for adaptation, resilience, and structural change [18].

8.2.3 Influences on the brain's capacity for change

It has been observed by the author that even though all humans share the biological mechanisms of neuroplasticity, each person's capacity for brain change and reorganization varies greatly depending on a variety of internal and external factors. Age is the most commonly recognized external influencer. We know that the brain undergoes "critical periods" in early childhood where it is primed for the learning of new skills such as language or vision, in a dynamic process allowing for rapid and efficient growth. However, although it may diminish, plasticity does not completely diminish with age; for example, the brain remains active in adapting, learning, and refining itself as an adult [19]. It has been observed that adolescence is marked by significant changes, including increased vascularization and the pruning of redundant connections, which contribute to the maturation of brain regions associated with decision-making. While certain plastic processes may slow in later years, the brain continues to demonstrate the capacity to form new connections and to compensate for the effects of injury or aging; this important function of brain plasticity conveys why learning can extend across the lifespan along with cognitive functioning until significantly impaired. Furthermore, each person's level of neuroplasticity can be influenced by their genetics and life experiences; for instance, certain genes influence how adaptable the brain is. The gene that produces the protein known as Brain-Derived Neurotrophic Factor (BDNF), which encourages neuron survival and new growth, is a good example. Genetic variations in BDNF levels will affect how well the brain learns or recovers. Environmental factors that have been demonstrated to interact with biological development go beyond biology. In addition to promoting physical exercise, meaningful social interaction, and cognitive challenges, a stimulating environment also promotes the growth of new neurons and fortifies existing ones [20]. Conversely, long-term stress, loneliness, and deprivation activate hormones that cause brain damage and waste the current neuronal network, inhibiting the same brain processes that promote neuron health. Therefore, it is possible to view acquiring new skills, engaging in physical activity for health, and maintaining fulfilling relationships with others as not only beneficial behaviors but also as significant support for the brain's inherent capacity for growth and repair. The future development of the brain's ability to change and evolve over the course of a lifetime will ultimately depend on a combination of age, gene expression, and life experiences [21].

8.3 BRAIN RECOVERY: AN OVERVIEW

It has been highlighted that when the brain sustains an injury, it does not passively accept the resulting deficits. Instead, it initiates a complex and coordinated process of repair and adaptation. This discussion focuses on identifying the types of injuries most responsive to neuroplastic interventions, while also distinguishing between the brain's inherent recovery mechanisms and those enhanced through therapeutic neurorehabilitation. It may be useful to think of the brain as an extensive and circuitous network of neural pathways, highways that have been established over time related to everything from movement to memory, and an injury caused by a stroke or traumatic injury is like a massive earthquake that destroys or disrupts neural pathways and severs communication or blocks critical pathways [22].

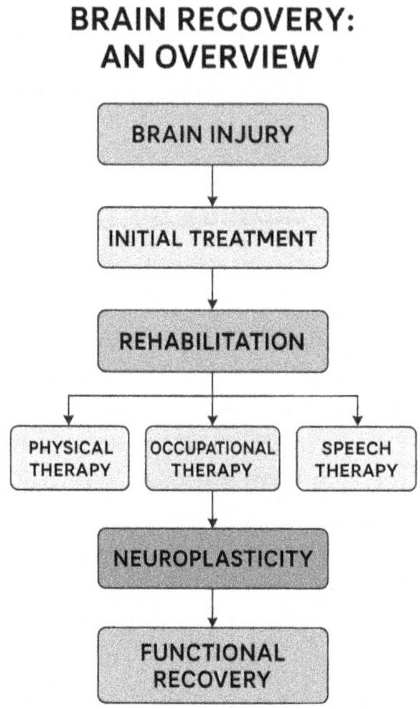

Figure 8.2 Brain recovery: an overview.

However, the first response from the brain is not "give up." The brain reacts by initiating a complex series of protective responses: stopping inflammation, clearing damaged cells, and readjusting for the difficult work ahead of it to recover. Following the principles of neuroplasticity, the brain begins to investigate pathways – seeking out new ways of reconnecting and communicating. There are two broad ways of repairing communication. The first example is when the brain finds its own ways around the damaged area; it will try different ways that may not always be the way it is supposed to, but sometimes the individual has to fail to find a different way [23]. The second example is the therapist facilitating the remapping – the therapist is a special kind of tour guide who helps guide the remapping process so the brain knows the most effective routes for recovery. With these controlled interventions helping with the remapping, these guided interventions will hopefully result in more stable and longer lasting paths/routes that will change what was once an injury-torn landscape into a reorganized, resilient landscape that can function again [24]. Figure 8.2 represents the flowchart that represents the process of brain recovery after injury, beginning with treatment and rehabilitation. Through therapies and neuroplasticity, the brain adapts and rebuilds, leading to functional recovery and improved quality of life.

8.3.1 Conditions ripe for neuroplastic recovery

It has been highlighted in research that there are possibilities for the brain to change and reorganize itself, known as neuroplasticity, across a wide range of neurological conditions. Some conditions are especially amenable to organized therapeutic treatments. Stroke is perhaps one of the best documented examples. Whether it's speech, movement, or coordination, a stroke will impair a particular brain function in that region [25]. However, the right brain, which is close by, is always prepared to step in and compensate for some of the loss, at least in part. With the help of directed therapy and practice that includes the development of motor skills and communication, a patient can regain neural pathways and regain more fundamental functions if rehabilitation is properly planned. This recuperation process demonstrates how certain neural circuits can be strengthened and used to support severely compromised functions. TBI is also a case where neuroplasticity is important for recovery, although often the areas of the brain affected are more broad, complex, and much more diffusely impaired. In these cases, rehabilitation focuses on optimizing what is left of the brain's resources and facilitating it to develop new networks to make up for impaired areas. The brain learns to make new pathways in the same speed as traffic systems adapt to make alternative routes when roads are blocked [26]. Other neurological conditions such as multiple sclerosis, cerebral palsy, and even recovery following brain tumor surgery also benefit from neuroplastic processes. These conditions potentially disrupt the brain's structure and signaling pathways; however, it may still be possible through the right interventions and perseverance for the nervous system to adapt and reorganize, creating new hope for recovery and enhanced quality of life.

8.3.2 Spontaneous vs. therapy-induced recovery

It has been described that following an injury, the brain engages in two interconnected recovery trajectories: one that occurs naturally and immediately, and another that is intentionally guided. The first, termed spontaneous recovery, reflects the brain's self-initiated attempt to repair itself without external intervention. In the hours and days following a brain injury, the brain methodically works to decrease swelling, clear away damaged cells, and restore blood flow. On a cellular basis, the brain might also strive to activate unused pathways or randomly generate new connections to compensate for lost functions. All these changes indicate that the brain has a natural response to injury; however, it is often an unplanned and incomplete response at best [27]. The response is important; however, it is not always functional or influential in promoting healing and, in the worst-case scenario, it may create inefficient or maladaptive pathways that disrupt recovery. The second trajectory, therapy-induced recovery, is an intentional and systematic approach to recovery based on neuroplasticity principles. Once the brain has initiated the process of healing itself, therapy provides a road map or guide for channeling the brain's adaptability in a functional and meaningful manner. Through exercises that are intentional and systematic yet repetitive, engaging, and relevant to the person, therapy helps the brain create stronger and more efficient neural connections and eliminate inefficient connections [28]. A useful analogy mentioned by the author for understanding the difference between spontaneous recovery and therapy is to think about spontaneous recovery as a neighborhood constructing random and winding footpaths to avoid sudden barriers, while therapy has delivery methods that provide step-by-step organized planning and infrastructure for a freeway. With effective stimulation and structured practice, the interventions in therapy work with, rather than against, the potential of strengthening useful connections (LTP) and eliminating unnecessary connections (LTD), and ensure it can facilitate recovery that is usable and permanent. The differences between the two paths show how natural brain recovery has potential but with skilled guidance, a collaborative engagement of the brain's natural processes leads to an integrated way of recovery through function and quality of life improvement [29]. Two different approaches to brain recovery are compared in Table 8.1, which tells about the two distinct modes of brain recovery.

Table 8.1 Comparative analysis of two distinct modes of brain recovery

Aspects	*Spontaneous recovery*	*Therapy-induced recovery*
Nature	Automatic, passive, undirected	Active, guided, purposeful
Driver	Natural biological healing (e.g., reduced swelling)	Specific, repetitive, targeted therapeutic tasks
Neural Process	Haphazard, nonspecific, potentially maladaptive	Organized, efficient, and functional rewiring
Timing	Primarily in the immediate days/weeks post-injury	A longer-term process driven by consistent effort
Analogy	Random footpaths forming around a roadblock	Engineers paving a new, efficient highway

8.4 COGNITIVE REHABILITATION STRATEGIES

It has been noted that there is no universal method for achieving effective cognitive rehabilitation. Instead, it is understood as an empirically supported process that systematically works to restore cognitive abilities essential for everyday functioning, tailored to the individual's specific needs and circumstances. Daily life relies heavily on cognitive processes such as attention, memory, and executive function, which form the basis for interaction with the external world. When these processes are disrupted, both emotional and physical consequences may arise. For example, difficulties with attention extend beyond test scores to include challenges such as following a conversation or retaining critical information. Similarly, impairments in memory are reflected not only in forgetting names but also in missing important commitments or becoming disoriented in familiar environments. This realization emphasizes that general exercises are insufficient for effective cognitive rehabilitation. It is an organized method of retraining the brain that is unique to each person and identifies which "cognitive muscles" require strengthening through particular exercises. The goal is not just academic remediation; it is also a rehabilitation process that helps people regain their independence and confidence in their capacity to carry out daily and interpersonal tasks. A field of proven techniques for addressing cognitive function following brain injury has emerged as a result of years of scientific study. One of the best techniques is Attention Process Training (APT), in which attention is treated like a muscle that can be strengthened by performing increasingly challenging exercises. APT aims to help patients learn to multitask, resist distraction, and maintain focus [30]. Two categories of strategies which are mentioned by the author are typically employed for memory rehabilitation. Restorative strategies, such as spaced retrieval, build the brain's natural abilities to encode and retrieve memory with longer intervals between recall attempts. Compensatory strategies are ways to work around memory loss. They can be internal, such as mnemonics and visualization, or external, such as calendars and alarms. Executive function training (like Goal Management Training (GMT)) teaches patients to think more deliberately through structured steps (i.e., goal management) in the development of cognitive function. The process teaches the patient how to break tasks into steps and monitor those steps in the achievement of goals. In this way, high-level cognitive control is rebuilt, and patients become more deliberate in their thinking when it comes to their goals. Central to cognitive retraining is personalization. A generic brain-training tool cannot replicate the depth of a therapy program entering on the person's brain function, lived experience, and identified desires. The process begins with a thorough neuropsychological evaluation that determines the patient's strengths and weaknesses and develops a cognitive profile that directs the treatment. From there, the patient and

Table 8.2 Cognitive rehabilitation strategies

Cognitive domain	*Daily life impairment examples*	*Rehabilitation approach/technique*	*Description of technique*
Attention	Inability to follow a conversation; difficulty retaining all information.	Attention process training (APT)	Treats attention like a muscle, using increasingly challenging exercises to help patients learn to multitask, resist distraction, and maintain focus
Memory	Forgetting important commitments; getting lost in familiar location	Restorative strategies	Builds the brain's natural ability to encode and retrieve memory by using progressively longer intervals between recall attempts
Executive function	Difficulty with planning, organization, and goal achievement	Goal management training (GMT)	Teaches patients to think more deliberately by breaking tasks into structured steps and monitoring those steps to achieve a goal, thereby rebuilding high-level cognitive control

therapist work together to establish engagement-based, practical goals, such as remembering family birthdays or cooking a meal they have never prepared before. The therapy is sparked by these personal goals, which also influence the therapeutic activities [31]. The Table 8.2 lists particular rehabilitation strategies for addressing deficits in important cognitive areas such as executive function, memory, and attention.

8.5 THE ROLE OF DATA AND AI IN PERSONALIZING NEUROREHABILITATION

Aside from the concentrated efforts with virtual reality (VR) and transcranial magnetic stimulation (TMS), they are enthusiastic about the future of neurorehabilitation and the potential to leverage data analytics and the fundamental principles of Internet of Things (IoT) to advance to new heights. The next phase is to establish a continuous, comprehensive feedback loop between the clinical team and the patient's life that is centered on data. This will help the patient recover without the need for constant clinical services. Patients will wear sensors that function as Internet of Things devices that can collect information on motor functioning, including how much they slept, how much they worked out at home, and other factors. In contrast to the hour they spend with patients during weekly appointments, this gives clinicians unmatched objective data about the course of a patient's recovery. Furthermore, it may be possible to use these massive datasets to produce useful clinical recommendations if AI-powered data analytics are applied [32]. They could transform neurorehabilitation into a completely new kind of "predictive maintenance for the brain" if data from thousands of patients are collected and machine learning models are developed to find almost undetectable patterns and recovery predictors. For example, creating an algorithm that can predict when a patient is likely to experience difficulties or, more crucially, when to increase the intensity of treatment. Our ultimate goal of providing hyper-individualized care is in line with this evidence-based strategy. With a dynamic, living protocol rather than a set rehabilitation plan that can be adjusted in real time to each patient's unique neural and functional recovery, the therapeutic efficacy of each intervention can be maximized [33]. The first step in gathering this kind of patient data is collecting large volumes of data, but efficient analysis and visualization of the data are crucial to maximizing any therapeutic benefits. Raw data streams from therapy apps or wearable sensors are frequently too complicated to be useful. The transformation of patient data into user-friendly dashboards that are displayed to patients and clinicians is a necessary step in the future development of digital therapeutics. Well-designed dashboards could instantly show therapists patterns in motor performance, cognitive task success rates, and adherence to the home program, giving them the freedom to change the treatment plan as necessary. For the patient, the visual output serves as another motivating aspect linking the experience of self-efficacy to improvement. Much the same way a student uses a fitness app to see daily imprints of improvement, a rehabilitation dashboard can generate positive visual feedback on accomplishments. For patients, seeing a graph of their own functional progression, no matter how small, is an important source of engagement in a difficult and lengthy process of rehabilitation. This paradigm turns the patient from a passive recipient of care into an active, informed agent in their own recovery, directly stimulating the activation and commitment essential for stimulating neuroplastic change. This transforms the patient from a passive recipient of care into an active, informed agent in their own recovery, directly fueling engagement and determination that are critical for driving neuroplastic change. As artificial intelligence becomes more immersed in health care, its acceptance on the neurorehabilitation spectrum will depend on one essential factor – trust. Clinicians are highly unlikely to accept "black box" AI systems that issue recommendations without any opportunity for reasoned reflection. The fundamental importance of developing Explainable AI (XAI) in the future of the field is therefore obvious. An advanced XAI system would not simply recommend "increase the difficulty of the attention exercises for the patient" but would provide explainable reasoning such as, "the recommendation is based on a 95% success rate on current tasks and a 15% decrease in response time over the last 48 hours which corresponds with a small group of patients in my database that achieved successful outcomes" [34]. It has been mentioned by the author that such transparency allows the therapist to use the AI as a very sophisticated clinical collaborator. The therapist can compare and contrast the AI's rationale against their professional judgment and holistic understanding of the patient. In this way, XAI would provide the therapist critical thinking and reflection space while creating trust with the AI where the "thinking" of the AI is explicit. The clinician can feel confident that humans remain in charge of the therapeutic journey and, at best, technology supports and augments our expertise as therapists. One of the biggest hindrances to improving our knowledge of brain recovery is the inability to share sensitive patient information between research organizations. Federated Learning, a method from machine learning, has the potential to provide a truly innovative solution to this problem. Federated Learning is a collaborative method to train a powerful, predictive AI model across multiple hospitals without requiring any institution to share any of its private patient data. In this method, each hospital will independently have its own patients' data to train a local copy of the AI. The only thing sent to a central server would be the anonymous mathematical improvements from that training, without sharing the data itself. The insights from all the models will be combined to build a single global model that will ultimately benefit from the diverse datasets in each collaborating institution. In the field of neurorehabilitation, this would mean that everyone could develop accurate models that can accurately predict outcomes of stroke recovery or help identify the best therapeutic pathways for individual types of TBI, while also rigorously protecting patients' privacy. It might portend a time when international cooperation can be used to advance research without jeopardizing patient privacy [35]. Figure 8.3 illustrates the essential elements of developing a data-driven neurorehabilitation system, in which patient data and technology combine to enhance and customize recovery results.

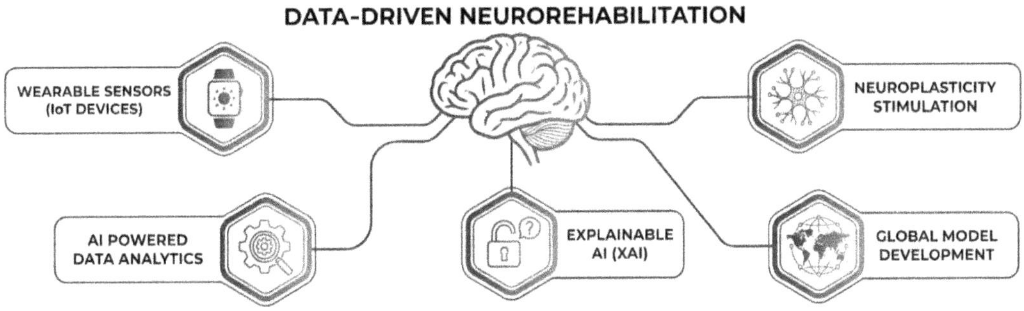

Figure 8.3 Data-powered recovery.

8.6 TECHNOLOGICAL ADVANCES IN NEUROREHABILITATION

It has been suggested that technology is now moving beyond the role of simply supplementing rehabilitation and is actively contributing to brain change. According to this perspective, digital formats are making long-term recovery more engaging and accessible, while emerging tools are creating powerful opportunities to stimulate neuroplasticity. By fostering brain adaptation and supporting meaningful recovery, these technological advancements are seen as transformative in the field of rehabilitation. While noninvasive brain stimulation will enable therapists to gently "turn on" previously underperforming brain regions for more effective therapy, technology will also enable patients to safely practice real-life scenarios in the virtual world. They can offer a variety of rehabilitation experiences. By converting ideas into actions in real time, brain-computer interfaces (BCIs) provide opportunities for individuals who have lost their ability to move. Patients can practice daily, pause, record, and review their patient reports over extended periods of time thanks to mobile and digital health solutions that integrate therapy into daily life. Together, they personalize rehabilitation and increase its appeal by possibly utilizing the brain's unique capacity for change and healing [36].

8.6.1 High-tech tools for rewiring the brain

Clinicians can now use techniques that directly facilitate brain reorganization in previously unheard-of ways thanks to a new suite of technologies.

Virtual reality: VR can be thought of as the brain's equivalent of a "flight simulator." Patients can practice real-world skills in a controlled, immersive, and safe setting with no real-world repercussions. Within the clinic, a person recovering from a stroke can practice crossing a busy street, and someone with a TBI can practice navigating a grocery store. With the use of rich sensory feedback, VR can transform monotonous tasks into captivating games that encourage the number of practice trials necessary to produce significant neuroplastic change. Focused magnetic pulses are used in TMS, a noninvasive technique that activates or deactivates brain regions [37]. It has been described that TMS is commonly utilized in rehabilitation to "prime" the brain for treatment. TMS can temporarily deactivate an overactive area of activity in the uninjured hemisphere to maximize the opportunity for the damaged or weaker hemisphere to engage and form new connections during therapy. This is analogous to preparing the soil to make it more amenable to planting new seeds of recovery.

BCIs: BCIs are direct communication linkages from brain activity to an external device. A patient suffering from severe paralysis may signal the intention to move a limb directly from their brainwaves, which the BCI will use to control a robotic arm (or cursor on a

Table 8.3 Overview of modern rehabilitation technologies

Feature	*Virtual reality (VR)*	*Transcranial magnetic stimulation (TMS)*	*Brain-computer interfaces (BCIs)*
Core principle	Immersive simulation	Noninvasive brain stimulation	Direct neural communication
Mechanism of action	Provides a safe, controlled environment for high-repetition practice to drive activity-dependent plasticity	Uses magnetic fields to increase or decrease the excitability of specific brain regions	Reads brain signals related to intent and translates them into commands for an external device
Primary goal	To rebuild functional skills (motor, cognitive, etc.) through safe and engaging practice	To "prime" the brain before or during therapy, making it more receptive to neuroplasticity change	To bypass damaged pathways and restore function by linking thought directly to an action or output
Patient interaction	Active: The patient physically and cognitively interacts with a simulated, digital environment	Passive: The patient receives magnetic pulses to the scalp; it is done *to* them	Active: The patient must consciously think or imagine actions to control a device.
Analogy	A flight simulator for the brain	Preparing the soil before planting seeds of recovery	A direct bridge from thought to action

screen). The feedback loop is immediate: the brain issues a command and the patient gets to see it enacted. This incredible coupling of intention with action is a powerful impetus for reorganizing the motor circuits of the brain, even if the body itself cannot move [38]. Table 8.3 presents a comparison of three of the most advanced neurorehabilitation technologies, VR, TMS, and BCIs. It notes that VR is immersive simulation of a physical space (or a virtual space) that patients actively practice (or attempt to improve) functional skills, such as a "flight simulator for the brain." TMS is described as a passive noninvasive brain stimulation method of stimulating the brain using magnetic fields to "prime" the brain to be more open to receiving therapy. Lastly, BCIs are described as direct neural communication methods that permit patients to actively control devices using their thoughts to bypass the pathways that are "damaged." The table also delineates these systems based on their principle, operation, and patient interaction methods.

The Rise of Digital Therapeutics: Rehab in Your Pocket, the effectiveness of rehabilitation hinges on consistency and long-term practice. Digital therapeutics and mobile health (mHealth) applications are revolutionizing this aspect of care by extending it beyond the clinic walls. These software-based platforms are very successful in promoting long-term recovery when they are distributed through computers, tablets, and smartphones. Their main benefit is that they are easily accessible. They significantly increase their practice's dosage and consistency by enabling patients to continue their rehabilitation program from home. A patient can perform quick, focused exercises several times a day in place of an hour-long therapy session three times a week. Additionally, these apps make the frequently difficult task of rehabilitation more motivating and less of a chore by implementing gamification principles (e.g., points, levels, and rewards). Additionally, they closely monitor performance data, providing therapists with a clear picture of a patient's development and difficulties in between sessions. This enables timely treatment plan modifications and remote monitoring. In the end, digital therapies are an essential instrument for sustained care. They ensure that recovery is a long-lasting, lifelong process of adaptation rather than a temporary event by helping to cement the progress made in formal therapy and offering continuing support [39].

8.7 CASE STUDIES: QUANTIFYING THE JOURNEY OF RECOVERY

Although knowing the tactics is crucial, the actual narrative of neuroplasticity is revealed by the quantified, practical recovery of individuals. The key evidence is provided by case studies, which convert therapeutic initiatives into measurable information on function, cognition, and quality of life. Now let's look at the data that has been made public regarding a male patient, age 22, who suffered a severe TBI. He had been in a vegetative state when he was admitted to a specialty rehabilitation hospital. His recuperation following a year of rigorous, multidisciplinary neurorehabilitation is shown in the following table. The process of rehabilitation started after the initial evaluation. The delivery of services was only one aspect of the therapeutic process; another was the active, cooperative relationship between the patient, his family, and the clinical team. Early objectives were fundamental and included things like enhancing visual tracking or generating reliable answers to basic instructions. As the patient emerged from the vegetative state, these goals evolved based on his progress. The team held weekly goal-setting meetings with his family. The objective was always to link clinical exercises to a meaningful, real-world outcome. A physical therapy goal to improve trunk control, for instance, was framed as the first step toward being able to sit at the dinner table with his family. A speech therapy goal to strengthen oral-motor skills was the precursor to his goal of eating his favorite food again. Maintaining motivation for the difficult task of rehabilitation required a patient-centered approach. The data's impressive outcomes are a clear indication that the principles of neuroplasticity are being used successfully. The thousands of repetitions of functional movements and everyday living activities that occupational and physical therapy practice are directly responsible for the increase in the Functional Independence Measure (FIM™) from 18 (total dependence) to 122 (modified independence). The brain was compelled to develop new, effective neural pathways for motor control as a result. Similarly, it was no coincidence that cognitive function recovered, as shown by the Repeatable Battery for the Assessment of Neuropsychological Status (RBANS) score of 85. It was the outcome of hierarchical, structured cognitive exercises that targeted weak points. This indicates that it is possible to systematically retrain abstract cognitive abilities like memory and attention. This case powerfully illustrates that when the brain is provided with the right type, intensity, and frequency of stimulation, it has a profound capacity to remap its own circuitry and restore critical functions that were once thought to be permanently lost.

The bar chart (Figure 8.4) compares a patient's assessment scores at admission versus discharge across four domains. The data shows a dramatic improvement in the patient's condition, with significant increases in the scores for (FIM™), (RBANS), and (QOLIBRI). Notably, the (DRS) score, where a lower number is better, plummeted from 21 to 1, indicating a profound recovery.

A crucial aspect of this case, highlighted in the data, is the patient's starting point of being "Not Testable" in the domains of Cognitive Function and Quality of Life. This clinical reality signifies more than just a low score; it represents a state where the individual is unable to participate in their own assessment, indicating a profound lack of awareness or ability to communicate. The journey from this nonresponsive state to achieving a "Low Average" score of 85 on the RBANS and a "Good" score of 78 on the QOLIBRI is arguably the most significant testament to the rehabilitation's success. It demonstrates that the initial therapeutic goal to reestablish awareness and basic communication was achieved, creating the foundation upon

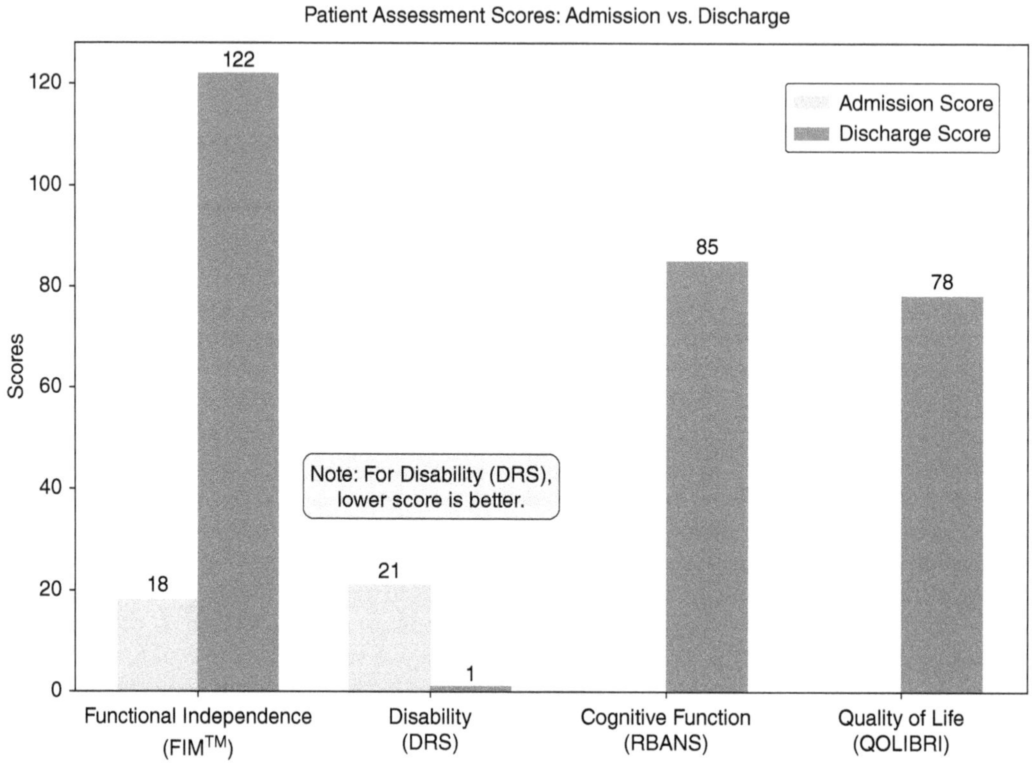

Figure 8.4 Patient recovery data.

which all other cognitive and functional skills could be rebuilt. Furthermore, the data, when viewed holistically, paints a picture of a truly comprehensive recovery rather than isolated gains. The remarkable leap to "Modified Independence" with a FIM™ score of 122 would be impossible without the parallel improvement in cognitive function. A person cannot manage their daily life without the ability to problem-solve, remember, and pay attention. The fact that the patient not only regained function but also self-reported a "Good" quality of life underscores the ultimate goal of rehabilitation. It shows that the intervention did not just restore a set of skills; it restored a person's ability to live a meaningful life with satisfaction and autonomy. Table 8.4 depicts a patient's profound recovery over one year by comparing their assessment scores at admission and discharge across four key domains. Initially, the patient presented in a "Vegetative State" with "Total Dependence," reflected by a Disability Rating Scale (DRS) score of 21 and a Functional Independence Measure (FIM™) score of 18. Following rehabilitation, their condition improved dramatically to only "Mild Disability" (DRS score of 1) and "Modified Independence" (FIM™ score of 122). Cognitively, the patient progressed from being "Not Testable" to achieving a "Low Average" score of 85, while their self-reported quality of life similarly improved from "Not Testable" to a "Good" score of 78.

The data reveals a life-changing and comprehensive recovery for the patient. While his functional independence jumped from complete dependence (FIM™ score of 18) to modified independence (FIM™ score of 122), his level of disability fell from a vegetative state (DRS score of 21) to only a mild disability (DRS score of 1). His cognitive ability improved from an untestable state to a score of 85 on the RBANS, and his self-reported quality of life reached a good score of 78 on the QOLIBRI, demonstrating that his remarkable progress extended to higher-order functions.

Table 8.4 Summary of patient assessment scores

Domain	*Assessment tool*	*Score at admission*	*Score at discharge (1 year)*	*Result: what this means*
Overall Disability	Disability Rating Scale (DRS)	21 (Vegetative State)	1 (Mild Disability)	A monumental shift from a nonresponsive state to having only minor, manageable disabilities.
Functional Independence	Functional Independence Measure (FIM™)	18 (Total Dependence)	**122** (Modified Independence)	The patient went from requiring complete care for all basic needs to being able to live on his own with minimal assistance.
Cognitive Function	Repeatable Battery for the Assessment of Neuropsychological Status (RBANS)	Not Testable	85 (Low Average)	From being unable to participate in testing to achieving cognitive scores within the low-average range for his age.
Quality of Life	Quality of Life after Brain Injury (QOLIBRI)	Not Testable	78 (Good)	The patient self-reported a good overall quality of life, indicating satisfaction with his health.

8.8 FUTURE TRENDS

The field of neurorehabilitation is about to undergo a significant transformation due to advancements in science and the evolving health system. It is anticipated that future therapies will be more potent, individualized, and readily available than in the past. Imagine a day when a straightforward gene test identifies a patient's natural ability to recover through neuroplasticity, enabling therapists to recommend the ideal "dose" of rehabilitation right away. What's more, mechanistic neuroimaging and blood-based biomarkers could be developed as a "weather forecast" for the brain, allowing clinicians to see how the brain is responding to therapy, even in real time, and modify therapy accordingly. A bit farther down the line, truly revolutionary tools, such as optogenetics, could provide unprecedented precision. Optogenetics is an animal-based strategy currently being used in brain research, in which specific neurons are turned on or off by light (if you think about it like a light switch in your house, it's like flipping the switch for some circuits in your brain). Although we are far away from using optogenetics in humans, one can only imagine the implications – clinicians could, in theory, "turn on" pro-recovery pathways and "turn off" "bad learning" pathways as desired, perhaps even creating a way to sculpt the rewiring of the brain. In order for these future innovations to reach patients, they must be endorsed by progressive health-care policies. The biggest hurdle is often reimbursement and access; many innovative treatments, from VR systems to digital therapeutics, remain to be consistently reimbursed by insurance, limiting access. In order to get past the access barriers, it is only when we understand how to analyze and visualize the data that we will realize the ultimate capabilities of digital therapeutics. The raw data streams from wearable sensors or therapy apps do not typically mean anything by themselves, as the streams of data are too complex for the user. The next iteration will translate the data into visually based, intuitive dashboards for both the clinician and patient. For a clinician, it will be an imaging dashboard that tells them what is changing in a person's movement performance, how well they are succeeding at a cognitive task, or if they are completing their home program. This visual aid turns into a powerful source of inspiration for the patient. A straightforward graph of one's own functional achievement may appear simple and unimportant, but it can strengthen one's resolve to persevere through the difficult rehabilitation process and likely increase patient engagement by empowering them to take an active role in their own treatment. An even smarter clinical associate has an opportunity as a result of this shift into environments that are becoming more and more data-rich: artificial intelligence. However, a crucial component of its application in neurorehabilitation will be trust. Clinicians will not employ "black box" AI systems that suggest novel treatments for patients unless they understand at least why the algorithm came up with the suggestion rather than merely the suggestion. For this reason, as a profession, we need to make sure that Explainable AI (XAI) is developed. A highly developed XAI system should ideally not only suggest a different course of treatment but also explain why based on objective data derived from the patient's performance. The therapist can evaluate the artificial intelligence's contribution to their professional interpretation and reflection thanks to this degree of transparency in the final result, since the human clinician still has the final say over all decisions, regardless of the technology at their disposal. In the end, this move to a data-driven approach has an impact beyond patient care. It will alter collaborative research as a whole. The inability to exchange sensitive patient data between institutions is one of the main obstacles to improving our understanding of brain recovery. Federated Learning, a privacy-preserving machine learning framework, is not a novel idea, but it offers exciting new opportunities. Federated learning enables institutions to work together across institutions to train a potent predictive AI without having to grant access to private, sensitive patient data. Federated learning would produce reliable models for forecasting recovery paths and identifying the best therapeutic avenues in fields like neurorehabilitation, enabling quicker discovery while maintaining patient privacy.

8.9 CONCLUSION

The most important development in contemporary neuroscience is the recognition that the brain is a living, adaptable system rather than a model of fixed, immutable structure. We are assisting the brain's remarkable capacity for adaptation and reconstruction rather than merely repairing the harm caused by neurological insults. Only recently has the theoretical idea of neuroplasticity gained acceptance, but it is now the cornerstone of contemporary rehabilitation techniques and is transforming hopelessness into possibility. These strategies rely on four key elements to be successful. First, to guarantee the legitimacy and effectiveness of neuroplastic therapies, research to advance their science must be continued. In order to mitigate harm to everyone, not just the privileged or wealthy, health system policies must facilitate access to those procedures. Third, to enable patients, families, and clinicians to exercise their agency during the healing process, education is essential. Lastly, bringing these therapies outside of the clinic and into the real world will require the use of VR platforms and customized digital therapeutic devices. The way we think about and react to brain compromise is going to change in the next ten years due to developments in the diagnosis, treatment, and recovery of the brain. For example, restoring lost abilities is becoming more important than making up for deficiencies. To make sure that people are neurologically resilient before a major neurological challenge arises, there will be a lot more attention paid to improving brain resilience throughout life. The combination of personalized medicine, tailored treatment plans, and sophisticated diagnostic tools will usher in a new era of neurology that is characterized by capabilities rather than deficiencies, by anticipation rather than reaction, and by rehabilitation rather than intervention. We are at the cusp of a more prosperous, inclusive, and empowering future for brain rehabilitation thanks to these technologies.

REFERENCES

1. Chen, W., & Liu, M. (2025). Digital therapies for poststroke cognitive dysfunction: Systematic review and network meta-analysis. *Journal of Medical Internet Research*, *27*(1), e73687. https://doi.org/10.2196/73687
2. Yu, X., Tang, Q., Lu, C., Wang, Q., Xu, Z., & Wang, H. (2025). Active and passive robotic gait training in stroke rehabilitation: A randomized controlled trial. *BMC Neurology*, *25*(1), 226. https://doi.org/10.1186/s12883-025-04226-0
3. Ben Ezzdine, O., Wang, M., & Smith, D. L. (2025). Exercise, cognitive training, and artificial intelligence in neurodegenerative disorders: Mechanisms and applications. *Frontiers in Neuroscience*, *19*, 1502417. https://doi.org/10.3389/fnins.2025.1502417
4. Kleim, J. A., & Jones, T. A. (2008). Principles of experience-dependent neural plasticity: Implications for rehabilitation after brain damage. *Journal of Speech, Language, and Hearing Research*, *51*(1), S225–S239. https://doi.org/10.1044/1092-4388(2008/018)
5. Hampstead, B. M., & Sarter, M. (2019). A framework for personalized cognitive rehabilitation. *Trends in Neurosciences*, *42*(12), 839–841. https://doi.org/10.1016/j.tins.2019.09.005
6. Sale, A., Berardi, N., & Maffei, L. (2014). Enrich the environment to empower the brain. *Trends in Neurosciences*, *37*(4), 193–202.
7. Martin, S. J., Grimwood, P. D., & Morris, R. G. (2000). Synaptic plasticity and memory: An evaluation of the hypothesis. *Annual Review of Neuroscience*, *23*, 649–711. https://doi.org/10.1146/annurev.neuro.23.1.649
8. Voss, M. W., Vivar, C., Kramer, A. F., & van Praag, H. (2013). Bridging animal and human models of exercise-induced brain plasticity. *Trends in Cognitive Sciences*, *17*(10), 525–544. https://doi.org/10.1016/j.tics.2013.08.001
9. Cramer, S. C., Sur, M., Dobkin, B. H., O'Brien, C., Sanger, T. D., Trojanowski, J. Q., ... & Vinogradov, S. (2011). Harnessing neuroplasticity for clinical applications. *Brain*, *134*(6), 1591–1609. https://doi.org/10.1093/brain/awr039
10. Zeiler, S. R., & Krakauer, J. W. (2013). The interaction between training and plasticity in the poststroke brain. *Current Opinion in Neurology*, *26*(6), 609–616. https://doi.org/10.1097/WCO.0000000000000025
11. Cicerone, K. D., Goldin, Y., Ganci, K., Rosenbaum, A., Wethe, J. V., Langenbahn, D. M., ... & Trexler, L. E. (2019). Evidence-based cognitive rehabilitation: Systematic review of the literature from 2009 through 2014. *Archives of Physical Medicine and Rehabilitation*, *100*(8), 1515–1533. https://doi.org/10.1016/j.apmr.2019.02.011
12. Faria, A. L., Pinho, M. S., & Badia, S. B. (2018). Capturing expert knowledge for the personalization of cognitive rehabilitation: Study combining computational modeling and a participatory design strategy. *JMIR Rehabilitation and Assistive Technologies*, *5*(2), e10714.
13. Massetti, T., Crocetta, T. B., da Silva, T. D., Trevizan, I. L., & de Araujo, L. V. (2016). Application of virtual reality in the rehabilitation of the upper limbs in patients with stroke: A systematic literature review. *Medical Express (São Paulo)*, *3*(4). https://doi.org/10.5935/MedicalExpress.2016.04.02
14. Kueider, A. M., Parisi, J. M., Gross, A. L., & Rebok, G. W. (2012). Computerized cognitive training with older adults: A systematic review. *PLOS One*, *7*(8), e43482. https://doi.org/10.1371/journal.pone.0043482
15. Alam, M., Ara, K., Javed, M. S., & Ansari, M. (2014, July). Detect and filter traffic attack through cloud Trace back and neural network, Imperial College. In *The 2014 International Conference of Data Mining and Knowledge Engineering (ICDMKE)*, London (pp. 2–4).

16. Ansari, M., & Alam, M. (2023). IoT-cloud-enabled smart framework for real-world applications. In *Intelligent Systems and Applications: Select Proceedings of ICISA 2022* (pp. 87–105). Springer Nature Singapore.
17. Ansari, M., & Alam, M. (2024). A novel hybrid time-series approach for IoT-cloud-enabled environment monitoring. *Journal of Supercomputing, 80*(7).
18. Noor, M. A. F., Khanum, S., Anwar, T., & Ansari, M. (2021). A holistic view on blockchain and its issues. In *Blockchain Applications in IoT Security* (pp. 21–44). IGI Global.
19. Ansari, M., & Alam, M. (2021). IoT-cloud enabled statistical analysis and visualization of air pollution data in India. In *Proceedings of Data Analytics and Management: ICDAM 2021, Volume 2* (pp. 125–139). Springer Singapore.
20. Ansari, M., Ali, S. A., Alam, M., Chaudhary, K., & Rakshit, S. (2023). Unlocking the power of explainable AI to improve customer experiences in E-Commerce. In *AI-Based Data Analytics* (pp. 31–48). Auerbach Publications.
21. Ansari, M., Ali, S. A., & Alam, M. (2022). Internet of things (IoT) fusion with cloud computing: current research and future direction. *International Journal of Advanced Technology and Engineering Exploration, 9*(97), 1812.
22. Ali, S. A., Ansari, M., & Alam, M. (2020). Resource management techniques for cloud-based IoT environment. In *Internet of Things (IoT) Concepts and Applications* (pp. 63–87). Springer International Publishing.
23. Ansari, M., & Alam, M. (2024). An intelligent IoT-cloud-based air pollution forecasting model using univariate time-series analysis. *Arabian Journal for Science and Engineering, 49*(3), 3135–3162.
24. Ali, S. A., Ansari, M., Alam, M., & Rakshit, S. (2023). Federated learning for business intelligence: Predictive maintenance in industry 4.0. In *AI-Based Data Analytics* (pp. 125–140). Auerbach Publications.
25. Khan, S., Liu, X., Ali, S. A., & Alam, M. (2019). Storage solutions for big data systems: A qualitative study and comparison. arXiv preprint arXiv:1904.11498.
26. Ansari, M., Chhabra, A., Hasan, N., & Alam, M. (2025). Quantum finance: Revolutionizing financial markets with quantum computing. In *AI-Driven Finance in the VUCA World* (pp. 96–105). Auerbach Publications.
27. Cassidy, J. M., & Cramer, S. C. (2016). Spontaneous & therapeutic-induced mechanisms of functional recovery after stroke. *Translational Stroke Research, 8*(1), 33–46. https://doi.org/10.1007/s12975-016-0467-5
28. Turolla, A., et al. (2018). Rehabilitation induced neural plasticity after acquired brain injury. *Neural Plasticity, 2018*, 6565418. https://doi.org/10.1155/2018/6565418
29. Yu, Y., Huang, W., Tuerxun, H., et al. (2025). Enhanced neuroplasticity and gait recovery in stroke patients: A comparative analysis of active and passive robotic training modes. *BMC Neurology, 25*, 239. https://doi.org/10.1186/s12883-025-04226-0
30. Chhabra, A., Hasan, N., Ansari, M., & Alam, M. (2025). Algorithmic trading and AI: A synergy for risk management. In *AI-Driven Finance in the VUCA World* (pp. 27–45). Taylor & Francis Publication.
31. Mughal, M. F., Ansari, M., & Alam, M. (2025). Revolutionizing financial forecasting: The rise of transformer models in long-term time series analysis. In *AI-Driven Finance in the VUCA World* (p. 142). Taylor & Francis Publication.
32. Hasan, N., Ansari, M., Chhabra, A., & Alam, M. (2025). Revolutionizing finance: The transformative impact of artificial intelligence on the finance sector. In *AI-Driven Finance in the VUCA World* (pp. 172–186). Auerbach Publications.
33. Chhabra, A., & Alam, M. (2025). RegGRU-Opt: A robust GRU-based RNN model for high-performance sentiment analysis and binary classification. *Arabian Journal for Science and Engineering*. https://doi.org/10.1007/s13369-024-09943-1
34. Chhabra, A., Chaudhary, K., & Alam, M. (2023). Strategic plans for market campaign using machine learning algorithms. In: Chaudhary, K., Alam, M., & Debnath, N. C. (eds) *Intelligent Data Analytics in Business. ICDABM 2022.* Lecture Notes in Electrical Engineering, vol 1068. Springer, Singapore. https://doi.org/10.1007/978-981-99-5358-5_15

35. Behboodi, A., Lee, W. A., Hinchberger, V. S., et al. (2022). Determining optimal mobile neurofeedback methods for motor neurorehabilitation in children and adults with non-progressive neurological disorders: A scoping review. *Journal of NeuroEngineering and Rehabilitation*, *19*, 104. https://doi.org/10.1186/s12984-022-01081-9
36. Szeto, S. G., Wan, H., Alavinia, M., et al. (2023). Effect of mobile application types on stroke rehabilitation: A systematic review. *Journal of NeuroEngineering and Rehabilitation*, *20*, 12. https://doi.org/10.1186/s12984-023-01124-9
37. Mohy Uddin, R., Ganapathy, K., Syed-Abdul, S., et al. (2024). Digital technology enablers of teleneurorehabilitation in pre- and post-COVID-19 pandemic era – A scoping review. *International Journal of Telerehabilitation*. https://doi.org/10.5195/ijt.2024.6611
38. Famale, R. D., et al. (2025). Federated learning in brain-computer interfaces: A systematic review on aggregating intrinsic data for enhanced performance. *Medico & Engineering Future*, *2*(1), 37–41. https://doi.org/10.5281/ZENODO.16018806
39. Sánchez Rodríguez, M. T., Collado Vázquez, S., Martín Casas, P., Cano de la Cuerda, R., et al. (2015). Neurorehabilitation and apps: A systematic review of mobile applications. *Neurología (English Edition)*. https://doi.org/10.1016/j.nrleng.2015.10.002

Chapter 9

AI-integrated diagnostic models for neurological disorders

Current trends, technologies, and future directions

Mohd. Maroof Siddiqui and Prajoona Valsalan

9.1 INTRODUCTION

Neurological disorders affect millions globally who suffer significant health-care burdens and significantly influence patients' quality of life. These diseases are a myriad of conditions, such as Alzheimer's disease (AD), Parkinson's disease (PD), epilepsy, multiple sclerosis (MS), or amyotrophic lateral sclerosis, referred to by the community almost pejoratively with its own set of symptoms and progression. The complexity of the brain and its nervous system means that diagnosing disorders can sometimes be extremely difficult [1–5]. Symptoms may be mild or masquerade as symptoms of another disease, causing misdiagnosis and delayed diagnosis, resulting in ineffective treatment. The critical aspect for neurodegenerative treatments is early and accurate detection, since it can largely slow down disease progression and improve patient's outcome or quality of life.

Traditionally, the diagnosis of neurologic conditions has relied heavily on clinical appraisal and is further plagued with semi-objectivity as well due to its manipulation by invasive methods (such as lumbar puncture), availability bias, and access restriction. Unfortunately, this is not very high; yet the rise of artificial intelligence (AI) in neurology has started to change recently by offering new and dynamic algorithms. Over the past few years, with advancements in AI-based techniques (i.e., machine learning and deep learning), approaches are increasingly being utilized to automatically identify patterns on neuroimaging data or genetic profiles such as electroencephalography (EEG) data. These platforms can identify early biomarkers of disease that are not visible to the naked eye, leading to more accurate and timely detection [6–8].

Through sifting through large amounts of data from various places, AI can detect patterns and correlations that could point toward the onset or progression of neural diseases. For instance, AI algorithms process subtle changes in brain structure and function to identify early indicators of neurodegenerative illnesses such as AD even decades before clinical symptoms. Moreover, predicting seizure activity in patients that live with epilepsy can also be made via AI mechanisms and systems, enabling real-time monitoring and early warning to prevent dangerous seizure states [9].

The emerging literature in AI-driven neurology also supports the promise that AI can both increase accuracy of diagnosis and create a more efficient health-care system by offloading humans from routine work, reducing clinician workloads, and allowing for personalized treatments. Despite this, implementing AI into the clinic presents its own set of hurdles. Data and privacy concerns, algorithmic transparency, and the requirement of strong validation for AI systems need to be resolved prior to safe implementation in neurological care [10].

The paper examines an extensive range of AI applications in the early detection and diagnosis process related to neurological disorders. It will feature the biggest success stories of

DOI: 10.1201/9781003685890-9

how AI has been promising thus far, e.g., in neuroimaging analysis and genetic data interpretation, as well as continuous monitoring via wearable devices. Besides, the paper will illuminate AI challenges in combination with neurological health-care ethical, data privacy, and regulatory frameworks. We will conclude by discussing the future of AI in neurology and how these developments could change the way we care for patients with neurological diseases [11].

9.2 LITERATURE SURVEY

The following literature survey Table 9.1 presents an overview of recent research studies on AI-based detection of neurological disorders:

9.3 AI IN NEUROIMAGING FOR DISEASE DETECTION

Neuroimaging is essential for neurological disease diagnosis. There have been great advances in the automation of neural imaging analysis, especially with AI techniques and deep learning algorithms, which aim at MRI (magnetic resonance imaging), CT (computed tomography), or PET scans (positron emission tomography).

Table 9.1 Recent advancements in AI methodologies applied to the detection and diagnosis of various neurological disorders [1–10]

Study	*Neurological disorder*	*AI methodology*	*Key findings*
Bosel et al., 2025	Various Neurological Disorders	Machine Learning and Deep Learning	AI influences diagnostics, prognostication, and therapy in neurology.
Ali et al., 2025	Various Neurological Disorders	Deep Learning on MRI Data	Systematic review showing improved diagnostic accuracy across neurological disorders.
Mohsen, 2025	Alzheimer's Disease	Deep Learning and Machine Learning	Comprehensive review highlighting early detection using AI models.
Qamar et al., 2025	Epilepsy	Deep Learning on Intracranial EEG	Advanced seizure detection and clinical application of AI techniques.
Dash et al., 2023	Various Neurological Disorders	Automated EEG Interpretation	High accuracy in interpreting clinical EEGs using AI.
Rabie & Akhloufi, 2025	Parkinson's Disease	Machine Learning and Deep Learning	Survey of AI approaches for early detection of Parkinson's disease.
Zhao et al., 2025	Parkinson's Disease	Multimodal AI Models	AI-enabled multimodal detection and assessment of Parkinson's disease.
Progress and trends, 2024	Various Neurological Disorders	Deep Learning	Trends in AI applications for neurological disorders using imaging and EEG data.
Naily et al., 2024	Epilepsy	Machine Learning on EEG & MRI	Novel techniques for seizure diagnosis with improved accuracy.
Dişli et al., 2025	Epilepsy	Wavelet CNN on EEG	Developed an AI model for epilepsy diagnosis from EEG signals.

9.3.1 Detection of Alzheimer's disease using AI

Introduction: AD is a neurodegenerative disorder that leads to cognitive impairment and deficits in memory and spatial orientation. Preventing AD is not yet possible and detection also remains elusive since traditional methods often detect AD only at an advanced stage. Early diagnosis is vital for controlling and managing the disease, and to date it has been rather challenging if not nearly impossible. Using MRI or PET scans, AI can detect biomarkers of AD such as amyloid plaques and tau tangles, which are 90% effective in revealing that a patient has the disease [12].

In other words, even if patients are asymptomatic, AI models can find these early signs more precisely than conventional methods. Deep learning algorithms can effectively distinguish between healthy, MCI (Mild Cognitive Impairment), and AD patients to enable earlier intervention for prevention [13–15].

9.3.2 Parkinson's disease and related disorders

Parkinson's disease is a type of progressive neurodegenerative disorder, which predominantly has an impact on movement and motor control. AI has been crucial in PD diagnosis based on analyzing neuroimaging data and using wearable sensors to detect movements. Researchers have now trained AI algorithms to pick up on subtle changes in gait, tremors, and motor skills that might precede diagnosis of PD years down the line [16–20].

Machine learning algorithms are also being employed to analyze MRI and diffusion tensor imaging (DTI) scans for aberrant changes in brain connectivity that occur prior to PD symptoms. The hope is for earlier diagnosis and intervention to stave off disease progression [21].

9.3.3 AI for detecting epilepsy

It is a condition defined by two or more unprovoked seizures separated by at least 24 hours. Epilepsy, for instance, can be diagnosed by studying the brainwave patterns caught on an electroencephalogram (EEG). AI has demonstrated significant capacity for analyzing EEG signals and detecting pathological electrical discharges, which may occur similarly to those before or during seizures [22].

These abnormalities can be very subtle, and deep learning algorithms are able to identify them in noisy data with high sensitivity, which is the basis of powerful AI-based systems for epilepsy diagnosis and seizure onset prediction. These could also be wirelessly tied into wearables that track patients 24/7 and provide fresh data, potentially heading off a seizure before it develops fully [23–26].

9.3.4 AI and genetic data for neurological disorders

After brain imaging, AI has started to analyze genetic data as it provides a blueprint of risk factors linked with neurological disorders [27–30]. By mining large genomic data sets, AI has been used to discover mutations and single nucleotide polymorphisms (SNPs) associated with Huntington's disease (HD), amyotrophic lateral sclerosis (ALS), MS, etc.

9.3.5 Artificial intelligence in Huntington's disease prediction

Huntington's disease is an autosomal dominant neural disorder due to a single gene mutation of the HTT gene. AI models are employed to study genetic data and predict when carriers may get sick. Obtaining genomic information along with clinical data can enable AI systems to tailor risk predictions and treatment recommendations for everyone [31–35].

9.3.6 AI and multiple sclerosis

Multiple sclerosis is an autoimmune disorder and a demyelinating condition of the central nervous system with resulting motor, sensory, or cognitive impairments. MRI scans have been analyzed by AI for MS lesions in the brain and spinal cord. In addition, AI models can interpret genetic data that could provide insight into risk factors for MS, which can allow early detection and individualized therapeutic plans [36].

The AI-based classification of sleep disorders begins with data collection from EEG, ECG, EMG signals, wearable devices (such as those tracking heart rate and motion), and patient sleep history mentioned in Figure 9.1. The collected data undergoes preprocessing, including noise filtering, normalization, segmentation, and feature extraction. Following this, AI models—using machine learning techniques like SVM, Random Forest, and KNN, or deep learning models such as CNN and RNN—are trained on the processed data. The models are then evaluated based on metrics like accuracy and sensitivity through cross-validation. Once validated, the system classifies sleep disorders such as insomnia, sleep apnea, narcolepsy, and

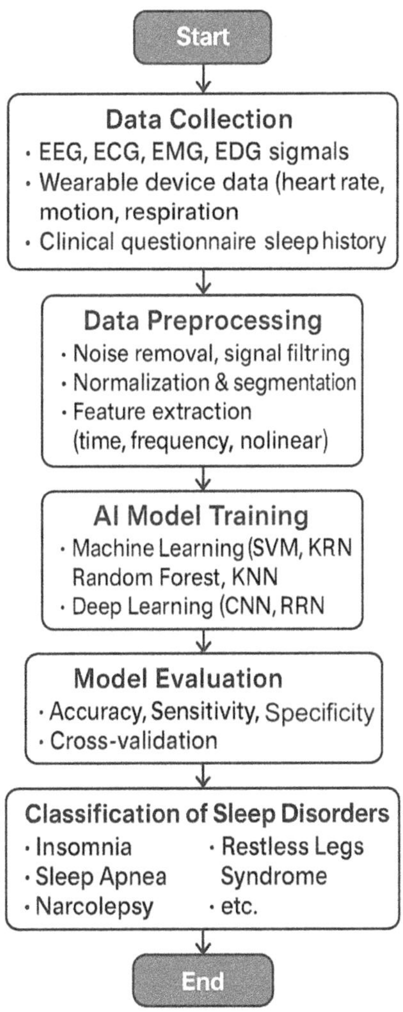

Figure 9.1 Classification of sleep disorders using AI.

others. Finally, a diagnostic report is generated, indicating the AI-based risk level and providing recommendations for consultation with sleep specialists [37–42].

9.4 CONTINUOUS MONITORING VIA WEARABLE DEVICES AND AI

Complemented by AI, wearable technology provides the ability to track brain health in different ways. The information gathered by these devices is then used to measure patient movement, physiological signals, and even brain activity over time, providing a dynamic window into the evolution of neurological diseases [43–46].

9.4.1 AI-based wearables for Parkinson disease

Wearable devices fitted with accelerometers and gyroscopes can monitor changes in motor function (tremors, gait abnormalities) associated with PD. AI routines that screen this information for the intensity of manifestations assist in fine-tuning treatment. These wearables provide patients with an opportunity to keep track of their own symptoms at home with less reliance on frequent clinical visits [47–49].

9.4.2 Epilepsy seizure prediction

AI-powered wearable devices show promise in predicting seizures for epilepsy patients by continuously analyzing EEG signals. Machine learning models can identify patterns in the vast amount of brain activity data researchers collect and can do it faster than a human doctor. Feeding back information an algorithm detects differences in one part of its calculation, for example before a seizure occurs, may not be as severe next time. This ongoing real-time monitoring improves the life expectancy of epilepsy patients and prevents complications related to seizures [50–53].

In this structure, you can easily compare research papers based on their focus, methods used, findings, and limitations. If you have specific papers in mind or want more categories, I can refine the table further.

9.5 ETHICAL CONSIDERATIONS AND CHALLENGES IN AI FOR NEUROLOGY

Although AI has many benefits in detecting and managing neurological disorders, it also raises some ethical problems. Such worries range from data privacy and algorithmic transparency to bias in AI models [54].

9.5.1 Data privacy and security

AI in neurology involves processing reams of sensitive patient data, including brain scans, gene-screening blood tests, and constantly recorded signals from wearable devices. It is important to keep this information secure, safe, and compliant with the requirements of privacy regulations for patients to trust AI systems [55–60].

9.5.2 Algorithmic bias and equity

AI is only as good or bad as the data it was trained on. In turn, it could lead to discriminatory AI systems that produce medical decisions based on biased results if the training data is not representative of diverse populations. This carries very worrying implications across neurology, with minority groups far more likely to be misdiagnosed or have conditions missed by AI-based systems. Therefore, it is vital to train AI models on diverse datasets and continuously monitor deployed machine learning systems for bias [61–65].

9.6 DISCUSSION

Central nervous system (CNS) diseases are still some of the most challenging conditions to diagnose and treat because their symptomatology is highly heterogeneous and their clinical presentations are often diverse. Traditional approaches, although prevalent, are subjective and may not allow timely intervention. AI provides a compelling alternative by permitting the analysis of multimodal data (such as neuroimaging, genetics, EEG, and electronic health records) more accurately and quickly.

Various machine learning and deep learning-based methods have already proved to be very promising. For instance, AI models are able to recognize early neurodegenerative changes in AD based on MRI biomarkers, forecast epileptic seizures from EEG patterns, and track PD evolution with data collected by wearable sensors. These developments underscore the promise of AI for transforming neurological care into early diagnosis and proactive intervention and personalized treatment planning.

Nonetheless, there are several limitations in clinical translation. Performance is only evident in the presence of large, diverse, and unbiased data sources, which remain limited. Ethical concerns concerning patient privacy, informed consent, and algorithm transparency must also be tackled. Furthermore, regulations surrounding health-care AI are still relatively new, hampering clinical acceptance.

Future developments will require interdisciplinary collaboration, common data-sharing standards, and the production of explainable AI models for safe and useful introduction into neurology.

9.7 CONCLUSION

Artificial intelligence transforms the way we detect and diagnose neurological disorders, enabling early detection of disease as well as personalized treatment monitoring around the clock. This includes everything from analyzing neuroimaging and genetic data to wearable devices allowing for real-time monitoring. AI is changing the way we understand complex conditions like AD, PD, or epilepsy. But ethical issues and robust data privacy measures to safeguard are signifiers of what needs to be addressed for AI–neurology interplay. With the further development of AI, it will be an even more important component in diagnosing and managing neurological disorders that translate into better patient outcomes.

REFERENCES

1. Bosel, J., Mathur, R., Cheng, L., Varelas, M. S., Hobert, M. A., & Suarez, J. I. (2025). AI and neurology. *Neurological Research and Practice*, 7(1), 11
2. Ali, S. S. A., Memon, K., Yahya, N., & Khan, S. (2025). Deep learning frameworks for MRI-based diagnosis of neurological disorders: A systematic review and meta-analysis. *Artificial Intelligence Review*, 58, 171–191.
3. Mohsen, S. (2025). Alzheimer's disease detection using deep learning and machine learning: A review. *Artificial Intelligence Review*, 58, 262.

4. Qamar, W. U. R., Lee, M.-H., & Abibullaev, B. (2025). Deep learning in intracranial EEG for seizure detection: Advances, challenges, and clinical applications. *Frontiers in Neuroscience*, 19, 1677898.
5. Dash, D., Dash, C., Primrose, S., et al. (2023). Automated interpretation of clinical electroencephalograms using artificial intelligence. *JAMA Neurology*, 80, 805–812.
6. Rabie, H., & Akhloufi, M. A. (2025). A review of machine learning and deep learning for Parkinson's disease detection. *Discover Artificial Intelligence*, 5, 24.
7. Zhao, A., Liu, Y., Yu, X., & Xing, X. (2025). Artificial intelligence-enabled detection and assessment of Parkinson's disease using multimodal data: A survey. *arXiv preprint:2502.10703*
8. Iqbal, M.S., Heyat, M.B.B., Parveen, S., Hayat, M.A.B., Roshanzamir, M., Alizadehsani, R., Akhtar, F., Sayeed, E., Hussain, S., Hussein, H.S., & Sawan, M. (2024). Progress and trends in neurological disorders research based on deep learning. *Computerized Medical Imaging and Graphics*, 116, 102400.
9. Naily, R., Yahia, S., & Zaied, M. (2024). Machine learning for epilepsy: Novel EEG and MRI techniques for seizure diagnosis. *Journal of Medical and Biological Engineering*, 44(6), 747–761.
10. Dişli, F., Gedikpinar, M., Fırat, H., & Şengür, A. (2025). Epilepsy diagnosis from EEG signals using wavelet CNN model. *Diagnostics*, 15(1), 84.
11. Ahmed, M., & Zaidan, A.A. (2020). A review of artificial intelligence applications in neuroscience with focus on brain-computer interface systems. *Journal of Neuroscience Methods*, 337, 108627. https://doi.org/10.1016/j.jneumeth.2020.108627
12. Armañanzas, R., & Ascoli, G.A. (2020). Towards the automatic classification of neurons. *Nature Reviews Neuroscience*, 21(4), 236–250.
13. Ashraf, A., Westerhuis, M.E., & van Gils, S.A. (2022). Artificial intelligence in early diagnosis of Alzheimer's disease. *Frontiers in Neuroscience*, 16, 927457.
14. Siddiqui, M.M., & Jain, R. (2021). Prediction of REM (rapid eye movement) sleep behavior disorder using EEG signal applied EMG1 and EMG2 channel. *Biomedical and Pharmacology Journal*, 14(1), 519–524.
15. Siddiqui, M.M., Srivastava, G., & Saeed, S.H. (2017). Alzheimer: A neurological disorder. *Research and Reviews: Journal of Medicine*, 7(1), 1–3.
16. Avanzo, M., Stancanello, J., & El Naqa, I. (2020). Beyond imaging: The promise of radiomics. *Physica Medica*, 38, 121–132. https://doi.org/10.1016/j.ejmp.2020.09.003
17. Choi, Y., & Lee, J. (2021). Artificial intelligence in neurological disorders: Past, present, and future. *Journal of Neural Engineering*, 18(3), 1–10.
18. Dastmardi, M., & Gandomi, A.H. (2021). Machine learning models for predicting seizure onset in epilepsy. *IEEE Transactions on Neural Systems and Rehabilitation Engineering*, 29(1), 132–141.
19. Siddiqui, M.M. (2024). Digitalize the system to diagnosis of neurological disorder (sleep disorder). *2024 Second International Conference on Emerging Trends in Information Technology and Engineering (ICETITE)* (pp. 1–6). IEEE.
20. Ali, R., & Khan, M.S. (2021). A novel deep learning framework for the early detection of Parkinson's disease. *IEEE Transactions on Neural Networks and Learning Systems*, 32(12), 5368–5379.
21. Belle, A., & Kon, M. (2015). Big data analytics in healthcare. *BioMed Research International*, 2015, 1–16. https://doi.org/10.1155/2015/370194
22. Siddiqui, M.M., & Valsalan, P. (2024). AI-Based Human Face Recognition System. *Journal of Electrical Systems*, 20, 357–362. https://doi.org/10.52783/jes.7174
23. Chen, R., & Guestrin, C. (2019). AI-driven diagnosis of neurological diseases: A comparative study. *Artificial Intelligence in Medicine*, 96, 34–43. https://doi.org/10.1016/j.artmed.2019.04.005
24. Srivastava, E. (2016). Carbon nano tubes & its application in medical field & communication. *International Journal of Advanced Research in Computer and Communication Engineering (IJARCCE)*, 5, 170–173.
25. Siddiqui, M.M., Valsalan, P., Qureshi, A.A., & Al Mashani, K.S. (2024) AI-based real-time heart rate (ECG) monitoring system. *Nanotechnology Perceptions*, 20(7), 1575–1586.
26. De Lange, S.C., & van den Heuvel, M.P. (2019). Integrating brain networks in neurodegenerative diseases using AI. *Nature Reviews Neurology*, 15(7), 385–396.

27. Esteva, A., & Robicquet, A. (2019). A guide to deep learning in healthcare. *Nature Medicine*, 25(1), 24–29. https://doi.org/10.1038/s41591-018-0316-z
28. Gao, J., & Zhang, Z. (2019). Artificial intelligence in the diagnosis of Parkinson's disease: A review. *IEEE Access*, 7, 108560–108571. https://doi.org/10.1109/ACCESS.2019.2932256
29. Gulshan, V., & Peng, L. (2021). Development and validation of deep learning algorithms for detecting neurological disease from medical imaging. *Lancet Digital Health*, 3(1), e20–e28.
30. Hinton, G., & Sejnowski, T. (2020). Deep learning and artificial intelligence in neuroscience. *Neuron*, 107(1), 1–4.
31. Siddiqui, M.M., et al. (2023). Detection of rapid eye movement behaviour sleep disorder using time and frequency analysis of EEG signal applied on C4-A1 channels. *Physical Sciences Reviews*, 8(10), 3269–3285. https://doi.org/10.1515/psr-2022-0990
32. Ismail, Z., & Herrmann, N. (2020). Role of artificial intelligence in Alzheimer's disease diagnosis and management. *Alzheimer's & Dementia*, 16(7), 1263–1271. https://doi.org/10.1002/alz.12115
33. Johnson, K.W., & Shameer, K. (2018). Artificial intelligence in clinical neuroscience: Expanding the horizons. *Nature Reviews Neurology*, 14(1), 9–11. https://doi.org/10.1038/nrneurol.2017.180
34. Keshavan, A., & Miller, J.B. (2022). Artificial intelligence applications in epilepsy: A review. *Journal of Clinical Neurophysiology*, 39(2), 101–110. https://doi.org/10.1097/WNP.0000000000000824
35. Litjens, G., & Sánchez, C.I. (2017). A survey on deep learning in medical image analysis. *Medical Image Analysis*, 42, 60–88. https://doi.org/10.1016/j.media.2017.07.005
36. Liu, Y., & Wang, C. (2021). Artificial intelligence applications in neuroimaging for diagnosing neurological disorders. *European Journal of Radiology*, 137, 109584. https://doi.org/10.1016/j.ejrad.2021.109584
37. Monteiro, A., & Oliveira, E. (2020). AI in neurology: Opportunities and challenges. *Journal of Neurology*, 267(5), 1235–1240.
38. Siddiqui, M.M., Qureshi, A.A., Al Mashani, K.S., & Shajanah, H.A. (2024, September). IoT-based ECG monitoring system for ICU. *IEEE International Conference on Signal Processing and Advance Research in Computing (SPARC-2024)*, Volume 1 (pp. 1–5). IEEE.
39. Siddiqui, M.M., Valsalan, P., Shajanah, H.A., Al Baraami, M.S., & Al Baraami, H.S. (2024, September). IoT-based RFID attendance system. *IEEE International Conference on Signal Processing and Advance Research in Computing (SPARC-2024)*, Volume 1 (pp. 1–5). IEEE.
40. Peterson, L. (2019). Integrating AI into healthcare for neurological disorders: A paradigm shift. *JAMA Neurology*, 76(1), 1–8.
41. Ramesh, A., & Bhaskar, H. (2021). AI for neuroimaging: A future roadmap. *IEEE Transactions on Medical Imaging*, 40(2), 617–635.
42. Vieira, S., & Pinaya, W.H. (2020). AI in neurodegenerative disorder research. *NeuroImage*, 221, 117151. https://doi.org/10.1016/j.neuroimage.2020.117151
43. Trapp, B.D., & Nave, K.A. (2008). Multiple sclerosis: An immune or neurodegenerative disorder? *Annual Review of Neuroscience*, 31, 247–269.
44. Kassani, S.H., & Kassani, P.H. (2021). Machine learning algorithms to diagnose Alzheimer's disease using fMRI data. *Journal of Healthcare Informatics Research*, 5(2), 213–227.
45. Pasa, F., & Boscolo, C. (2021). AI-based electroencephalogram analysis for epilepsy detection. *Journal of Neuroscience Methods*, 347, 108956.
46. Namdev, K., & Siddiqui, M.M. (2015). Different types of electrical signals produced by human body. *International Journal of Advanced Research in Science and Engineering*, 4(02), 232–239.
47. Rieke, N., & Hancox, J. (2020). Improving radiomics with artificial intelligence. *Nature Machine Intelligence*, 2(3), 165–177.
48. Harrer, S., & Shah, P. (2019). Artificial intelligence for clinical trial design. *The Lancet Digital Health*, 1(9), e415–e416.
49. Siddiqui, M.M., et al. (2024). Analysis of EEG data using different techniques of digital signal processing. *Biomedical and Pharmacology Journal*, 17(1), 135–139.
50. Hu, W., & Ge, Y. (2022). Role of artificial intelligence in brain imaging analysis: Current status and future perspectives. *Frontiers in Neuroscience*, 16, 923451.

51. Jamieson, T., & Williams, D.R. (2020). Machine learning in neurological disorder prediction and classification. *IEEE Reviews in Biomedical Engineering*, 13, 321–336.
52. Siddiqui, M.M., Srivastava, G., & Saeed, S.H. (2019). *Diagnosis of Sleep Disorders Using EEG Signal*. LAP LAMBERT Academic Publishing.
53. Farina, D., & Vujaklija, I. (2020). Machine learning in neuroprosthetics: Hybrid approaches for improving performance. *Trends in Neurosciences*, 43(11), 915–928. https://doi.org/10.1016/j.tins.2020.09.008
54. Raghavendra, U., Acharya, U.R., & Adeli, H. (2020). Artificial intelligence techniques for automated diagnosis of neurological disorders. *European Neurology*, 82(1–3), 41–64.
55. Surianarayanan, C., Lawrence, J.J., Chelliah, P.R., Prakash, E., & Hewage, C. (2023). Convergence of artificial intelligence and neuroscience towards the diagnosis of neurological disorders—A scoping review. *Sensors*, 23(6), 3062.
56. Patel, U.K., Anwar, A., Saleem, S., Malik, P., Rasul, B., Patel, K., et al. (2021). Artificial intelligence as an emerging technology in the current care of neurological disorders. *Journal of Neurology*, 268, 1623–1642.
57. Alam, M., Ara, K., Javed, M.S., & Ansari, M. (2014, July). Detect and filter traffic attack through cloud Trace back and neural network, Imperial College. *The 2014 International Conference of Data Mining and Knowledge Engineering (ICDMKE)*, London (pp. 2–4).
58. Ansari, M., & Alam, M. (2023). IoT-cloud-enabled smart framework for real-world applications. In *Intelligent Systems and Applications: Select Proceedings of ICISA 2022* (pp. 87–105). Springer Nature Singapore.
59. Ansari, M., & Alam, M. (2024). A novel hybrid time-series approach for IoT-cloud-enabled environment monitoring. *Journal of Supercomputing*, 80(7), 9019–9053.
60. Noor, M.A.F., Khanum, S., Anwar, T., & Ansari, M. (2021). A holistic view on blockchain and its issues. In *Blockchain Applications in IoT Security* (pp. 21–44). IGI Global.
61. Ansari, M., & Alam, M. (2021). IoT-cloud enabled statistical analysis and visualization of air pollution data in India. In *Proceedings of Data Analytics and Management: ICDAM 2021*, Volume 2 (pp. 125–139). Springer Singapore.
62. Ansari, M., Ali, S.A., Alam, M., Chaudhary, K., & Rakshit, S. (2023). Unlocking the power of explainable AI to improve customer experiences in E-Commerce. In *AI-Based Data Analytics* (pp. 31–48). Auerbach Publications.
63. Ansari, M., Ali, S.A., & Alam, M. (2022). Internet of things (IoT) fusion with cloud computing: Current research and future direction. *International Journal of Advanced Technology and Engineering Exploration*, 9(97), 1812.
64. Ali, S.A., Ansari, M., & Alam, M. (2020). Resource management techniques for cloud-based IoT environment. In *Internet of Things (IoT) Concepts and Applications* (pp. 63–87). Springer International Publishing.
65. Ansari, M., & Alam, M. (2024). An intelligent IoT-cloud-based air pollution forecasting model using univariate time-series analysis. *Arabian Journal for Science and Engineering*, 49(3), 3135–3162.

Chapter 10

Deep learning-based approaches for enhancing battery prognostics and health management

V. Lakshmi Chetana, Balasubbareddy Mallala, and P. V. R. Ravindra Reddy

10.1 INTRODUCTION

Battery prognostics and health management (BPHM) [1] plays a pivotal role in predicting the state of health (SoH) and the remaining useful life (RUL) of batteries, thereby ensuring their safe, reliable, and efficient performance. In this context, BPHM integrates advanced data analysis techniques with intelligent algorithms to assess battery behavior and anticipate future operational states. Figure 10.1 presents the different approaches for battery SoH and RUL.

By continuously monitoring critical parameters such as temperature, voltage, and capacity, BPHM facilitates the early detection of potential issues. Consequently, timely maintenance or replacement actions can be taken, which not only extends the operational lifespan of batteries but also significantly enhances their reliability. Such improvements are especially

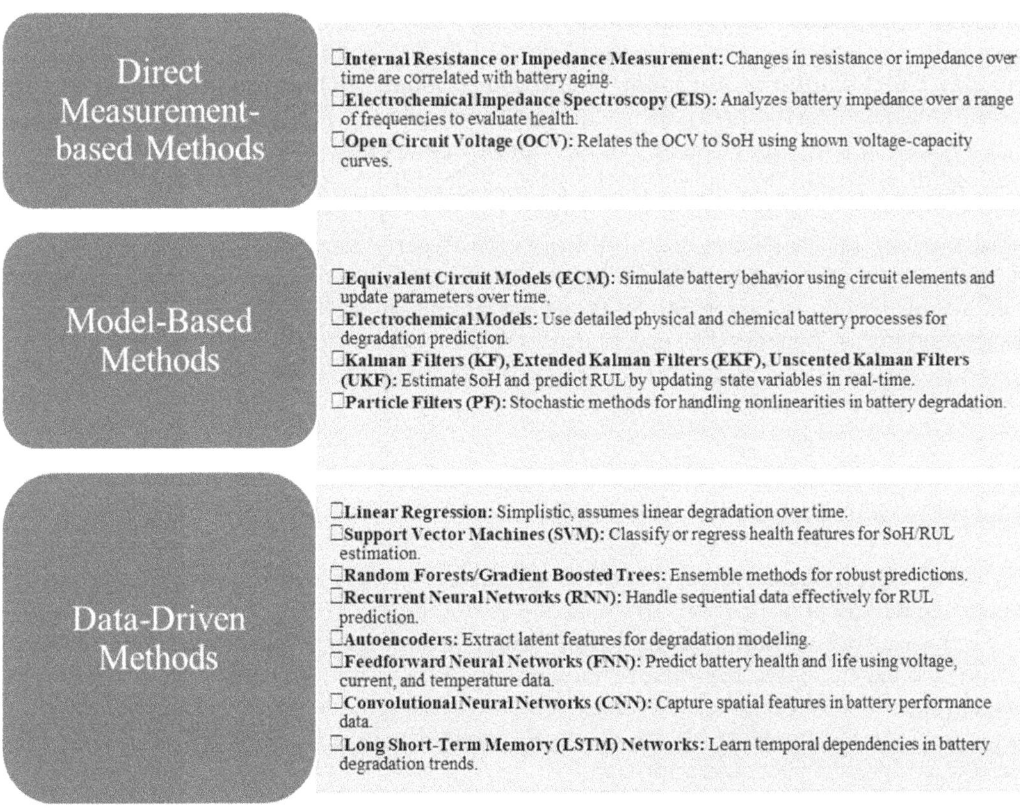

Figure 10.1 Different approaches for battery SoH and RUL..

DOI: 10.1201/9781003685890-10

crucial in sectors where uninterrupted power supply is vital, including electric vehicles, aerospace, and renewable energy systems. Therefore, with the rapid growth of these industries, the development of advanced and robust BPHM techniques has become more important than ever. Lithium-ion batteries [2], in particular, have emerged as the preferred energy storage technology owing to their superior energy density and extended cycle life. Nevertheless, their performance is subject to gradual degradation due to aging mechanisms, temperature variations, and diverse operating conditions. To address these challenges, researchers and engineers are actively developing innovative BPHM strategies specifically tailored for lithium-ion batteries [3–5]. These approaches are designed to enable real-time monitoring of battery health and performance, thereby supporting preventive maintenance and ensuring optimal operation throughout their service life. Furthermore, BPHM frameworks not only contribute to improved operational reliability but also provide valuable insights for the design of next-generation battery materials and system architectures. Hence, they serve as a foundation for advancing efficiency, durability, and overall dependability in future energy storage systems.

10.2 LITERATURE SURVEY

In recent years, a growing body of research has concentrated on the application of deep learning methods for estimating the SoH of lithium-ion batteries, with particular emphasis on capacity prediction. In this regard, several studies [6–19] have explored diverse deep learning architectures, including feedforward neural networks (FNNs), often integrated with optimization strategies and battery management frameworks, to enhance predictive accuracy. Furthermore, complementary works [20–30] highlight the integration of IoT-enabled cloud platforms, advanced forecasting techniques, large-scale data storage solutions, and neural network-based tracking systems, thereby broadening the technological ecosystem that supports effective battery health monitoring. Taken together, these contributions demonstrate the significant potential of deep learning in improving the accuracy, robustness, and scalability of Li-ion battery capacity estimation. However, despite these advances, several challenges remain unresolved. In particular, issues related to the heavy dependence on large datasets, limited generalizability across varying operating conditions, and the inherent lack of interpretability in deep learning models persist. Therefore, while the current literature provides a strong foundation, there remains substantial scope for further investigation into developing more adaptive, transparent, and data-efficient approaches for battery health prognostics.

10.3 METHODOLOGY

Figure 10.2 presents an overview of the proposed framework, which is structured into three main stages: data preprocessing, training, and estimation.

The process begins with data cleaning, followed by min-max normalization [31–32]. After preprocessing, the dataset is divided into training, validation, and test sets. In the second stage, the validation set is used to determine the most suitable model. Finally, in the third stage, the selected models [33] are employed to predict battery capacity and evaluate the performance of the proposed approach.

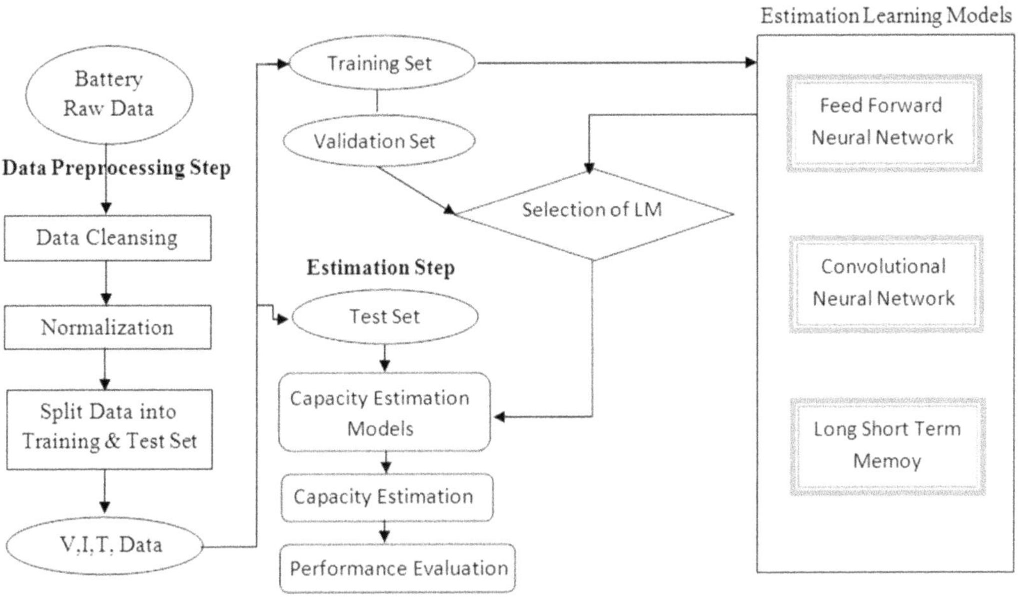

Figure 10.2 A comprehensive framework for the proposed capacity estimation.

10.3.1 Data preprocessing

Prior to employing the battery dataset [34–35] in this study, a comprehensive data preprocessing stage was carried out to ensure both accuracy and consistency. This stage involved the removal of anomalies and the verification of the completeness of all relevant parameters. The batteries chosen for analysis, namely #5, #6, #7, and #18, as designated in the online repository, were selected due to the availability of complete operational records. The detailed specifications of these batteries, along with the corresponding charging and discharging conditions, are summarized in Table 10.1.

The proposed models were developed using features extracted from uniformly sampled raw battery data. Each feature vector was constructed with 30 dimensions, derived from ten evenly spaced samples of voltage (V), current (I), and temperature (T) profiles during the charging process. This sampling strategy effectively captures the temporal dynamics of battery behavior while maintaining a balance between model accuracy and computational complexity. To minimize the influence of abrupt fluctuations in the raw data, average values were computed within each sampling interval. Furthermore, a min-max normalization technique [36–37] was applied, which scales all data points to a normalized range of 0 to 1. This preserves the underlying data distribution while ensuring consistency and stability during the training phase of the models.

10.3.2 Training

Figure 10.3 exhibits notable alterations in charging as a battery deteriorates over time. Contrasted with a new cell, the aged cell achieves the peak voltage (4.2 V) sooner, undergoes a decline in current from the constant current stage earlier, and reaches maximum temperature faster. These changes stem from repeated charging, resting, and discharging cycles. Typically, the temperature at the discharge's conclusion surpasses that at the commencement, affecting

Table 10.1 Battery specifications data repository

Battery no	*Charging*			*Discharging*		*Operating Conditions*		
	Constant current (A)	*Upper voltage limit (V)*	*Cut-off current (mA)*	*Constant current (A)*	*Cut-off voltage (V)*	*Operating temperature*	*Initial capacity (Ah)*	*End of Life criteria (Ah)*
5	1.6	4.2	20	2	2.7	RT	1.86	1.40
6	1.6	4.2	20	2	2.5	RT	2.04	1.40
7	1.6	4.2	20	2	2.2	RT	1.89	1.40
8	1.6	4.2	20	2	2.5	RT	1.86	1.40

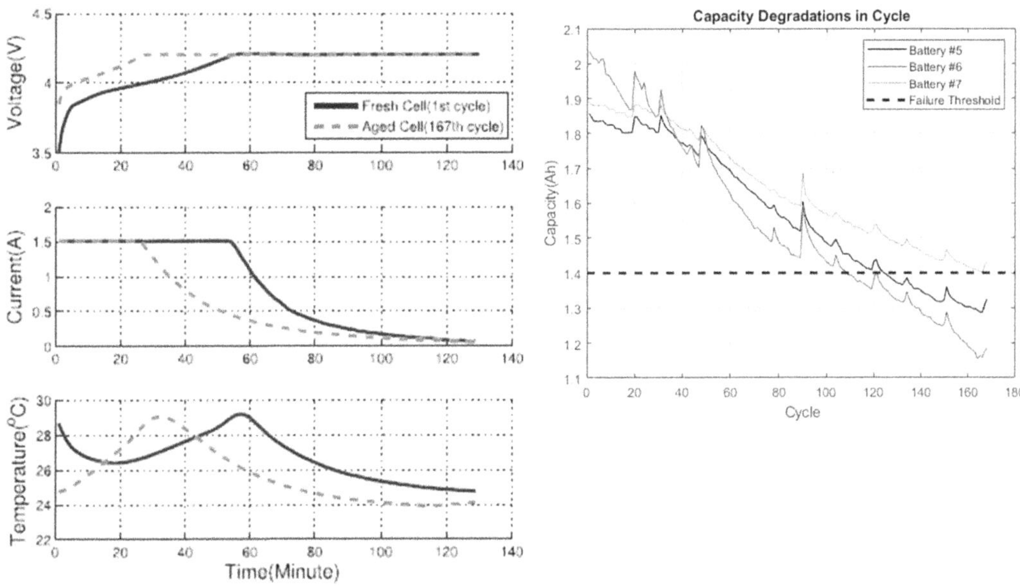

Figure 10.3 New and deteriorated cell characteristics.

the subsequent charging cycle's initial temperature. To quantify battery aging, a well-defined SoH is required. While there's no single universal definition, it's commonly expressed in terms of capacity, as shown by

$$\mathrm{SoH}(\%) = \mathrm{Ck} / \mathrm{C0} \times 100$$

For effective training, we utilize min-max normalization. This method maintains the original data distribution while introducing a scaling factor, guaranteeing all values lie within the 0 to 1 range. This normalization simplifies the training of the deep learning model.

10.3.2.1 Feedforward neural network (FNN)

In this initial strategy, we utilized an FNN, a widely recognized machine learning method distinguished by the lack of feedback connections among neurons. FNNs consist of numerous layers, with neurons in each layer applying a nonlinear activation function to weighted inputs from the previous layer. To adapt to the input data's inherent variability, we selected a conventional FNN design with a solitary hidden layer.

10.3.2.2 Convolutional neural network (CNN)

Moreover, we investigate the application of convolutional neural networks (CNNs), a well-known deep learning structure recognized for utilizing convolution operations rather than conventional matrix multiplication in at least one layer. Convolution, within the realm of two-dimensional data, can be elucidated as follows. A standard CNN structure [38] comprises convolutional layers [39] that extract features via convolution operations, followed by the application of an activation function to the output. Subsequently, pooling layers decrease feature map dimensions, bolstering the network's resilience to input fluctuations. The refined data proceeds to the subsequent layer. Through successive iterations of convolutional and pooling layers, the network discerns increasingly complex features.

10.3.2.3 Long short-term memory (LSTM)

Recurrent Neural Networks (RNNs) [30] are designed to work with sequential data by using feedback connections that allow them to remember information over time. However, standard RNNs [40] suffer from the vanishing gradient problem, which makes it difficult for them to learn long-term dependencies. In such cases, error signals weaken as they pass through many time steps, limiting the network's ability to capture long-range patterns. To address this limitation, Long Short-Term Memory (LSTM) networks [31] were developed. LSTMs include a memory cell and three control gates input, output, and forget that decide what information should be stored, updated, or removed. By managing the flow of information in this way, LSTMs [41] can retain knowledge over long sequences, making them effective at handling long-term dependencies in sequential data.

10.4 EXPERIMENTAL RESULTS

10.4.1 Dataset

Our research utilizes battery datasets procured from a data repository. These datasets cover eight lithium-ion batteries subjected to three distinct operational profiles: charging, discharging, and resting periods, all executed at room temperature. Throughout charging, the CCCV (constant current, constant voltage) principle was applied. The cell voltage hit its upper threshold of 4.2 V, succeeded by a constant voltage application until the current diminished to 20 mA. The discharging phase employed a constant current of 2 A until the cell voltage reached predetermined cut-off values specific to each battery (#5: 2.7 V, #6: 2.6 V, #7: 2.2 V, and #18: 2.5 V). These trials persisted until the batteries encountered a 30% capacity reduction, corresponding to 1.4 Ah.

10.4.2 Evaluation metrics

The mean absolute percentage error (MAPE) was utilized as a representative error index to evaluate the estimation accuracy.

$$\text{MAPE}(\%) = \frac{100}{K}\sum_{k=1}^{K}\frac{\left|l(k)-\hat{l}(k)\right|}{l(k)}$$

$$\text{MAE} = \frac{1}{K}\sum_{k=1}^{K}\left|l(k)-\hat{l}(k)\right|,$$

$$\text{RMSE} = \sqrt{\frac{1}{K}\sum_{k=1}^{K}l(k)-\hat{l}(k)^2}.$$

Table 10.2 Estimation errors using voltage, current, and temperature profiles

Model	*RMSE*	*MAE*	*MAPE (%)*
FNN-10	0.0379	0.0329	1.9800
FNN-40	0.0298	0.0242	1.7300
CNN-1	0.0584	0.0443	2.8961
CNN-2	0.0443	0.0364	2.3731
LSTM	0.0246	0.0159	1.0320

10.4.3 Results and discussions

Given the limited dataset availability, maximizing existing data utilization is essential. Hence, a k-fold cross-validation technique was adopted. The four datasets were divided into three training sets and one testing set, repeating this process four times with varying folds. This ensures each data point appears in the testing set once. Ultimately, performance assessment involves averaging results across all four folds, yielding a more robust model generalizability evaluation. Table 10.2 shows the estimation errors using voltage, current, and temperature profiles.

Battery capacity during charging was estimated using the proposed method. Each channel undergoes uniform sampling at ten data points, and capacities estimated using the proposed methods are plotted for each charging cycle. Figures 10.4–10.6 show the capacity estimation for different profiles for batteries #5, #6, and #7 respectively.

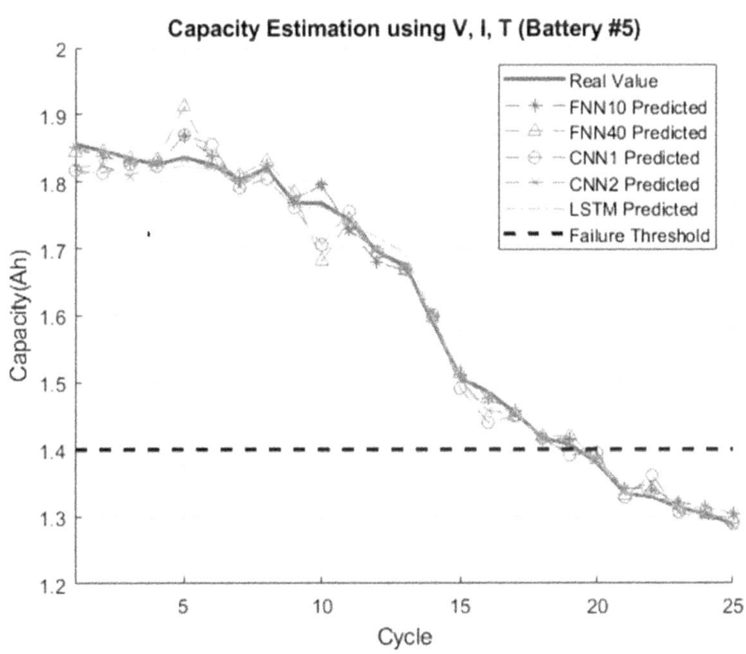

Figure 10.4 Capacity estimation using voltage, current, temperature profiles (Battery #5)..

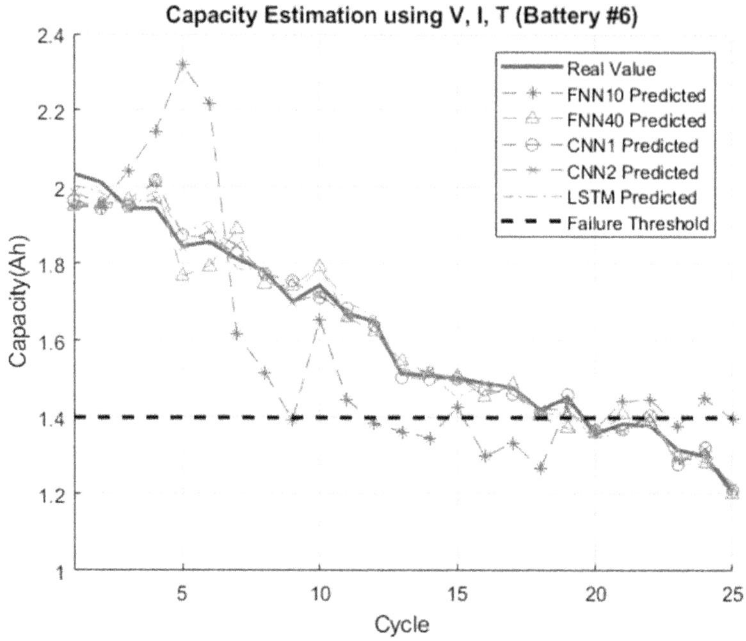

Figure 10.5 Capacity estimation using voltage, current, temperature profiles (Battery #6).

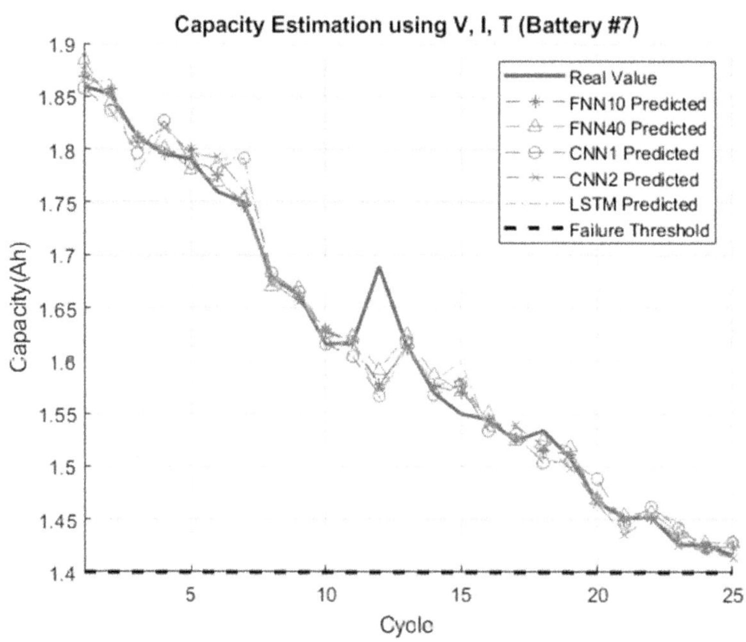

Figure 10.6 Capacity estimation using voltage, current, temperature profiles (Battery #6).

10.5 CONCLUSION

This paper evaluates an innovative framework for lithium-ion battery capacity estimation employing a blend of convolution networks. This framework utilizes multi-channel data encompassing parameters to attain precise estimations. The proposed method's performance was assessed using a lithium-ion battery dataset from NASA, with evaluation based on deviation per cycle. Our results underscore the pivotal role of diverse data channels in achieving high estimation accuracy. In future endeavors, we intend to enhance this method by devising an online approach capable of dynamically updating in real time, thereby capturing the evolving degradation process during battery operation.

REFERENCES

[1] Y. Choi, S. Ryu, K. Park, and H. Kim, "Machine learning-based lithium-ion battery capacity estimation exploiting multi-channel charging profiles," *IEEE Access*, vol. 7, pp. 75143–75152, 2019, https://doi.org/10.1109/ACCESS.2019.2920932.

[2] Z. Liu, Z. Jia, C.-M. Vong, J. Han, C. Yan, and M. Pecht, "A patent analysis of prognostics and health management (PHM) innovations for electrical systems," *IEEE Access*, vol. 6, pp. 1808818107, 2018.

[3] B. Saha, K. Goebel, S. Poll, and J. Christophersen, "Prognostics methods for battery health monitoring using a Bayesian framework," *IEEE Transactions on Instrumentation and Measurement*, vol. 58, no. 2, pp. 291–296, Feb. 2009; Condition Monitoring and Diagnostics of Machines Prognostics Part1: General Guidelines, Int. Standards Organisation, Geneva, Switzerland, Nov. 2004.

[4] J. Meng, D.-L. Stroe, M. Ricco, G. Luo, and R. Teodorescu, "A simplified model-based state-of-charge estimation approach for lithium-ion battery with dynamic linear model," *IEEE Transactions on Industrial Electronics*, vol. 66, no. 10, pp. 7717–7727, Oct. 2019.

[5] D.-I. Stroe, V. Knap, M. Swierczynski, A.-I. Stroe, and R. Teodorescu, "Operation of a grid-connected lithium-ion battery energy storage system for primary frequency regulation: A battery lifetime perspective," *IEEE Transactions on Industry Applications*, vol. 53, no. 1, pp. 430–438, Jan./Feb. 2017.

[6] B. Bennehalli, L. Singh, S. Stephen, P. V. Prasad, B. Mallala, A. Purna, and C. Rao, "Machine learning approach to battery management and balancing techniques for enhancing electric vehicle battery performance," *Journal of Electrical Systems*, vol. 20, no. 2, pp. 885–892, 2024.

[7] Sanjay Dubey, "End-of-discharge and end-of-life predic- tion in lithium-ion batteries with electrochemistry-based aging models," *Proceedings, 2023 IEEE International Conference on Paradigm Shift in Information Technologies with Innovative Applications in Global Scenario: 28–29 December 2023*, SAGE University, Indore. IEEE Computer Society, 2023.

[8] A. Harshavardhan, A. Alam, R. S. Kumar, and B. Mallala, "Bayesian asymmetric quantized neural networks for MRI/mammography-based breast cancer identification in PTEN hamartoma syndrome," *Biomedical Materials & Devices*, 2025, https://doi.org/10.1007/s44174-025-00456-5

[9] K. P. Joshua, J. Ranga, P. V. Prasad, B. Mallala, M. Rajendiran, and R. Maranan, "Optimized scheduling of electric vehicles charging in smart grid using deep learning," *Proceedings – 2024 International Conference on Expert Clouds and Applications, ICOECA 2024*, pp. 408–412, 2024, https://doi.org/10.1109/ICOECA62351.2024.00079

[10] V. P. Papana, B. Mallala, V. K. Reddy Chinthalacheruvu, and K. Palle, "Evaluating the impact of distribution static compensators on power quality enhancement with finite element interpolated neural networks in modern power systems," *Expert Systems with Applications*, vol. 297, pp. 2026, https://doi.org/10.1016/j.eswa.2025.129214

[11] A. Paul, K. T. Thilagham, K. G. Jaya Christiyan, P. R. Reddy, S. Ravishankar, and S. Boopathi, "Multi-criteria optimization on friction stir welding of aluminum composite (AA5052-H32/B4C) using titanium nitride coated tool," *Engineering Research Express*, vol. 6, no. 1, 2024, https://doi.org/10.1088/2631-8695/ad248d

[12] P. Rajendran, J. Ranga, P. V. Prasad, B. Mallala, G. Senthilkumar, and L. Natrayan, "Development of intelligent power quality management in renewable energy system in smart grid using deep learning," *7th International Conference on Inventive Computation Technologies, ICICT 2024*, pp. 1178–1182, 2024, https://doi.org/10.1109/ICICT60155.2024.10544835

[13] B. V. S. Rao, G. Chandra, M. Reddy, G. Krishna, M. Rao, and P. R. Reddy, "Numerical analysis on the effect of various parameters on fracture limit for deep drawn cups," n.d., www.sciencedirect.comwww.materialstoday.com/proceedings2214-7853

[14] D. Santhakumar, D. Mishra, B. Mallala, S. Sridhar, and T. Dhanadhya, "Optimizing smart grid energy and load prediction with Namib beetle algorithm and self-adaptive physics-informed neural networks," *3rd International Conference on Automation, Computing and Renewable Systems, ICACRS 2024 – Proceedings*, pp. 367–373, 2024, https://doi.org/10.1109/ICACRS62842.2024.10841617

[15] D. Santhakumar, K. K. Sankar, B. Mallala, S. Sridhar, and S. Kadam, "Enhancing energy management in photovoltaic systems with battery storage through artificial rabbits optimization in connected and disconnected modes," *3rd International Conference on Automation, Computing and Renewable Systems, ICACRS 2024 – Proceedings*, pp. 374–379, 2024, https://doi.org/10.1109/ICACRS62842.2024.10841665

[16] S. Thankaraj Ambujam, B. Mallala, P. K. Aggarwal, and P. V. Prasad, "Enhanced IoT-enabled community microgrid energy management with hybrid COA-HQNN approach with battery degradation consideration," *Journal of Energy Storage*, vol. 107, 2025, https://doi.org/10.1016/j.est.2024.114739

[17] S. Thankaraj Ambujam, B. Mallala, P. K. Aggarwal, and P. V. Prasad, "Enhanced IoT-enabled community microgrid energy management with hybrid COA-HQNN approach with battery degradation consideration," *Journal of Energy Storage*, vol. 107, 2025, https://doi.org/10.1016/j.est.2024.114739

[18] B. V. S. Rao, and P. V. R. Ravindra Reddy, "Influence of drawing ratio on thickness variation along the walls of deep drawn cups," *Journal of Manufacturing Engineering*, vol. 11, no. 2, 2016, www.smenec.org

[19] N. Venkateswaran, J. Ramprabu, R. Marimuthu, N. V. Ramana, M. Kotha, and B. Mallala, "Enhanced demand-side management for electric vehicle charging stations using pyramidal dilation attention neural networks," *6th International Conference on Mobile Computing and Sustainable Informatics, ICMCSI 2025 – Proceedings*, 572–577, 2025, https://doi.org/10.1109/ICMCSI64620.2025.10883404

[20] M. Alam, K. Ara, M. S. Javed, and M. Ansari, "Detect and filter traffic attack through cloud Trace back and neural network, Imperial College," *The 2014 International Conference of Data Mining and Knowledge Engineering (ICDMKE)*, London, pp. 2 1, 2014.

[21] M. Ansari, and M. Alam, "IoT-cloud-enabled smart framework for real-world applications," *Intelligent Systems and Applications: Select Proceedings of ICISA 2022* (pp. 87–105). Singapore: Springer Nature Singapore, 2023.

[22] M. Ansari, and M. Alam, "A novel hybrid time-series approach for IoT-cloud-enabled environment monitoring," *Journal of Supercomputing*, vol. 80, no. 7, 2024.

[23] M. A. F. Noor, S. Khanum, T. Anwar, and M. Ansari, "A holistic view on blockchain and its issues," *Blockchain Applications in IoT Security* (pp. 21–44). IGI Global, 2021.

[24] M. Ansari, and M. Alam, "IoT-cloud enabled statistical analysis and visualization of air pollution data in India," *Proceedings of Data Analytics and Management: ICDAM 2021, Volume 2* (pp. 125–139). Singapore: Springer Singapore, 2021.

[25] M. Ansari, S. A. Ali, M. Alam, K. Chaudhary, and S. Rakshit, "Unlocking the power of explainable AI to improve customer experiences in E-Commerce," *AI-Based Data Analytics* (pp. 31–48). Auerbach Publications, 2023.

[26] M. Ansari, S. A. Ali, and M. Alam, "Internet of things (IoT) fusion with cloud computing: Current research and future direction," *International Journal of Advanced Technology and Engineering Exploration*, vol. 9, no. 97, p. 1812, 2022.
[27] S. A. Ali, M. Ansari, and M. Alam, "Resource management techniques for cloud-based IoT environment," *Internet of things (IoT) concepts and applications* (pp. 63–87). Cham: Springer International Publishing, 2020.
[28] M. Ansari, and M. Alam, "An intelligent IoT-cloud-based air pollution forecasting model using univariate time-series analysis," *Arabian Journal for Science and Engineering*, vol. 49, no. 3, pp. 3135–3162, 2024.
[29] S. A. Ali, M. Ansari, M. Alam, and S. Rakshit, "Federated learning for business intelligence: Predictive maintenance in industry 4.0," *AI-Based Data Analytics* (pp. 125–140). Auerbach Publications, 2023.
[30] S. Khan, X. Liu, S. A. Ali, and M. Alam, "Storage solutions for big data systems: A qualitative study and comparison," *arXiv preprint arXiv:1904.11498*, 2019.
[31] M. Rashid, and A. Gupta, "Effect of relaxation periods over cycling performance of a Li-ion battery," *Journal of the Electrochemical Society*, vol. 162, no. 2, pp. A3145–A3153, Jan. 2015.
[32] L. He, G. Meng, Y. Gu, C. Liu, J. Sun, T. Zhu, Y. Liu, and K. G. Shin, "Battery-aware mobile data service," *IEEE Transactions on Mobile Computing*, vol. 16, no. 6, pp. 1544–1558, Jun. 2017.
[33] A. Jain, K. Nandakumar, and A. Ross, "Score normalization in multimodal biometric systems," *Pattern Recognition*, vol. 38, no. 12, pp. 2270–2285, Dec. 2005.
[34] I. Goodfellow, Y. Bengio, and A. Courville, *Deep Learning*. Cambridge, MA: MIT Press, 2016, http://www.deeplearningbook.org
[35] H. Choi, S. Ryu, and H. Kim, "Short-Term load forecasting based on ResNet and LSTM," *Proceedings IEEE International Conference Communication, Control, Computer Technology Smart Grids (SmartGridComm)*, Oct. 2018, pp. 1–6.
[36] S. Hochreiter and J. Schmidhuber, "Long short-term memory," *Neural Computation*, vol. 9, no. 8, pp. 1735–1780, 1997.
[37] M. Abadi, P. Barham, J. Chen, Z. Chen, A. Davis, J. Dean, M. Devin, S. Ghemawat, G. Irving, M. Isard, and M. Kudlur, "TensorFlow: A system for large-scale machine learning," *Proceedings 12th USENIX Symposium on. Operating Systems Design and Implementation (OSDI)*, Nov. 2016, pp. 265–283.
[38] Ying Zhang, and Yan-Fu Li, Prognostics and health management of Lithium-ion battery using deep learning methods: A review, *Renewable and Sustainable Energy Reviews*, vol. 161, p. 112282, 2022, https://doi.org/10.1016/j.rser.2022.112282
[39] D. P. Kingma and J. Ba, "Adam: A method for stochastic optimization," *Proceedings International Conference Learning Representations (ICLR)*, May 2015, pp. 1–13.
[40] A. L. Maas, A. Y. Hannun, and A. Y. Ng, "Rectifier nonlinearities improve neural network acoustic models," *Proceedings of the International Conference on Machine Learning (ICML)*, Jun. 2013, vol. 30, no. 1, pp. 1–6.
[41] T. Wang, J. Yu, D. Siegel, and J. Lee, "A similarity-based prognostics approach for remaining useful life estimation of engineered systems," *2008 International Conference on Prognostics and Health Management*, Oct. 2008, pp. 1–6.

Chapter 11

A review on applications of artificial intelligence in hydrogen energy

P. V. R. Ravindra Reddy, Balasubbareddy Mallala, and P. Prabhakar Reddy

11.1 INTRODUCTION

Hydrogen energy is globally recognized as a major alternative to fossil fuels in the transition toward low-carbon and net-zero energy systems. Its high gravimetric energy density and compatibility with intermittent renewable generation sources make it an inevitable solution for decarbonization [1, 2]. As per predictions of the International Energy Agency, 10% of global energy will be taken care of by hydrogen supply by 2050, assuming technological barriers in production, distribution, storage, and infrastructure challenges are overcome [3]. This could be achieved through optimization of production processes [4], enhancement of safety with early detection of leaks and risk assessment [5], improvement of logistics through demand forecasting and network optimization [6], and supporting intelligent transportation integration with fuel-cell systems [7], thus transforming various stages of the hydrogen value chain with the prevailing developments in artificial intelligence (AI), machine learning (ML), deep learning (DL), predictive modeling, and autonomous robotics. Collectively, progress in AI technologies is becoming a transformative enabler for scalable, safe, and economically viable hydrogen energy deployment in the coming decades [8]. Hydrogen value chain consists of production, storage, distribution, conversion, and utilization, in which the first one, production, can be carried out through various established processes as presented in Figure 11.1.

The production processes are classified as thermochemical processes, electrochemical processes, biological processes, photochemical processes, and hybrid processes.

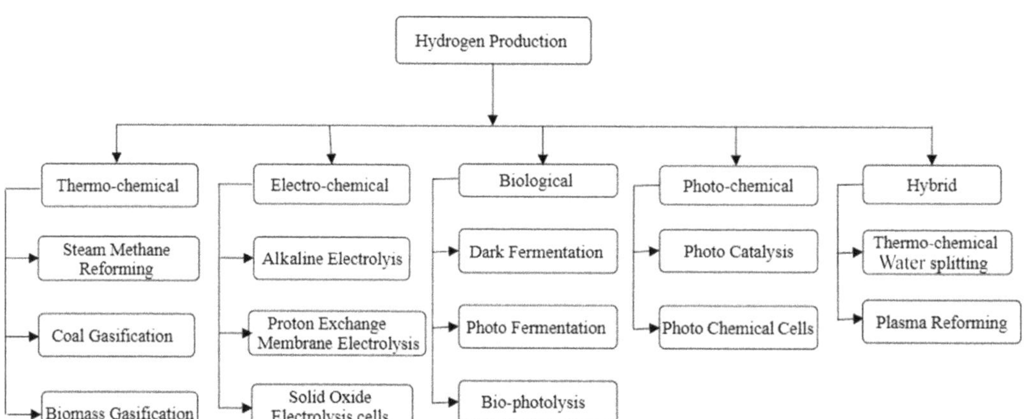

Figure 11.1 Classification of hydrogen production methods.

DOI: 10.1201/9781003685890-11

1. **Thermochemical Processes:** Hydrogen is extracted by heating fossil and biomass materials in this process. In steam methane reforming methane is reacted with steam at a high temperature. In coal gasification coal is partially oxidized with oxygen and steam under high temperature and pressure. In biomass gasification biomass is thermally converted to hydrogen.
2. **Electrochemical Methods:** In these methods, water is split into hydrogen and oxygen. These methods are classified based on the way the water is split: in alkaline electrolysis with alkaline electrolyte, in proton exchange membrane electrolysis with Nafion, and in solid oxide electrolysis cells by ceramic cells.
3. **Biological Methods:** These methods involve microorganisms or enzymes to generate hydrogen under mild conditions. In dark fermentation anaerobic bacteria ferment organic matter to release hydrogen. Photo fermentation uses photosynthetic bacteria, sunlight, and organic matter to release hydrogen. Biophotolysis uses solar energy and hydrogenase/nitrogenase enzymes to split water.
4. **Photochemical and Photoelectrochemical Methods:** Light-activated catalysts are used to split water. In photocatalysis semiconductor materials absorb sunlight for water splitting. In photoelectrochemical cells photovoltaic and electrolysis processes are carried out in one device.
5. **Hybrid Approaches:** It combines multiple methods. In thermochemical water-splitting cycles high-temperature cycles using metal oxides are employed. In plasma reforming hydrogen is produced from hydrocarbons or biomass at lower bulk temperatures.

11.2 AI IN HYDROGEN PRODUCTION

Artificial intelligence techniques are increasingly being applied to optimize hydrogen production processes, particularly in water electrolysis, where high capital and operating costs stem mainly from the expensive proton exchange membranes and the substantial energy required per kilogram of hydrogen produced. Moreover, operational inefficiencies remain common in electrolytic systems, further impacting overall cost-effectiveness. Computational fluid dynamics (CFD) modeling in combination with ML algorithms could significantly improve electrolyzer design and operational control, achieving reductions in operating costs of up to 24% while maintaining prediction accuracies exceeding 95% for flow distribution and polarization curve estimation [9]. In terms of design, CFD could identify problems of uneven flow distribution, high-pressure drops, gas bubble formations, and local overheating through simulation of fluid flow and heat transfer [10], while AI algorithms, serving as surrogate models, use CFD-generated data and experimental results to quickly predict performance of new designs, saving time by reducing the number of simulations and enabling the evolution of optimized designs with minimum pressure drops, better flow paths, and lower overpotential losses. Operating performance was improved by adaptive control of operational parameters through AI-driven algorithms for maximization of hydrogen yield for a given power input. The durability of the cell could be improved by identifying regions prone to corrosion and membrane degradation through CFD simulations and enhanced predictive maintenance by AI models through sensory data.

Even though most AI research in hydrogen production focuses toward enhancement of design and operation of proton exchange membranes and alkaline electrolyzers, considerable research is evident in high-temperature systems such as solid oxide electrolytic cells (SOEC), which are sensitive to thermal stresses; biomass systems, which are highly sensitive to variation of feedstock quality and reaction conditions; and solar-to-hydrogen photochemical systems, where the fluctuations in light intensity and catalytic conditions are inevitably common. By leveraging predictive analytics and multi-objective optimization, AI models can dynamically adjust input parameters such as temperature, pressure, and current density to maximize hydrogen yield while minimizing degradation rates [11].

By integrating multiphysics simulations of steam and hydrogen flow inside cathode and anode channels using Navier–Stokes equations, simulation of heat transfer modes through ceramic layers and gas channels, simulation of electrode reactions through Butler–Volmer kinetics and Nernst potential calculations, and thermal stress analysis for prediction of cracking and delamination coupled with AI surrogate models, researchers can optimize steam utilization and temperature uniformity in SOEC and also identify hot spots and areas of uneven current distribution to prevent faster degradation of the cell. Subsequently, ML models such as artificial neural networks (ANNs) and support vector machines (SVMs) could be trained with the simulation and experimental data augmented by AI surrogate models for the prediction of hydrogen output rate and efficiency under different operating conditions and response to fluctuating energy power supply, especially in the case of renewable energy, and for adaptively controlling the system for higher hydrogen yield and reduced wear and tear, thus leading to more resilient and economically viable SOEC systems.

Biomass-to-hydrogen conversion encounters nonlinear complexities such as heterogeneous feedstock, variation in moisture content, and ash composition variability that challenge stable reactor performance under dynamic operating conditions. ML models, ANNs, and Gaussian process regressors (GPRs) identify the relations of hydrogen yield with operating variables under variable conditions and dynamically adjust operating parameters such as steam–biomass ratio, gasification temperature, and catalyst loading to maximize hydrogen output while suppressing tar formation and unwanted byproducts. In addition, AI-enabled fault detection systems identify early deviations in reactor behavior such as suboptimal carbon burnout and heat profile inconsistencies, enabling predictive maintenance and reducing downtime.

Furthermore, hybrid AI-quantum computing frameworks are accelerating materials discovery for photocatalytic water splitting. In particular, density functional theory (DFT) simulations integrated with DL and inverse design algorithms have been applied to identify novel catalyst materials such as titanium dioxide (TiO_2) and bismuth vanadate ($BiVO_4$), enabling faster screening of candidate materials with enhanced light absorption and catalytic efficiency [12]. These advancements highlight the dual role of AI not only as a process optimizer but also as an innovation driver in materials science, effectively linking fundamental research with large-scale production of hydrogen in a cost-effective manner.

Summary of AI technologies in the production of hydrogen along with their advantages is presented in Table 11.1.

Table 11.1 Comparison of AI techniques in hydrogen production

S. No.	*Model*	*Application*	*Area*	*Advantages*
1.	Machine learning surrogate models with Computational Fluid dynamics	Design and Operation of Electrolyzer	Prediction of flow distribution	Lesser number of trials with faster optimization
2.	Artificial Neural Networks along with Support Vector Machines	Solid Oxide Electrolyzer cells	Tar prevention and consistent yield	Variability of feedstock can be handled
3.	Gaussian Process Regression with Artificial Neural Networks	Hydrogen from Biomass	Tar prevention and consistent yield	Variability of feedstock can be handled
4.	Combination of Deep Learning, Density Functional Theory and Inverse Design	Photocatalysis	Discovery of new catalyst materials	Mineral innovation

11.3 SAFETY ENHANCEMENT VIA AI

Safety is the most important concern over productivity and cost-effectiveness in hydrogen technologies due to high flammability, low self-ignition temperature of hydrogen, and its tendency to leak through minute gaps. Conventional leak detection systems normally rely on single-point sensors—electrochemical, catalytic bead, or semiconductor detectors—which show considerable constraints such as delayed leak detection in critical environments due to their localized sensing and sensitivity to environmental conditions, compromising system reliability. The limited spatial coverage of point detectors limits leak detection in complex installations. In order to combat these limitations, AI techniques offer transformative opportunities in enabling faster, more accurate, and proactive hydrogen safety systems by leveraging advanced data analytics, pattern recognition, and real-time decision-making. This section reviews AI-enabled implementations in safety systems across four major domains: sensor network integration, ANNs, advanced localization methodologies, and predictive risk modeling.

11.3.1 AI-enhanced sensor networks and data fusion

AI algorithms can detect anomalies more reliably through continuous, distributed monitoring by integrating wireless sensor networks (WSNs) with AI-enabled analytics [13]. In such systems, spatially distributed sensors capture multi-parameter data such as hydrogen concentration, temperature, humidity, and flow dynamics and transmit them in real time. AI models employing convolutional neural networks (CNNs), long short-term memory (LSTM) networks, or hybrid DL architectures predict trends, perform pattern recognition, and filter signals to distinguish leak signatures from transient environmental conditions. Within the hydrogen energy literature, this AI-WSN integration has been identified as a pivotal advancement over conventional safety systems [4], due to the transformation of a reactive, single-point leak-sensing paradigm into a proactive, high-resolution monitoring network. Multi-sensor data fusion methodology not only enhances detection accuracy but also reduces false alarms, thereby supporting faster resolution and mitigation strategies [14, 15].

Wireless near-field hydrogen sensors have also achieved breakthroughs in responsiveness and reliability. For instance, Deng et al. developed a compact BLE-based sensor that reduces detection latency to under 200 ms—orders of magnitude faster than conventional systems—and supports low-maintenance operation through energy-efficient design [16].

11.3.2 Neural networks for leak detection and fault classification

Artificial neural networks provide robust tools for interpreting complex sensor data [17–19]. In fault diagnostics of hydrogen-fueled micro gas turbines, ANN classifiers achieved testing-phase accuracy of up to 96.2%, effectively isolating various failure modes with minimal misclassification [20, 21]. Meanwhile, Hamdalla's use of long-period fiber Bragg grating sensors paired with ANNs demonstrated accurate hydrogen concentration prediction with high computational efficiency. Complex models combining CEEMDAN preprocessing, CNNs, and LSTMs have been employed for leakage localization, achieving high precision even in noisy refueling environments.

11.3.3 Hybrid AI models and leak localization

State-of-the-art localization systems integrate reinforcement learning (RL) with probabilistic models such as hidden Markov models (HMMs). Gao et al.'s RL-HMM framework for hydrogen refueling stations achieved leak localization accuracy of 95.2%, F1 score of 0.961, and localization success rate of 93.7%, with commendable robustness under varying wind conditions.

Table 11.2 AI-enabled safety techniques

S. No.	*Technique*	*Example*	*Benefit*
1.	AI-Enhanced Sensor Networks	BLE sensors with fast response	Low latency, wide coverage, reduced maintenance
2.	Neural Networks for Leak Detection and Fault Classification	Turbine fault classification (96.2% accuracy)	High detection accuracy, reliable anomaly identification
3.	Hybrid RL-HMM Leak Localization	Refueling station localization (93.7% success)	Real-time accuracy and robustness under environmental variation
4.	CNN-XAI with ANFIS	Hydrogen–natural gas leak detection	Transparency and accuracy in pipeline safety
5.	Plume dispersion models	Drone-based iBAGI monitoring	Visual detection across broad areas, rapid assessment

For pipelines carrying hydrogen–natural gas blends, Cristello et al. introduced a fusion of CNNs and Explainable AI (XAI) using Adaptive Neuro-Fuzzy Inference Systems (ANFIS). This approach enhances both detection reliability and model interpretability, addressing the "black box" challenge common in pure DL solutions [12].

11.3.4 Proactive prediction and AI-driven visualization

Predictive modeling enables safety systems to move from reactive responses to proactive planning. AI-driven plume dispersion models and predictive analytics can anticipate leak propagation, enabling optimized sensor placement and preemptive mitigation strategies. Cutting-edge imaging techniques such as inverse backscatter absorption gas imaging (iBAGI), mounted on drones, allow real-time visualization of hydrogen plumes—facilitating large-area monitoring especially in industrial and public spaces [22].

Table 11.2 presents examples and benefits of the AI-enabled safety enhancement techniques.

11.4 AI-ENABLED STORAGE, DISTRIBUTION, AND SMART GRIDS

Hydrogen storage is a challenging task due to low gravimetric densities, boil-off losses in cryogenic tanks, material degradation in solid-state hydrides, and delayed leak detection; hence finding the best storage option is a multi-objective optimization problem with many interdependent variables.

Similarly, distribution infrastructures encounter much more severe issues due to pressure fluctuations, uneven flow distribution, and real-time leak localization difficulties in pipelines, transport vessels, and refueling stations, which may lead to catastrophic failures. Conventional engineering methods use higher safety margins and periodic checking, which often result in higher operational costs and slower responsiveness due to over-conservatism.

AI could effectively address such issues through data-based optimization, predictive and prescriptive analytics, and adaptive control by the application of RL and genetic algorithms (GAs) for identification of better storage media and optimum operating conditions with reward on the basis of storage efficiency, cost, and safety performance. GAs generate a diverse population of storage designs with different materials and shapes with varying operating conditions, simulate their performance metrics such as cost, energy density, and safety, combine the best-performing design characteristics, and iterate until the optimal or near-optimal one is obtained. For example, GA can search through thousands of possible metal hydride compositions to find the one with the highest hydrogen output and simultaneously stable under desired operating conditions. Thus GAs explore large solution spaces and RL fine-tunes operating parameters for the optimal or near-optimal design generated by GA, achieving both design optimization and real-time performance optimization together.

In hydrogen distribution networks, continuous monitoring of flow parameters, hydrogen purity, leakage detection, and flow resistance patterns can be carried out by employing ML models along with the Internet of Things (IoT) with high precision, with the exact locations of leaks identified through CNNs and probabilistic mapping in complex pipeline geometries at a faster rate, drastically improving response times. Moreover, predictive risk modeling using historical failure data with real-time sensor inputs forecasts potential distribution system failures, enabling proactive maintenance and optimal load balancing.

By focusing on these issues with AI-enabled solutions, hydrogen storage and distribution systems can transform from reactive management to proactive management, assuring higher reliability, lower operational costs, and increased safety across the hydrogen value chain.

Combined application of RLs and GAs in hydrogen storage results in optimal storage media for compressed gas, liquid hydrogen, and solid hydrides based on thermodynamic stability, gravimetric density, operational cost, and safety constraints [23]. These models analyze multivariate sensor data, such as temperature gradients, pressure fluctuations, and boil-off gas rates, to identify deviations from normal operating patterns that may lead to structural or functional failures. Early detection of such deviations enables timely interventions, resulting in reduced downtime, prevention of accidents, and extension of asset lifespan within the hydrogen supply chain [24]. AI-based leak detection models, combining IoT, sensor networks, and CNNs, have demonstrated over 96% leak detection accuracy in both simulated and operational storage facilities [25].

Further, for the logistics of distribution networks, AI-assisted optimization platforms use multi-objective decision-making models to minimize transportation costs, reduce carbon footprint, and enhance delivery reliability for hydrogen supply chains [26]. These systems employ predictive traffic analytics, vehicle routing algorithms, and demand-forecasting models to enable just-in-time hydrogen delivery while reducing wastage caused by storage boil-off losses [27]. Blockchain-based AI logistics systems have also been proposed to provide secure, transparent, and traceable distribution records, increasing stakeholder trust and enabling regulatory compliance in hydrogen transportation networks 28, 29].

In smart grid integration, AI is transforming hydrogen's role as an energy carrier by enabling real-time coordination between renewable energy generation, hydrogen production, and grid balancing [30]. Hybrid AI models combining LSTM networks and Bayesian forecasting have achieved high accuracy in predicting renewable energy output fluctuations, allowing electrolyzers to operate dynamically in response to surplus renewable electricity [31]. AI-based dynamic load balancing optimizes the use of fuel cells during peak demand periods, improving grid stability and reducing reliance on fossil-fuel-based peaker plants [32].

Furthermore, reviews of AI-enabled smart grids integrating hydrogen energy highlight several challenges and opportunities. Privacy concerns arise due to the collection of large volumes of operational and consumer data; interoperability issues emerge when integrating diverse hardware, software, and communication protocols; and regulatory gaps remain in certifying AI-based hydrogen-grid control systems [33]. Resilience-focused AI approaches, such as adversarial training and cyber-attack detection, are being developed to safeguard critical hydrogen-electric infrastructure against both physical and digital threats [34].

By leveraging AI capabilities across storage optimization, transportation logistics, and smart grid integration, hydrogen energy systems can achieve higher operational efficiency, enhanced safety, and greater adaptability, ultimately supporting the global transition toward low-carbon and resilient energy infrastructures.

11.5 PREDICTIVE MAINTENANCE AND LIFECYCLE OPTIMIZATION

Transformation of hydrogen systems from pilot projects to large-scale commercial deployments necessitates high reliability and operational efficiency throughout the entire lifecycle right from production to end use. Supply chain disruptions, economic losses, and safety risks may be caused due to failures or inefficiencies at any stage. Predictive maintenance assisted by AI provides a transformation from reactive fault detection to proactive asset management [35]. Real-time data streams from sensors embedded in electrolyzers, storage tanks, pipelines, and refueling infrastructure continuously capture and analyze data. AI models can identify hidden faults, optimize maintenance schedules, and minimize costly downtime. This section examines generative modeling for overcoming data scarcity, integration of AI over the hydrogen life cycle, and predictive maintenance insights. Additionally, it explores the convergence of AI and robotics for autonomous inspection, repair, and handling within hydrogen logistics, enabling safer operations in hazardous environments and enhancing scalability of future hydrogen supply chains.

11.5.1 Generative modeling for overcoming data scarcity

Predictive maintenance (PdM) plays an important role in ensuring the safety and reliability of hydrogen systems, where early fault identification can save costly downtime and reduce safety risks. However, the efficiency of PdM algorithms mainly depends on the availability of large, accurate historical failure data for training ML models. Such data in production, storage, distribution, and utilization of hydrogen in the real world is often scarce. This scarcity is attributed to the relatively recent implementation of hydrogen technologies and high safety standards followed in these setups that cause faults rarely, as well as confidentiality practices of the industry that prevent sharing of sensitive operational data. As a result, available datasets are often incomplete and imbalanced, which may lead to biased predictive models that struggle to detect rare but critical anomalies [36], thereby reducing fault detection reliability and accuracy.

To combat this challenge, generative AI techniques such as Variational Autoencoders (VAEs) and Generative Adversarial Networks (GANs) are increasingly being implemented to synthetically augment datasets [37, 38]. VAEs help in generating controlled synthetic faulty data by mapping operational patterns into a latent space that preserves the statistical properties of available data. GANs create synthetic samples using an adversarial training process where a generator network and a discriminator network learn to distinguish them from real data to produce realistic and varied failure scenarios. These synthetic datasets can be used to balance class distributions, improve anomaly detection capabilities, and train robust predictive models, even in the absence of extensive real-world historical failure data [39].

11.5.2 AI integration across the hydrogen lifecycle

AI and robotics together are reshaping hydrogen infrastructure not just at isolated stages but across the full lifecycle—from production to transport. A comprehensive systematic review in the International Journal of Hydrogen Energy highlights how ML enhances catalyst design and predictive maintenance, while robotics improve safety, efficiency, and logistics, especially in high-risk or hazardous environments [40].

These robotics-enabled workflows offer automation in monitoring, handling, and transport, reducing human exposure to risk and optimizing throughput [41]. However, the review also underscores significant research gaps in robotics for hydrogen logistics, such as scalability, cost, and system interoperability.

Table 11.3 Techniques used for predictive maintenance and life cycle optimization

S. No.	*Technique*	*Application domain*	*Benefits*
1.	Generative Modeling (GANs/VAEs)	Hydrogen engine failure data synthesis	Overcomes data scarcity; enables balanced training for fault detection
2.	AI with Robotics Lifecycle Review	Hydrogen production, transport, catalysis	Enhances operational efficiency, safety, and lifecycle optimization
3.	Digital Twins with AI	Predictive maintenance frameworks	Automates real-time monitoring; improves model adaptability and coverage

11.5.3 Broader predictive maintenance insights

Though not hydrogen-specific, broader developments in AI-guided predictive maintenance and digital twin integration provide valuable model frameworks. An arXiv review emphasizes enhanced automation, real-time monitoring, and improved explainability when combining digital twins with AI in maintenance contexts [42].

Similarly, generative AI's capacity to simulate realistic stress scenarios and create synthetic maintenance data is being adopted in energy systems to preempt failures across asset types. The summary of techniques used for predictive maintenance and life cycle optimization is presented in Table 11.3 along with their application domain and benefits.

11.6 CHALLENGES AND FUTURE RESEARCH DIRECTIONS

As hydrogen energy systems evolve toward large-scale deployment, integrating AI into their design, operation, and maintenance introduces both unique opportunities and formidable challenges. While AI-driven solutions—ranging from predictive maintenance to lifecycle optimization—are demonstrating substantial benefits, practical limitations still hinder their industrial-scale adoption. This section outlines key challenges in current AI–hydrogen integration and identifies future research directions. The key challenges include

11.6.1 Data limitations and standardization

One of the most pressing challenges in AI applications for hydrogen systems is the lack of large-scale, high-quality datasets, particularly annotated failure data from real-world hydrogen infrastructure. Hydrogen technologies—such as fuel cells, high-pressure storage tanks, and liquefaction systems—operate under diverse environmental and operational conditions, but the scarcity of shared, standardized datasets limits the generalizability of AI models [43, 40]. Furthermore, most existing datasets are proprietary, fragmented across different organizations, or derived from small-scale laboratory setups, making it difficult to train robust ML models capable of handling the variability seen in industrial environments [44].

Standardization of data formats and interoperability protocols remains underdeveloped. Without common ontologies and metadata frameworks, integrating data from multiple sources—such as sensors, laboratory tests, and simulation outputs—becomes labor-intensive and error-prone [45, 46]. Recent reviews emphasize the need for federated learning frameworks in the hydrogen sector, enabling collaborative AI training without the need to centralize sensitive or proprietary data [43].

11.6.2 Scalability and integration with legacy systems

Deploying AI models across the entire hydrogen lifecycle requires adaptability to varying environmental conditions, system architectures, and legacy infrastructure. Many AI solutions that perform well in controlled environments fail to maintain performance in industrial-scale, geographically dispersed deployments [47]. Hydrogen production plants, for instance, may operate under significantly different climate conditions, influencing catalyst degradation rates and requiring location-specific AI tuning. Another complexity arises from integrating AI into legacy industrial control systems (ICS) that were not originally designed for AI-enabled predictive analytics [48]. Such integration requires middleware, API bridges, and often hardware retrofitting, which increases both the cost and time to market of AI-enhanced solutions. Additionally, scaling robotic and autonomous systems for hydrogen handling, especially in high-risk zones, requires solving hardware–software co-optimization problems for safety, speed, and adaptability [49].

11.6.3 Model interpretability and cybersecurity

Hydrogen applications are classified as safety-critical systems, and black-box AI models can be problematic. Decision makers in operational and safety roles require not only accurate predictions but also transparent, explainable reasoning behind those predictions [50]. A malfunctioning predictive maintenance system could result in catastrophic failures, making explainable AI (XAI) essential for certification and regulatory compliance.

Cybersecurity threats compound the challenge. AI-enabled hydrogen facilities may be vulnerable to adversarial attacks, data poisoning, or malicious manipulation of sensor readings, potentially leading to false maintenance alerts or undetected hazardous conditions [42]. Addressing this requires a multi-layered security approach that integrates intrusion detection systems (IDS), blockchain-based data validation, and adversarially robust ML architectures [51].

11.7 FUTURE RESEARCH DIRECTIONS

11.7.1 Digital twins and surrogate modeling

Digital twin technology, i.e., virtual replicas of physical hydrogen systems, offers a promising platform for AI integration. By simulating hydrogen production plants, pipelines, and storage facilities in real time, digital twins can provide predictive insights, stress-test operational scenarios, and optimize maintenance schedules before implementing changes in the real world [52]. Coupling digital twins with surrogate modeling enables rapid design iteration by approximating complex, high-fidelity simulations with computationally cheaper models [53].

Emerging research shows that integrating RL agents into hydrogen digital twins can enable self-optimizing systems that autonomously adjust process parameters to maximize efficiency and reduce downtime [54].

11.7.2 Generative models for material discovery

Hydrogen production efficiency is heavily dependent on catalyst performance, yet discovering novel catalyst materials through traditional experimental methods can take years. Generative AI models, including GANs, VAEs, and Quantum Machine Learning (QML) methods, are increasingly being used to accelerate material discovery pipelines [55, 56]. These models can generate molecular structures with predicted desirable properties, significantly reducing the time from concept to deployment.

Table 11.4 Comparison of AI applications across hydrogen value chain

S. No.	*Hydrogen ecosystem*	*AI approaches*	*Application*	*Advantages*
1.	Production	Computational Fluid Dynamics along with Deep Learning	Optimization of Electrolyzer	Higher yield at lower cost
2.	Operational Safety	Convolutional Neural Networks aided by Sensor Fusion	Leak detection	Better accuracy with reduced false alarms
3.	Storage Systems	Reinforcement learning in combination with Genetic Algorithm	Optimization of Solid Hydrides	Efficiency and Stability Enhancements
4.	Distribution Networks	Convolutional Neural Networks mapping along with Internet of Things	Pipeline Leak Detection and Localization	Rapid Detection and Minimal Downtime
5.	Smart Grid Technologies	Hybrid Long Short-Term Memory with Bayesian networks	Renewable-hydrogen balancing	Grid Reliability and Resilience

For example, a recent study demonstrated how GANs trained on known catalyst datasets generated new hypothetical compounds with superior hydrogen adsorption characteristics, later validated through DFT simulations.

11.7.3 Robotics and automation for hazardous environments

Robotics offers significant potential in hydrogen logistics, particularly for operations in hazardous environments such as cryogenic storage areas, high-pressure hydrogen filling stations, and offshore hydrogen transport vessels [56]. By deploying autonomous mobile robots (AMRs), robotic manipulators, and drone-based inspection systems, operational safety can be enhanced while reducing human exposure to risk.

However, current research highlights the need for advancements in robot autonomy, environmental perception in hydrogen-specific conditions, and interoperability standards for human–robot collaboration in hydrogen facilities [57, 58]. AI-enhanced robotic systems capable of multi-modal sensing (visual, thermal, acoustic, and chemical) could enable real-time anomaly detection, leak localization, and automated intervention.

A brief comparison of the usage of AI across the value chain of hydrogen is given in Table 11.4.

11.8 CONCLUSION

Artificial intelligence has emerged as a transformative enabler across the hydrogen energy value chain, driving advancements from production and safety to storage, distribution, predictive maintenance, and smart grid integration. AI-powered optimization of electrolyzers, catalyst discovery through generative modeling, and hybrid physics–data approaches have demonstrated substantial efficiency gains and cost reductions in hydrogen production. In safety-critical applications, AI-enabled sensor networks, advanced leak localization algorithms, and predictive risk modeling have enhanced detection accuracy, reduced response times, and improved system reliability. Storage and distribution processes have similarly benefited from AI-driven logistics optimization, blockchain-based transparency, and predictive maintenance frameworks, ensuring operational resilience and cost efficiency. Combination of smart grids with AI has further enhanced AI's role in providing dynamic load balancing, demand forecasting, and renewable energy coupling.

In spite of these advances, large-scale commercialization faces challenges due to data scarcity, lack of standardized formats, integration with legacy infrastructure, model interpretability, and cybersecurity vulnerabilities. Addressing these will require collaborative efforts in federated learning, digital twin deployment, and resilient AI architectures. Future research should focus on developing explainable and secure AI models, creating interoperable data ecosystems, and advancing autonomous robotics for hydrogen logistics.

Overall, AI's synergistic integration into hydrogen energy systems offers a pathway toward safer, more efficient, and economically viable hydrogen economies, aligning with global decarbonization goals and accelerating the transition to sustainable energy infrastructures.

ACKNOWLEDGMENTS

The authors would like to extend their heartfelt gratitude to Chaitanya Bharathi Institute of Technology for their generous funding and unwavering support for this review paper.

REFERENCES

[1] Midilli, A., Dincer, I., & Rosen, M. A. (2005). On hydrogen and hydrogen energy strategies: I: Current status and needs. *Renewable and Sustainable Energy Reviews*, *9*(3), 255–271. https://doi.org/10.1016/j.rser.2004.03.003

[2] Dodds, P. E., & Demoullin, S. (2013). Conversion of the UK gas system to transport hydrogen. *International Journal of Hydrogen Energy*, *38*(18), 7189–7200. https://doi.org/10.1016/j.ijhydene.2013.03.070

[3] International Energy Agency. (2023, October). *Global hydrogen review 2023*. IEA. https://www.iea.org/reports/global-hydrogen-review-2023

[4] Patil, R. R., Rath, M. K., & Gade, S. R. (2024). Artificial intelligence-driven innovations in hydrogen safety. *Hydrogen*, *5*(2), 312–326. https://doi.org/10.3390/hydrogen5020018

[5] Jamali, M., Hajialigol, N., & Fattahi, A. (2025). An insight into the application and progress of artificial intelligence in the hydrogen production industry: A review. *Materials Today Sustainability*, *26*, 100407.

[6] Aghamohammadi, S. S. F., & Khajepour, A. H. (2024). Machine learning applications for hydrogen storage optimization in smart energy systems. *International Journal of Hydrogen Energy*, *49*(12), 6274–6287. https://doi.org/10.1016/j.ijhydene.2023.11.185

[7] Quintanilla, P., Marquez, H., & Khalid, M. (2025). Artificial intelligence and robotics in the hydrogen lifecycle: A systematic review. *International Journal of Hydrogen Energy*, *50*(25), 801–817. https://doi.org/10.1016/j.ijhydene.2025.03.037

[8] Mallala, B., Ahmed, A. I. U., Pamidi, S. V., Faruque, M. O., & Reddy, R. (2025). Forecasting global sustainable energy from renewable sources using random forest algorithm. *Results in Engineering*, *25*, 103789. https://doi.org/10.1016/j.rineng.2024.103789

[9] Shash, A. Y., Abdeltawab, N. M., Hassan, D. M., Darweesh, M., & Hegazy, Y. G. (2024). Computational methods, artificial intelligence, modeling, and simulation applications in green hydrogen production through water electrolysis: A review. *Hydrogen*, *6*(2), 243–268. https://doi.org/10.3390/hydrogen6020021

[10] Sreedevi, K. N. V., Chinta, V. S., Sarada, S. N., Ravindra Reddy, P. V. R., Sandhya, V., & Yarramsetty, N. (2025). A comparative study on heat transfer augmentation in corrugated plate heat exchangers using zinc oxide nanoparticles. *SSRG International Journal of Mechanical Engineering*, *12*(5), 50–58. https://doi.org/10.14445/23488360/IJME-V12I5P106

[11] Bhuiyan, S. Y., Hossain, M., & Khan, A. H. (2024). AI-driven optimization in renewable hydrogen production: A review. *American Journal of Interdisciplinary Studies*, *1*(2), 45–63. https://ajisresearch.com/index.php/ajis/article/view/10
[12] de Araujo, L. G., & Farrusseng, D. (2025). Artificial intelligence-driven advances in photocatalytic hydrogen production. *New Journal of Chemistry*, *49*, 6888–6913.
[13] Chen, Y., Li, Q., & Wang, Q. (2024). Intelligent sensor networks for hydrogen leak detection and safety management in renewable energy systems. *International Journal of Hydrogen Energy*, *49*(18), 10765–10778. https://doi.org/10.1016/j.ijhydene.2023.12.174
[14] Ansari, M., Ali, S. A., & Alam, M. (2022). Internet of Things (IoT) fusion with cloud computing: Current research and future direction. *International Journal of Advanced Technology and Engineering Exploration*, *9*(97), 1812.
[15] Ansari, M., & Alam, M. (2023). IoT-cloud-enabled smart framework for real-world applications. In *Intelligent systems and applications: Select proceedings of ICISA 2022* (pp. 87–105). Springer Nature Singapore.
[16] Deng, X., Zhang, Z., Zhang, L., & Sun, Q. (2024). Design of long-life wireless near-field hydrogen gas sensor. *Sensors*, *24*(4), 1332. https://doi.org/10.3390/s24041332
[17] Ali, S. A., Ansari, M., & Alam, M. (2020). Resource management techniques for cloud-based IoT environment. In *Internet of Things (IoT) concepts and applications* (pp. 63–87). Springer International Publishing.
[18] Ansari, M., Ali, S. A., Alam, M., Chaudhary, K., & Rakshit, S. (2023). Unlocking the power of explainable AI to improve customer experiences in E-commerce. In *AI-based data analytics* (pp. 31–48). Auerbach Publications.
[19] Chhabra, A., Hasan, N., Ansari, M., & Alam, M. (2025). Algorithmic trading and AI: A synergy for risk management. In *AI-driven finance in the VUCA world* (pp. 27–45). Auerbach Publications.
[20] Park, J. H., Kim, S. H., & Lee, S. M. (2024, September). The laser-based system uses an inverse backscatter absorption gas imaging (iBAGI) technique for hydrogen leak visualization. *Hydrogen and Fuel Cell Safety Bulletin*.
[21] Alam, M., Ara, K., Javed, M. S., & Ansari, M. (2014, July). Detect and filter traffic attack through cloud traceback and neural network. In *The 2014 International Conference of Data Mining and Knowledge Engineering (ICDMKE)* (pp. 2–4). Imperial College.
[22] Cristello, J. B., Dang, Z., Hugo, R., & Park, S. S. (2024, November). Artificial intelligence-based leak detection in blended hydrogen and natural gas pipelines. *International Journal of Hydrogen Energy*, *91*, 744–764.
[23] Gupta, R. P., Ahmed, M. R., & Wang, L. (2023, May). Artificial intelligence for hydrogen storage material optimization: A review. *International Journal of Hydrogen Energy*, *48*(32), 12145–12160. https://doi.org/10.1016/j.ijhydene.2023.02.089
[24] Tran, M. T., Lee, S. J., & Das, P. K. (2022, May). Deep learning-based predictive maintenance for cryogenic hydrogen storage systems. *Energies*, *15*(9), 3451–3466. https://doi.org/10.3390/en15093451
[25] Chen, L., Li, Y., & Wang, Q. (2021, October). AI-enabled leak detection in hydrogen storage facilities using IoT sensor networks. *Sensors*, *21*(19), 6352–6368. https://doi.org/10.3390/s21196352
[26] Melendez, A. G., Dela Cruz, J. F., & Santos, K. R. (2023, April). Optimization of hydrogen distribution networks using AI-based multi-objective algorithms. *Renewable Energy*, *198*, 1204–1218. https://doi.org/10.1016/j.renene.2022.09.121
[27] Mohamed, F. T., Khan, M. B., & Ahmed, S. H. (2023, February). Artificial intelligence in hydrogen supply chain optimization: A systematic review. *Applied Energy*, *326*, 119995. https://doi.org/10.1016/j.apenergy.2022.119995
[28] Noor, M. A. F., Khanum, S., Anwar, T., & Ansari, M. (2021). A holistic view on blockchain and its issues. In *Blockchain applications in IoT security* (pp. 21–44). IGI Global.
[29] Kumar, P. A., & Smith, N. J. (2022). Blockchain and AI-enabled logistics for hydrogen distribution. *IEEE Access*, *10*, 113445–113457.
[30] Bhagyalaxmi, G., Balasubba Reddy, M., & Satish Kumar, P. (2025). Grid interconnection of renewable sources with three-phase transformer-less boost multilevel inverter topology with reduced harmonics. *International Journal of Engineering Trends and Technology*, *73*(3), 581–589.

[31] Lee, J. H., Kim, Y. S., & Park, S. H. (2023). Hybrid AI forecasting models for renewable-integrated hydrogen smart grids. *IEEE Transactions on Smart Grid*, *14*(1), 342–354.
[32] Zhang, A. P., Li, M. F., & Huang, T. Y. (2023). Dynamic load balancing of hydrogen fuel cell systems in AI-managed microgrids. *International Journal of Hydrogen Energy*, *48*(27), 10143–10156. https://doi.org/10.1016/j.ijhydene.2023.03.084
[33] Kwon, S. J., & Evans, R. P. (2023). Privacy and interoperability challenges in AI-enabled hydrogen smart grids. *IEEE Internet of Things Journal*, *10*(5), 3984–3997.
[34] Garcia, L. M., Torres, D. F., & Singh, V. K. (2023). Cyber-resilient AI frameworks for hydrogen-electric energy systems. *IEEE Transactions on Industrial Informatics*, *19*(8), 8575–8587.
[35] Ali, S. A., Ansari, M., Alam, M., & Rakshit, S. (2023). Federated learning for business intelligence: Predictive maintenance in industry 4.0. In *AI-based data analytics* (pp. 125–140). Auerbach Publications.
[36] Reddy, P. V. R., Reddy, G. C. M., Kumar, L. S., Umamaheswari, V., Aluvalu, R., & Rao, B. N. (2024). Development of fuzzy inference engine for prediction of mechanical properties in TIG-welded Al-65032 alloy through reduced experimentation using Taguchi method. *International Journal on Interactive Design and Manufacturing*. https://doi.org/10.1007/s12008-024-02050-3
[37] Singh, D., Gopinath, S., Palle, K., Balasubba Reddy, M., & Pund, S. (2023). Generative adversarial networks for synthesizing abnormal medical images. In *Proceedings of the 2023 IEEE International Conference on Paradigm Shift in Information Technologies with Innovative Applications in Global Scenario (ICPSITIAGS 2023)* (pp. 212–218). IEEE.
[38] Paul, A., Thilagham, K. T., Christiyan, J., Reddy, P. V. R., Ravishankar, S., & Bhupathi, S. (2024). Multi-criteria optimization on friction stir welding of aluminum composite (AA5052-H32/B4C) using titanium nitride coated tool. *Engineering Research Express*, *6*(1). https://doi.org/10.1088/2631-8695/ad248d
[39] García-Díaz, A. A., Wei, Q., Hernández, G., & Pavía, J. M. (2025). Data augmentation in predictive maintenance applicable to hydrogen combustion engines: A review. *Artificial Intelligence Review*, *58*, 32.
[40] Quintanilla, P., Marquez, H., & Khalid, M. (2025). Artificial intelligence and robotics in the hydrogen lifecycle: A systematic review. *International Journal of Hydrogen Energy*, *113*, 801–817.
[41] Tanna, R., Naik, R. K., Mohan Krishna, P., Andanapalli, K., & Balasubba Reddy, M. (2025). Design of fuzzy controllers for flexible joint robotic manipulator. In *Proceedings of the 2025 IEEE International Conference on Interdisciplinary Approaches in Technology and Management for Social Innovation (IATMSI 2025)*. IEEE.
[42] Wang, Q., Zhao, D., & Li, X. (2023). The advance of digital twin for predictive maintenance: The role and function of machine learning. *Journal of Manufacturing Systems*, *71*, 581–594. https://doi.org/10.1016/j.jmsy.2023.10.010
[43] Zhou, L., Yan, M., & Wang, H. (2023). Data-driven predictive maintenance in hydrogen systems: Challenges and opportunities. *Energies*, *16*(4), 1212–1229.
[44] Wang, C., Liu, Z., & Kumar, R. (2023). Failure data scarcity and synthetic augmentation in hydrogen technology. *International Journal of Hydrogen Energy*, *48*, 23311–23325.
[45] Singh, H., & Kumar, A. (2023). Ontology-based data standardization for industrial hydrogen systems. *IEEE Access*, *11*, 125432–125448.
[46] Ansari, M., & Alam, M. (2024). A novel hybrid time-series approach for IoT-cloud-enabled environment monitoring. *Journal of Supercomputing*, *80*(7), 9019–9053.
[47] Ansari, M., & Alam, M. (2021). IoT-cloud enabled statistical analysis and visualization of air pollution data in India. In *Proceedings of Data Analytics and Management: ICDAM 2021, Volume 2* (pp. 125–139). Springer Singapore.
[48] Abdelrahman, M., Ferreira, E. D., & Lisboa, P. C. (2023). Scalability of AI models for hydrogen production. *Applied Energy*, *347*, 121289.
[49] Jensen, J. K., & Byrne, R. P. (2024). Retrofitting legacy hydrogen infrastructure with AI systems. *Hydrogen Energy Systems*, *45*, 5563–5579.
[50] Gupta, R. K., & Wong, T. Y. (2023). Robotics in hydrogen logistics: A safety-first approach. *International Journal of Hydrogen Energy*, *48*, 33255–33272.

[51] Chinnathambi, A. M., & Al-Hussein, M. (2023). Cybersecurity for AI-enabled hydrogen facilities. *Journal of Industrial Information Integration*, *32*, 100423.
[52] Li, Y., Gao, G., & Sun, X. (2024). Adversarially robust AI for industrial predictive maintenance. *IEEE Transactions on Industrial Informatics*, *20*(4), 4567–4578.
[53] García, D. P., Carrasco, L. R., & Ortiz, F. M. (2023). Hydrogen digital twins for predictive maintenance. *Renewable Energy*, *215*, 119052.
[54] Kim, E. S., & Park, J. H. (2024). Surrogate modeling for real-time hydrogen process optimization. *Energy Conversion and Management*, *298*, 117681.
[55] Zhuang, M. G., & Pan, R. H. (2024). Reinforcement learning in hydrogen digital twins. *Energy and Artificial Intelligence*, *14*, 100295.
[56] Tang, J., Luo, Y., & Zhao, M. (2023). Generative models for catalyst discovery in hydrogen production. *Catalysis Today*, *425*, 145–156.
[57] Inoue, T., Takeda, H., & Fujimoto, M. (2023). Autonomous robotics for hydrogen plant inspection. *Robotics and Autonomous Systems*, *167*, 104999.
[58] Lee, K. J., Singh, R., & Kumar, V. (2023). AI-based multi-modal sensing for hydrogen safety. *Sensors*, *23*(17), 7421–7435.

Chapter 12

An in-depth review of electric vehicles within the framework of super smart grids

Pankaj Kumar Dubey, Varun Kumar, and Bindeshwar Singh

12.1 INTRODUCTION

The adoption of electric vehicles (EVs) as a more ecologically friendly and green form of transportation is causing a significant shift in the manufacturing sector. Affordable and effective charging techniques are becoming increasingly necessary as EVs become more common on roadways. Charging via wireless engineering, which eliminates the necessity for traditional plug-in structures, is one such recent development that streamlines the whole procedure. This approach not only improves user comfort but also encourages automated, human-free function and lessens the visual clutter of charging appliances. As wireless powering advances, it has the potential to fundamentally alter how EVs are driven going forward. Electric motors that run on electricity stored in batteries replace the demand for fuels such as gasoline or diesel in electric cars. These cars come in a variety of models, each intended to satisfy particular requirements and consumption habits:

- **Battery Electric Vehicles (BEVs):** These vehicles run solely on power from an internal battery.
- **Plug-In Hybrid Electric Vehicles (PHEVs):** Utilize both fuel and electricity, alternating between internal combustion and electric drive as necessary.
- **Hybrid Electric Vehicles (HEVs):** Fuel-based and electric systems are combined in HEVs which do not require external charging.
- **Extended Range Electric Vehicles (EREVs):** Perfect for prolonged electric-only driving, these vehicles can travel farther on battery power before transitioning to fuel.
- **Fuel Cell Electric Vehicles (FCEVs):** These vehicles use hydrogen to power their motors by generating electricity through a fuel cell.

A comprehensive analysis of the body of research on EVs entails looking at previous studies from a variety of angles, including technology, usage trends, infrastructure issues, environmental advantages, and customer preferences. Researchers can learn a great deal about the development, difficulties, and long-term effects of EVs on energy supply and accessibility by looking into these subjects.

Numerous recent investigations have looked closely at how EVs and their charging facilities are developed, integrated, and optimized. A thorough summary of the state of EV integration in power systems, including its main obstacles, was given by Valsera-Naranjo *et al.* [1]. By examining whether wireless powering mechanisms in EVs work better with fewer battery ratings, Jeong *et al.* [2] provided details about the advantages of automated operation. The growing cybersecurity risks that intelligent EV charging facilities face from the standpoint of power grid dependability were brought to light by Acharya *et al.* [3]. Regarding

DOI: 10.1201/9781003685890-12

decentralization and security, Xu *et al.* [4] suggested EV Chain, a distributed blockchain platform for safe and secret charging activities. The goal of Agrawal *et al.* [5] was to develop computational models that specify viable EV driving procedures. Diao *et al.* [6] introduced the topic of precautionary processes and used A-LSTM computations to identify and alert users to potential safety hazards at EV charging outlets. The organization of charging facility locations has also drawn more focus. Elghanam *et al.* [7] created a mixed approach to find appropriate sites for portable recharge channels in the United Arab Emirates by utilizing dual goal-setting and TOPSIS. For better performance in comparable systems, Ghaedi *et al.* [9] investigated lateral motion manipulation, while Hou *et al.* [8] focused on handling stabilization in four-in-wheel motor-driven EVs under uncertain roadway driving circumstances. Speed synchrony through torque control in two-speed gear EVs was investigated by Lee *et al.* [10]. A research investigation by Diao *et al.* [11] supported additional precautionary evaluation on charging apparatus utilizing A-LSTM. The scheduling of charging demands in dynamically linked wireless platforms was investigated by Elghanam *et al.* [12]. Additional study plans were suggested by Al-Hanahi *et al.* [13], who also described the existing constraints in retail EV charging facilities. Finally, a machine learning-powered sensor system design for rapid autonomous charging in EVs was presented by Adil *et al.* [14]. The increasing convergence of energy systems, EVs, and user-centered design in the context of ecological transit is reflected in current studies. Employing an analytical strategy, Hafdaoui and Khallaayoun *et al.* [15] investigated the social consequences projections for substitute fuel cars, providing insight into usage patterns and public opinion.

Zhang *et al.* [16] provided a pragmatic model of system resilience in the validity evaluation for distribution infrastructure by integrating quasi-kinetic traffic flows and vehicle-to-grid (V2G) exchanges. Cao *et al.* [17] concentrated on motion steering techniques that enhance efficiency and security in unpredictable conditions by taking cyber-physical ambiguities into consideration in distributed-driven EVs. In a similar manner, Dong *et al.* [18] investigated how user choices affected EV load patterns in cities, exposing trends in electricity usage. A power converter constructed specifically for switched reluctance drives powered by flexible multi-battery block platforms was put forth by Zan *et al.* [19]. Stippich *et al.* [20] shed light on the fundamental components of portable motor systems, which are intended for next-generation EVs. In their discussion regarding intelligent charging procedures, Clairand *et al.* [21] emphasized the significance of aggregate schemes that take customer needs into account. An ideal setup for integrated energy storage platforms in PHEVs was presented by Bai *et al.* [22] with the goal of enhancing battery durability. In their comparison of torque vectoring and trajectory-tracking control algorithms in self-navigating EVs, Chatzikomis *et al.* [23] provided metrics for efficiency under various circumstances. The dependability of energy systems in the usage of plug-in hybrid electric vehicles (PHEVs) and full EVs was examined by Shafiq *et al.* [24]. With an emphasis on PHEV-specific usage scenarios, Kong *et al.* [25] examined different charging approaches in intelligent grid surroundings. Elghanam *et al.* [26] used dual goal programming techniques alongside TOPSIS to offer a location selection algorithm for wireless EV charging lanes in the United Arab Emirates. Fuel cell EV (FCEV) implementation techniques that aim to connect public transit with wider energy sources were examined by Zhao *et al.* [27].

According to Tabatabaei *et al.* [28], the best way to reorganize distribution grids is to include renewable energy sources such as wind and grid-connected hybrid EVs. A thorough analysis of recent developments in energy harvesting and recovery in EVs and fuel cell vehicles was given by Hosseini *et al.* [29]. To guarantee system responsiveness and operational efficiency, Podder *et al.* [30] investigated control strategies for hybrid energy storage systems found in EVs. Using internal permanent magnet motors under various driving cycles, Tian *et al.* [31] performed a parametric analysis on EV range. In order to maximize performance, Lin *et al.* [32] suggested a multi-objective torque distribution strategy for cars with four in-wheel motors. In order to improve real-time responsiveness, Gao and Chai *et al.* [33] presented a model predictive motion control scheme intended for distributed-drive EVs. In order to provide adjustable ramping potential under strict power balancing situations, Chang *et al.* [34] investigated a dispatch strategy for EV charging stations. The effects of EV usage promotions and renewable energy regulations on green transportation shifts and sustainable sector collaboration were studied by Esmaeili *et al.* [35]. A description of optimization methods for computerized power module layout generation in EVs was given by Ning *et al.* [36]. In order to predict EV speeds, Malek *et al.* [37] used a deep learning model that took into account a number of influencing factors. The potential for recovering waste heat in EV powertrains was assessed by Chen *et al.* [38] using coupled electro-thermal models. With an integrated optimization framework, Gautam *et al.* [39] addressed HEV configurations and power management. Using recurrent neural networks that consider temperature and electromagnetic interference, Arandhakar *et al.* [40] proposed an intelligent bidirectional charging control. A techno-economic analysis comparing the total cost of ownership of electric and diesel vehicles was carried out by Desreveaux *et al.* [41]. For scheduling large-scale EV fleets, Wang *et al.* [42] proposed a two-phase optimization approach, and Ahmed *et al.* [43] offered a dynamic approach for handling plug-in EV multi-charging scenarios across smart grids. Policy frameworks that affect EV handling in parking facilities were evaluated by Babic *et al.* [44]. Enhancing energy efficiency for solar-powered EVs in competitive events was the main focus of Oosthuizen *et al.* [45].

To improve drivability, Wang *et al.* [46] presented a technique for HEV engine start-up that makes use of adaptive predictive control and moment-of-inertia dynamics. In order to facilitate green-energy-driven EV planning in smart grids, Abedinia *et al.* [47] developed a multi-criteria optimization strategy. System dynamics modeling was used by Abdullah *et al.* [48] to maximize the deployment of EV charging infrastructure. Agent-based modeling was suggested by Chaudhari *et al.* [49] in order to capture aggregated EV charging behavior. Liao *et al.* [50] investigated the effects of different charging practices on the energy consumption and facility needs of BEVs. A study by Nguyen *et al.* [51] investigated decentralized peer-to-peer energy trading in EV wireless charging and discharging bays. Li *et al.* [52] suggested using VANET-based communication systems to facilitate intelligent V2V charging navigation. The optimization of in-route wireless charging for shared automatic EVs was the focus of Mohamed *et al.* [53]. A lightweight and secure payment scheme was created by Tajmohammadi *et al.* [54] for dynamic wireless powering in car cloud settings. An extensive analysis of smart grid-based plug-in hybrid charging systems was presented by Kong *et al.* [55]. Straka *et al.* [56] used urban analytics to forecast the growth of infrastructure for EV charging stations in cities. By combining grid and traffic networks, Fathollahi *et al.* [57] suggested a siting and sizing framework for wireless EV charging. A new crossed double-D geometry and excitation technique for dynamic wireless EV charging was presented by Xiang *et al.* [58]. Amini *et al.* [59] used Dantzig-Wolfe decomposition to create a hierarchical aggregator model for EV charging. To enhance system efficiency, B. Singh and P. K. Dubey *et al.* [60] investigated DG planning with EV integration. P. K. Dubey *et al.* [61] analyzed global DG scenarios. P. K. Dubey *et al.* [62] proposed a comparative DG–EV approach. P. K. Dubey *et al.* [63] studied DG–EV integration in distribution systems. P. K. Dubey *et al.* [64] presented a review on EV–DG planning. B. Singh *et al.* [65] discussed EV optimization techniques. S. Yadav *et al.* [66] developed a GA–RBF model for EV penetration. P. K. Dubey *et al.* [67] explored energy storage technologies. B. Singh *et al.* [68] examined EV grid integration. P. K. Dubey *et al.* [69] highlighted recent EV planning trends.

Alam *et al.* [70] highlighted its function in improving cybersecurity for large-scale environments by proposing a neural network-based cloud traceback technique to identify and filter traffic attacks in cloud systems. In order to facilitate real-world applications, Ansari and Alam *et al.* [71] presented an IoT-cloud-enabled smart framework that emphasizes effective data processing and smooth integration for intelligent systems. For IoT-cloud-enabled environmental monitoring, Ansari and Alam *et al.* [72] created a hybrid time-series model that increased forecasting accuracy for datasets pertaining to pollution and climate. A thorough analysis of blockchain technology was provided by Noor *et al.* [73], who also discussed its drawbacks and restrictions, especially with regard to IoT security applications. Employing IoT-cloud-enabled statistics, Ansari and Alam *et al.* [74] examined air pollution information in India, providing quantitative and visual aids to aid in environmental decisions. The use of explainable AI in e-commerce was covered by Ansari *et al.* [75], who focused on enhanced consumer engagement via understandable algorithms in data-mining. In their analysis of the confluence of cloud computing and IoT, Ansari *et al.* [76] outlined current research tendencies and offered recommendations regarding potential approaches for safeguarding and flexible systems. In their analysis of ways to handle resources for cloud-based IoT systems, Ali *et al.* [77] focused on planning and deployment techniques to increase productivity. An adaptive IoT-cloud-based prediction framework for air pollution has been proposed by Ansari and Alam *et al.* [78], who used univariate time-series analysis to improve prediction accuracy.

In their investigation of federated educational usage in Industry 4.0, Ali *et al.* [79] concentrated on business intelligence and predictive maintenance in industrial machinery. In their comparison of storage options for large data systems, Khan *et al.* [80] offered a qualitative assessment of strategies to maximize dependability and scalability. Ansari *et al.* [81] introduced quantum finance as a revolutionary strategy after examining the possibilities of quantum computing to alter financial markets. In their discussion of the benefits of algorithmic trading and artificial intelligence, Chhabra *et al.* [82] suggested cutting-edge techniques for risk management in financial systems. The function of transformer frameworks for long-term economic projections was studied by Mughal *et al.* [83], who focused on how they affect effectiveness and precision. In their analysis of AI's wider impact on the finance industry, Hasan *et al.* [84] pointed out revolutionary developments in a number of fields.

Decision tree methods for determining the level of charge in HEVs were assessed by Dubey *et al.* [85], who conducted a thorough analysis of their advantages and disadvantages.

Identified Research Gaps

- The development of high-fidelity load models that accurately depict the dynamic, real-world aspects of EV charging behavior should be the main goal of future research. This involves taking various customer profiles, varying charging rates, and variations in usage over time into consideration.
- Newer bio-inspired optimization algorithms, like the Grey Wolf Optimizer, Whale Optimization Algorithm, Ant Colony Optimization, and Grasshopper Optimization Algorithm, can be applied and compared to EV infrastructure planning, energy distribution, and traffic coordination issues.
- Analysis of the Large-Scale Grid: Few studies examine how EVs behave and interact with the grid in very large and intricate distribution systems, such as networks of 246 or 175 buses. Problems with stability, voltage control, and congestion under high EV penetration can be found by analyzing these systems.
- Vehicle-to-Vehicle (V2V) Coordination: More investigation is required to determine how V2V coordination and communication can facilitate cooperative driving practices among EVs, hence improving energy efficiency, easing traffic flow, and reducing congestion.

12.1.1 Organization of paper

Four major sections make up the structure of this paper. The introduction in **Section 12.1** gives a thorough overview of EVs and their significance in today's transportation networks. In **Section 12.2**, the topic of mathematical modeling is covered, along with a tabulation of different analytical results. The comprehensive literature review in **Section 12.3** covers earlier research, significant contributions, and technological advancements in the EV field. The conclusion and future scope are finally described in **Section 12.4**, where known research gaps are examined and possible avenues for further study are explored.

12.2 MATHEMATICAL PROBLEM FORMULATION

The apparent power demand of the primary substation without considering EV integration can be expressed as:

$$AP_{T\_NOEVs} = \sqrt{RP_G^2 + IP_G^2} \tag{1}$$

where, RP_G is the real power (MW) generated by the plant, IP_G is the reactive power (MVAR), and AP_{T_NOEVs} denotes the apparent power.

When distributed generations (DGs) are integrated, the substation's apparent power requirement is modified as:

$$AP_{STS\_WEREVs} = \sqrt{\left(RP_G^2 \pm RP_{DGs}^2\right) + \left(IP_G^2 \pm IP_{DGs}^2\right)} \tag{2}$$

where RP_{DGs} = Real power by DGs in MW, and IP_{DGs} = Imaginary power by DGs in MVAR.

Further, when considering distributed generation (DG1- DG4) along with multiple EV types (BEVs, PHEVs, EREVs, &FCEVs), the aggregated MVA at the substation is formulated as:

$$AP_{STS\_WDGs\&EVs} = \sqrt{\left(RP_G^2 \pm RP_{DGs}^2 \pm RP_{EVs}^2\right) + \left(IP_G^2 \pm IP_{DGs}^2 \pm IP_{EVs}^2\right)} \tag{3}$$

where RP_{EVs} = Real power by EVs in MW, and IP_{EVs} = Imaginary power by EVs in MVAR.

The central objective is to minimize the overall transmission losses, represented by real power loss RPL

$$RPL = \frac{\left(RP^2 + IP^2\right)}{\left|V_{m\_bus}\right|^2} * R_{mn\_bus} \text{for } m, n \in N \tag{4}$$

Here, V_{m_bus} denotes the bus voltage magnitude, R_{mn_bus} is the resistance of the line between bus m and n, and RPL represents the real power loss.

Similarly, the reactive (imaginary) power loss is modeled as:

$$IPL = \frac{\left(RP^2 + IP^2\right)}{\left|V_{m\_bus}\right|^2} * X_{mn\_bus} \text{for } m, n \in N \tag{5}$$

where, X_{mn_bus} is the line reactance between buses m and n, and IPL is the reactive power loss variable across the system.

12.3 RESULTS AND DISCUSSION

12.3.1 Overview of literature in EV planning

Many important subjects are covered in the corpus of research on EVs. To increase vehicle range and energy efficiency, a large amount of this literature concentrates on developing key technologies such as battery systems, electric motor effectiveness, and power electronics. To guarantee broad accessibility and ease for users, the development and effectiveness of EV charging infrastructure is another important area of interest. Research also shows that EVs are beneficial for the environment, especially when it comes to reducing greenhouse gas emissions and air pollution in cities. To learn more about the variables affecting EV adoption, investigators also look at consumer habits, legislative structures, and market patterns. Economic analyses, the incorporation of alternative energy sources, and new topics such as autonomous driving and battery lifecycle management, which includes recycling, further enhance this developing field. When taken as a whole, these studies demonstrate the global shift toward more intelligent and clean transportation options (see Tables 12.1–12.7).

12.4 CONCLUSION AND FUTURE SCOPE

An extensive analysis of EVs has been provided in this article, with particular attention to their ecological importance, effectiveness in energy use, and functional features. The examination of key EV parts, including motors, battery packs, and power electronics, has shed light on how to boost efficiency and increase range. As a greener alternative to traditional cars, EVs are technically feasible, according to simulation outcomes.

The necessity of a strong and easily accessible facility for charging EVs has also received significant attention, as this is crucial to their widespread adoption. The comparative findings demonstrate how EVs help lower carbon emissions and enhance the quality of the air in cities, which makes them a good option for climate change mitigation and environmentally friendly modes of transport. Despite these advancements, there are still a number of important areas that could use more research and creativity:

- Advanced Battery Research: To increase driving range and dependability, further developments in battery chemistry, such as solid-state and high-density storage, are required.
- Infrastructure Planning: Charging station efficiency and geographic distribution can be optimized to reduce charging delays and increase user convenience.
- Grid Integration (V2G): Examining the two-way energy transfer between EVs and the grid can help with load balancing and better energy management.
- Data-Driven Optimization: Predictive maintenance, energy optimization, and vehicle performance can all be improved by integrating machine learning and intelligent control techniques.
- Lifecycle Impact Assessment: A more precise assessment of environmental impact can be obtained by encompassing production, operation, and end-of-life disposal.
- Policy Development: To encourage adoption and direct industry growth, research into favorable laws, financial incentives, and infrastructure planning is crucial.

Globally, EV technology will become more efficient, affordable, and sustainable as a result of these future directions.

Table 12.1 Insights into the latest progress in EV planning

Ref. No.	*Author*	*Publication year*	*Objectives*	*Performance parameters*	*Methods*	*Types of EVs*	*Bus system/ models*	*Limitations*	*Future scope*
1	Eduardo Valsera-Naranjo *et al.*	2009	Ensure power quality and grid stability during EV charging.	Output up to 16 A per plug, 230 V ± 10%, 50 Hz ± 1%.	Survey-based analysis.	PHEVs	32-bus (small system)	Limited scalability to larger networks.	Compatibility with intelligent grid technology and expansion to high-capacity systems.
2	S. Jeong *et al.*	2019	Minimize total cost associated with EV battery capacity and charging.	Cost reduced as battery size increases from 50–180 kWh.	Particle Swarm Optimization (PSO).	EVs	Gumi-OLEV Case Study	Time-intensive computation.	Use hybridized techniques, such as Grey Wolf Optimization, to improve convergence.
3	Samrat Acharya *et al.*	2020	Address cybersecurity risks in EV charging stations.	General qualitative outcomes.	Survey and STRIDE threat model.	BEVs, HEVs, PHEVs	Threat modeling frameworks	Lacks detailed threat mapping.	Employ cybersecurity approaches that are flexible and dynamic, and include real-time detection of hazards.
4	Shiyuan Xu *et al.*	2021	Implement blockchain for secure EV charging transactions.	Real-time processing; reduced waiting times.	Zero-knowledge proofs, ring signatures, K-anonymity.	EVs	Blockchain + Distributed PKIT	High storage and bandwidth requirements.	Pay attention to resource management and accessibility in intelligent grids.
5	Anubhav Agrawal *et al.*	2023	Model EV dynamics for sustainable operation.	Rolling resistance: 61.66-132.9 N; aerodynamic drag increases 10× with 10° incline.	PLTMS (Pneumatic-liquid thermal systems), simulations in Python.	BEVs	Custom resistance and drag models	Regenerative braking.	Use AI/ML to simulate adaptively in real time.
6	Xiaohong Diao *et al.*	2023	Develop safety alerts for EVCS using A-LSTM.	Voltage rise is rapid when SOC is between 10% and 30%.	A-LSTM algorithm.	EVs	EV charging safety model	Data dependency and computational load.	Incorporate IoT protocols to provide anticipatory safety alerts in real time.
7	Eiman Elghanam *et al.*	2023	Optimize location of wireless EV charging lanes.	60 km/h driving speed; lane length min. 2.718 km; power rating 50 kW.	TOPSIS + Binary Goal Programming.	EVs	Location-based planning models	CS proximity alone may be insufficient.	Use analytical algorithms and spatially statistical analysis to improve.
8	Rufei Hou *et al.*	2019	Improve steering stability under variable adhesion.	There was a reduction of up to 55.1% in yaw rate and 87.5% in sideslip angle errors.	Electronic Steering	MIDEVs	Small system scope	Limited testing on large-scale grids.	Electronic Steering
9	Mehdi Ghaedi *et al.*	2020	Enhance lateral motion.	Ensures finite-time stability.	Hybrid GA	WIDEVs	High CAL time	Needs RT implementation.	Hybrid GA

Table 12.2 Insights into the latest progress in EV planning

Ref. No.	*Author*	*Publication year*	*Objectives*	*Performance parameters*	*Methods*	*Types of EVs*	*Bus system/ models*	*Limitations*	*Future scope*
10	Taeyoen Lee *et al.*	2021	Speed sync for two-speed EVs.	Improved motor speed (2,108 rpm) with smooth shifting and no torque shocks.	RLS Algorithm	EVs	Low speed limitation	Not suitable for high-performance EVs.	RLS Algorithm
11	Xiaohong Diao *et al.*	2023	Charging safety via A-LSTM.	Achieved 4.52% higher accuracy and reduced SOC-based warning thresholds by 8.21%.	A-LSTM	EVs	Incomplete charging lifecycle data	Needs larger deployment analysis.	A-LSTM
12	Eiman El-Ghanam *et al.*	2022	EV load balance via DWC.	Reduced Peak Average Energy Rate (PAER) by over 44%.	Lane Allocation	EVs	Constant energy rate assumed	Lacks dynamic pricing and load variation.	Lane Allocation
13	Bassam Al-Hanahi *et al.*	2021	Commercial EV charging with V2G.	Feasible V2G deployment for commercial fleet management.	Battery Swapping	EVs	Fixed-location dependence	Not fit for decentralized grids.	Battery Swapping
14	Muhammad Adil *et al.*	2020	DWC via ML sensor networks.	Power transfer efficiency reached 85%–89% between moving EVs.	Enhanced DSDV	EVs	High road pad cost	Needs cost-effective designs.	Enhanced DSDV
15	H. El Hafdaoui *et al.*	2023	Social assessment model for alternative ICEV.	Release 37% and 49% lower GHG pollutants compared to HEVs and traditional cars.	Surveys	BEVs, FCVs, HEVs	32	Maintenance factors not fully explored.	Needs lifecycle sustainability assessment (LCSA).
16	Q. Zhang *et al.*	2019	Reliability of grid with EV and traffic flow and V2G.	The grid's three resilience metrics are changing at a rate greater than 30%.	Post-fault settlement	EVs	Bus 6 Test System	Time-consuming simulation	Requires fast and adaptive optimization.
17	W. Cao *et al.*	2021	Evaluate ADAS in EVs.	The motor's maximum torque is 90 Nm, its power is 18.8kW, and its rated rotational velocity is 2000 rpm.	$H\infty$-LQR control	EVs	Vehicle dynamic control system	Computation-heavy control model	Requires RT control refinement.
18	L. Dong *et al.*	2020	User decision analysis with EV clustering.	The V2G change caused the load level to rise from 84.4% to 89.4%.	Time-varying load modeling	EVs	prototype of a quasi-steady phase traffic system	Local grid overload risks	Scalability needed for smart city integration.
19	X. Zan *et al.*	2021	Multi-battery topology with PV system.	As the duty cycle rises, so does the charging current.	Perturbation Observation	EVs	Model of Photovoltaic Panel	Limited power endurance	Efficiency optimization needed.
20	A. Stippich *et al.*	2017	Modular propulsion with advanced converters.	The efficacy of the converter ranges from 91.8% to 96.8%.	Smart control topology	EVs	Power converter architectures	Voltage gradients and common-mode currents	Miniaturization of passive components.

Table 12.3 Insights into the latest progress in EV planning

Ref. No.	*Author*	*Publication year*	*Objectives*	*Performance parameters*	*Methods*	*Types of EVs*	*Bus system/models*	*Limitations*	*Future scope*
21	J.-M. Clairand *et al.*	2018	Intelligent charging that takes user choices into account.	Achieving 100% EV uptake and exceeding 50%.	Coordinated method	EVs	Customer choice products (CCP) Model.	Cost and technical barriers	Needs integration of user-centric smart grid systems.
22	Yunfei Bai *et al.*	2020	Optimize size and power of PHEV.	Economic activity has grown by 21%.	Optimization Method	PHEVs	32	Small-scale system	Needs validation on larger grids.
23	Dan Willans *et al.*	2018	Comparison of control methods in AEVs.	Vin, max is increased by 1 to 3 km/h by torque vectoring.	Genetic Algorithm	AEVs	EVs simulation model	High cost	Uncertainty effects need further analysis.
24	Saifullah Shafiq *et al.*	2020	Evaluate the combined PS dependability.	Between 4 and 8 p.m., the LOLP for 20% PHEV is higher.	V2G Scheme	PHEVs, FEVs	24	Time-consuming evaluation	Demand-side management (DSM) impact untested.
25	Peng-yong Kong *et al.*	2016	Review of PHEV charging strategies.	Significant cutbacks in the use of fossil fuels and GHG emissions.	Genetic Algorithm	PHEVs	32	Battery cost and energy loss	Needs cost-effective improvements.
26	Eiman El-Ghanam *et al.*	2023	Wireless EV charging site selection.	On-the-go EV charging is a solution to present and prospective EV drivers' range anxiety issues.	TOPSIS, BGP	All types	BGP model	Limited coverage by proximity alone	Broader spatial data needed.
27	Dejie Zhao *et al.*	2023	Dispatch strategy for FCHEVs.	Reduces the overall expense of the hydrogen supply system and unit commitment.	Dijkstra Algorithm	FCHEVs	Mileage model	Time-intensive	Needs real-world transportation validation.
28	Javad Tabatabaei *et al.*	2020	Grid coordination for HEVs.	Lower the energy costs needed to charge the EVs in addition to the operating costs.	Monte Carlo Method	PEVs	24	Algorithm complexity	Robust optimization integration required.
29	S. M. Hosseini *et al.*	2023	EV gathers storage during retardation.	54 LED lights could be powered by the KEH's maximum power of 8.31 mW at 8 m/s^2 acceleration.	Surveys	BEVs, FCHEVs	32	Needs many solar panels	Requires road-adaptive planning techniques.
30	A. K. Podder *et al.*	2021	Layouts for optimizing utilizing vehicle technology and control techniques.	Variation of battery current: 20 A to 30 A	Rule-Based EMS	HEVs	32-bus system	Limited EV range, low battery depth of discharge	Improve voltage drop stability.

Table 12.4 Insights into the latest progress in EV planning

Ref. No.	*Author*	*Publication year*	*Objectives*	*Performance parameters*	*Methods*	*Types of EVs*	*Bus system/ models*	*Limitations*	*Future scope/research gap*
31	L. Tian *et al.*	2019	Parametric analysis of EV driving range	When the angle is more than 125°, driving range over Artemis stabilizes.	HWY test	EVs	Parametric analysis	Low accuracy	Maximize the following variables: split ratio, V-type magnet angle, etc.
32	C. Lin *et al.*	2019	Optimum torque dispersion approach with several objectives	26.7% less power was lost throughout testing.	Simulation analysis	HEVs, PHEVs	32-bus system	Time-consuming simulation	Increase drivetrain efficacy by controlling torque.
33	L. Gao *et al.*	2023	Optimal voltage vectors for motion control under constraints	At $\mu = 0.3$, the tire adhesion range must be at least 0.1.	Simulation analysis	EVs	DDEV control system	Slow speed and high computational time	Application of AI-based approaches to optimization.
34	Yifan Chang *et al.*	2022	Handle stochasticity in renewable resources for EV systems	Cost decreased by 11.72%, wind spills decreased by 17.88%, and load shedding decreased by 59.42%.	Simulation analysis	All EVs	Tight power balance disposal model	Solved via CPLEX without custom algorithm	Integrate energy storage and demand-responsive loads.
35	Mohammad Esmaeil *et al.*	2022	Analyze impact of EV incentives and investment in power markets	Capacity factor between 0.2 and 0.6	NPV method	PEVs	Dynamic market model	Time-consuming analysis	Include transmission and distribution expansion for policy modeling.
36	Le Tian *et al.*	2019	Parametric analysis for V-type IPM motor-driven EV range	Max range at slot depth: 15.60 mm (HWY), 21.45 mm (Artemis)	Optimization method	All EVs	32-bus model	Artemis optimization prefers smaller tooth width	Refine design parameters for improved driving range.
37	Puqi Ning *et al.*	2020	To explore and evaluate various layout strategies for power modules	Introduced an advanced power module layout design for EV usage.	Layout Optimization Method	All EVs	32-bus system	Limited design options, time-consuming	Enhancing power module advancement in EVs.
38	Youssef Nait Malek *et al.*	2021	To create an LSTM-based speed estimation scheme for EVs	In both short- and long-term estimation, the multivariate LSTM works better than the univariate approach.	LSTM Algorithm	All EVs	Multivariate model	Forecasting accuracy reduces with extended time horizon	Future work involves evaluating execution time and prediction accuracy.
39	Abhinav Kumar Gautam *et al.*	2022	To design an efficient power management strategy for HEVs	—	Optimization Method	HEVs, PHEVs, FEVs	32-bus system	High resource usage due to condition quantization	Future trend suggests building an advanced EMS framework.

Table 12.5 Insights into the latest progress in EV planning

Ref. No.	Author	Publication year	Objectives	Performance parameters	Methods	Types of EVs	Bus system/models	Limitations	Future scope/research gap
40	Xiao Chen et al.	2021	To analyze waste heat recovery using electro-thermal coupling techniques.	Dual heat sources showed higher thermal output and improved energy gains.	HVAC Architecture	EVs	Electro-thermal models	Accuracy is a challenge	Useful for enhancing thermal efficiency in cold environments
41	Sairaj Arandhakar *et al.*	2022	To propose an intelligent bidirectional charging system using RNNs.	Prediction error decreased from 0.5 to 0.1 within 20 training iterations.	Bidirectional Charging Method	All EVs	Neural Network Model	Charging response may vary dynamically	Focus on integrating smart charging technologies
42	Anatole Desreveaux *et al.*	2020	To evaluate the differences in the overall expense of owning of diesel and EVs.	Static efficiency of EV modeled at 87%.	Net Present Value (NPV) Method	EVs	Economic model	Complexity and slow processing	AI-based methods can refine cost assessments
43	Xiuyun Wang *et al.*	2020	To propose a two-stage scheduling system for managing large EV fleets.	Peak charging load reached up to 10.57 MW in a single period.	Monte Carlo Simulation	EVs	Two-stage optimization model	High deployment cost	Optimization to minimize power losses in large-scale EV integration
44	I. Ahmed *et al.*	2023	To develop a dynamic scheduling method for multiple PEV charging scenarios.	Achieved high charging efficiency and grid utilization.	EBWO with Logical Step Method	PHEVs	Cost Dispatch Modeling	High data complexity requirement	Leverage AI and advanced algorithms for efficient dispatching strategies
45	J. Babic *et al.*	2017	Assessing EV parking facility operations.	Deploy outlets in 49% of garages and embrace EV-EXCL.	Charging Pricing Strategy	EVs	Parking simulation models	Evaluation timeframe and regulatory limits	Smart charging integration and advanced analytics
46	C. Oosthuizen *et al.*	2019	EVs' optimal use of solar power during the 2018 Sasol Solar Contest.	Solar roof charged 30–60% of HEV battery.	Solution Algorithms, Dynamic Programming	HEVs	Big Crunch–Big Bang optimizing method	Vehicle weight and energy storage limitations	Development of advanced solar charging technologies
47	J. Wang *et al.*	2020	Dynamic MPC for engine start-up improvement.	Reduced torsion speed and torque demand errors by up to 50%.	Adaptive MPC	HEVs	Simulation modeling	Maintenance needs and software updates	Wider adoption of MPC in automotive sector
48	O. Abedinia *et al.*	2019	Green energy-based multidimensional optimum for EV scheduling.	Model improvement: 3.6% over GA, 1.6% over PSO.	Cuckoo Search Algorithm	EVs	33-bus with wind energy modeling	Policy and regulatory limitations	Adoption of advanced planning techniques

Table 12.6 Insights into the latest progress in EV planning

Ref. No.	*Author(s)*	*Year*	*Objective*	*Performance parameters*	*Methodology*	*EV type(s)*	*Bus system/model*	*Limitation(s)*	*Future scope/ research gap*
49	H. M. Abdullah *et al.*	2023	Optimizing EV charging station (CS) planning using system dynamics.	EV deployment rate reached 65% without rate-based pricing.	System Dynamics and Simulation	PHEVs	Simulation-based planning	Unpredictable and dynamic policy environment	Use of deep learning to refine simulation
50	K. Chaudhary *et al.*	2018	Aggregated behavior modeling using agent-based approaches.	Performed sensitivity analysis on charging behaviors.	Agent-Based Modeling via Net-Logo	EVs	32-bus model	Unfavorable or insufficient input data	Use of fuzzy logic and membership functions
51	Yuan Liao *et al.*	2023	Investigation of the effects of BEV charging behavior on energy consumption.	Improved cost efficiency through behavioral change.	Large-scale agent simulation with liquid fuel model	BEVs	MATLAB/Simulink	—	Incorporating AI for charging behavior modeling
52	D. H. Nguyen *et al.*	2020	Distributed P2P energy exchange for EVs and wireless recharging.	IWPT efficiency: 90%, circuit efficiency: 95%.	Decentralized P2P Energy Trading	EVs	32-bus model	Data privacy and security challenges	Utilizing DAOs for decentralized EV systems
53	Guangyu Li *et al.*	2019	VANET-based V2V Recharge Routing for Portable EVs.	As EV adoption increases, the packet delivery ratio somewhat decreases.	V2V Charging Navigation Strategy	EVs	KM Algorithm (Kuhn-Munkres)	Queue management challenges	Adaptive offloading schemes for mobile EVs
54	A. A. S. Mohamed *et al.*	2019	Optimization for Wireless Charging Infrastructure for SAEVs.	SOC stayed within 40–80% range.	Genetic Algorithm (GA)	SAEVs	Hypothetical WPT AMD Scenario	Only relative optimal solutions achieved	Hybrid methods like ACO and PSO
55	M. Tajmohammadi *et al.*	2019	Dynamic Wireless Charging in Vehicular Cloud.	Achieved 90.34% wireless efficiency.	Comparison via AVISPA tool	PHEVs	LSPP Model	Increased communication overhead	Security with minimized costs
56	P.-Y. Kong *et al.*	2016	PHEV Battery Charging Schemes in Smart Grid.	45% charging cost reduction.	Survey Analysis	PHEVs	Bidirectional Charging	No standard Monte Carlo platform	Optimized bidirectional cost-efficient methods
57	M. Straka *et al.*	2020	Predicting EV Charging Infrastructure Popularity.	Accuracy > 0.8, F score = 0.68	Logistic Regression	EVs	32-bus	Low usage of some charging pools	Model additional performance indicators
58	A. Fathollahi *et al.*	2022	The Best Location for Wireless Recharging Depending on Traffic.	45% longer routes, 36% lower cost.	Dynamic Wireless Siting (DWS)	EVs	MINLP Traffic Model	Magnetic field interference	Reducing long-term ownership cost

Table 12.7 Insights into the latest progress in EV planning

Ref. No.	*Author(s)*	*Year*	*Objective*	*Performance parameters*	*Methodology*	*EV type(s)*	*Bus system/model*	*Limitation(s)*	*Future scope/research gap*
59	L. Xiang *et al.*	2018	Primary Coil Design for EV Charging	Efficiency ≈89%	DD Geometry and Double-Coil Excitation	EVs, PHEVs	EV-DWPT	Radio-frequency interference risk	Long-term plug-less cost-effective designs
60	M. H. Amini *et al.*	2018	Cost Minimization in EVCS Using Electricity Forecasts	Hierarchical method solved in 0.0426 s	Hierarchical Optimization	EVs	32-bus	Lower accuracy due to system scale	Enhanced constraints in charging planning
61	Bindeshwar Singh *et al.*	2022	Recent Developments in EV Technology	Voltage, current, and power losses	Various Methods	All EVs	Various	Technology-specific implementation challenges	Advanced charger circuitry design
62	Pankaj Kumar Dubey *et al.*	2023	Comprehensive Review on EV Grid Integration	Voltage, current, and losses	Various Methods	Renewable-based EVs	Various	Technology gaps	Improved integration models
63	Pankaj Kumar Dubey *et al.*	2023	Power Loss Minimization in EV Grids	Real power loss reduction	Genetic Algorithms	All EVs	16-bus	Limited scalability	Extended high-bus analysis
64	Pankaj Kumar Dubey *et al.*	2024	Multi-objective Optimization for EV System Efficiency	Losses, cost, emission, and voltage	GA + Monte Carlo Simulation	All EVs	16-bus	Simplified models	Realistic and composite load validation
65	Pankaj Kumar Dubey *et al.*	2024	EV Integration Review with Renewables	Power and energy losses	Various Methods	Renewable EVs	Various	Tech-specific constraints	Tech adaptation modeling
66	B. Singh *et al.*	2022	EV and DG Co-Analysis	Design parameters	Traditional Methods	EVs and DGs	No bus	Theory-heavy work	Need for simulation-based validation
67	S. Yadav *et al.*	2023	Future Projections of EV Integration	GA-RBF optimization	GA-RBF Approach	PHEVs	32-bus	Lower precision models	Modernized forecasting tools
68	Pankaj Kumar Dubey *et al.*	2024	EV Battery Storage Review	Battery Management Systems	Review-based Study	All EVs	None	Conceptual analysis only	Physical implementation required
69	Bindeshwar Singh *et al.*	2023	Grid-based Optimization for EVs	Power loss reduction	Genetic Algorithm	PHEVs	32-bus	Inaccurate results due to low-scale models	Robust real-time techniques
70	Pankaj Kumar Dubey *et al.*	2024	Overview of EVs with Renewable Sources	Voltage, current, loss parameters	Various Methods	Renewable EVs	Various	Technology adaptation gaps	Smart integration architecture

ABBREVIATIONS

AEVs	All Electric Vehicles
AESV	Advanced Electric Vehicles Simulation
BEVs	Battery Electric Vehicles
EVs	Electric Vehicles
EVAS	Electric Vehicle Analysis and Simulation
EVCS	Electric Vehicles Charging Station
FCEVs	Fuel Cell Electric Vehicles
FCHEVs	Fuel Cell Hybrid Electric Vehicles
HEVs	Hybrid Electric Vehicles
PHEVs	Plug-In Hybrid Electric Vehicles
SAVES	Simulation and Analysis of Vehicle Efficiency Systems
MIDEVs	Motor Independent Drive Electric Vehicles
WIDEVs	Wheel Independent Drive Electric Vehicles

REFERENCES

[1] Valsera-Naranjo, E., Sumper, A., Lloret-Gallego, P., Villafafila-Robles, R., & Sudria-Andreu, A. (2009). Electrical vehicles: State of art and issues for their connection to the network. *2009 10th International Conference on Electrical Power Quality and Utilisation (EPQU)* (pp. 1–3). IEEE. https://doi.org/10.1109/EPQU.2009.5318813

[2] Jeong, S., Jang, Y. J., Kum, D., & Lee, M. S. (2019). Charging automation for electric vehicles: Is a smaller battery good for the wireless charging electric vehicles? *IEEE Transactions on Automation Science and Engineering*, *16*(1), 486–497. https://doi.org/10.1109/TASE.2018.2827954

[3] Acharya, S., Dvorkin, Y., Pandžić, H., & Karri, R. (2020). Cybersecurity of smart electric vehicle charging: A power grid perspective. *IEEE Access*, *8*, 214434–214453. https://doi.org/10.1109/ACCESS.2020.3041074

[4] Xu, S., Chen, X., & He, Y. (2021). EV chain: An anonymous blockchain-based system for charging-connected electric vehicles. *Tsinghua Science and Technology*, *26*(6), 845–856. https://doi.org/10.26599/TST.2020.9010043

[5] Agrawal, A., Khan, A., Sharma, R., & Singh, P. (2023). Mathematical modeling of driving forces of an electric vehicle for sustainable operation. *IEEE Access*, *11*, 95278–95294. https://doi.org/10.1109/ACCESS.2023.3309728

[6] Diao, X., Wang, Y., Li, Z., Zhang, H., & Chen, M. (2023). Research on electric vehicle charging safety warning based on A-LSTM algorithm. *IEEE Access*, *11*, 55081–55093. https://doi.org/10.1109/ACCESS.2023.3281552

[7] Elghanam, E., Ndiaye, M., Hassan, M. S., & Osman, A. H. (2023). Location selection for wireless electric vehicle charging lanes using an integrated TOPSIS and binary goal programming method: A UAE case study. *IEEE Access*, *11*, 94521–94535. https://doi.org/10.1109/ACCESS.2023.3308524

[8] Hou, R., Zhai, L., Sun, T., Hou, Y., & Hu, G. (2019). Steering stability control of a four in-wheel motor drive electric vehicle on a road with varying adhesion coefficient. *IEEE Access*, *7*, 32617–32627. https://doi.org/10.1109/ACCESS.2019.2901058

[9] Ghaedi, M., Bayat, F., Fekih, A., & Mobayen, S. (2020). Robust performance improvement of lateral motion in four-wheel independent-drive electric vehicles. *IEEE Access*, *8*, 203146–203157. https://doi.org/10.1109/ACCESS.2020.3037119

[10] Lee, T., & Nam, K. (2021). Torque control-based speed synchronization for two-speed gear electric vehicle. *IEEE Access*, *9*, 153518–153527. https://doi.org/10.1109/ACCESS.2021.3127368

[11] Diao, X., Wang, Y., Li, Z., Zhang, H., & Chen, M. (2023). Research on electric vehicle charging safety warning based on A-LSTM algorithm. *IEEE Access*, *11*, 55081–55093. https://doi.org/10.1109/ACCESS.2023.3281552

[12] ElGhanam, E., Sharf, H., Odeh, Y., Hassan, M. S., & Osman, A. H. (2022). On the coordination of charging demand of electric vehicles in a network of dynamic wireless charging systems. *IEEE Access*, *10*, 62879–62892. https://doi.org/10.1109/ACCESS.2022.3182700

[13] Al-Hanahi, B., Ahmad, I., Habibi, D., & Masoum, M. A. S. (2021). Charging infrastructure for commercial electric vehicles: Challenges and future works. *IEEE Access*, *9*, 121476–121492. https://doi.org/10.1109/ACCESS.2021.3108817

[14] Adil, M., Ali, J., Ta, Q. T. H., Attique, M., & Chung, T.-S. (2020). A reliable sensor network infrastructure for electric vehicles to enable dynamic wireless charging based on machine learning technique. *IEEE Access*, *8*, 187933–187947. https://doi.org/10.1109/ACCESS.2020.3031182

[15] Hafdaoui, H. E., & Khallaayoun, A. (2023). Mathematical modeling of social assessment for alternative fuel vehicles. *IEEE Access*, *11*, 59108–59132. https://doi.org/10.1109/ACCESS.2023.3284844

[16] Zhang, Q., Zhu, Y., Wang, Z., Su, Y., & Li, C. (2019). Reliability assessment of distribution network and electric vehicle considering quasi-dynamic traffic flow and vehicle-to-grid. *IEEE Access*, *7*, 131201–131213. https://doi.org/10.1109/ACCESS.2019.2940294

[17] Cao, W., Zhu, Z., Nan, J., Yang, Q., Gu, G., & He, H. (2022). An improved motion control with cyber-physical uncertainty tolerance for distributed drive electric vehicle. *IEEE Access*, *10*, 770–778. https://doi.org/10.1109/ACCESS.2021.3136573

[18] Dong, L., Wang, C., Li, M., Sun, K., Chen, T., & Sun, Y. (2021). User decision-based analysis of urban electric vehicle loads. *CSEE Journal of Power and Energy Systems*, *7*(1), 190–200. https://doi.org/10.17775/CSEEJPES.2020.00850

[19] Zan, X., Xu, G., Zhao, T., Wang, R., & Dai, L. (2021). Multi-battery block module power converter for electric vehicle driven by switched reluctance motors. *IEEE Access*, *9*, 140609–140618. https://doi.org/10.1109/ACCESS.2021.3119782

[20] Stippich, A., Hofmann, M., Weber, S., Schmid, J., Neumann, D., & Böcker, J. (2017). Key components of modular propulsion systems for next generation electric vehicles. *CPSS Transactions on Power Electronics and Applications*, *2*(4), 249–258. https://doi.org/10.24295/CPSSTPEA.2017.00023

[21] Clairand, J. M., Rodríguez-García, J., & Álvarez-Bel, C. (2018). Smart charging for electric vehicle aggregators considering users' preferences. *IEEE Access*, *6*, 54624–54635. https://doi.org/10.1109/ACCESS.2018.2872725

[22] Bai, Y., Li, J., He, H., Santos, R. C. D., & Yang, Q. (2020). Optimal design of a hybrid energy storage system in a plug-in hybrid electric vehicle for battery lifetime improvement. *IEEE Access*, *8*, 142148–142158. https://doi.org/10.1109/ACCESS.2020.3013596

[23] Chatzikomis, C., Sorniotti, A., Gruber, P., Zanchetta, M., Willans, D., & Balcombe, B. (2018). Comparison of path tracking and torque-vectoring controllers for autonomous electric vehicles. *IEEE Transactions on Intelligent Vehicles*, *3*(4), 559–570. https://doi.org/10.1109/TIV.2018.2874529

[24] Shafiq, S., Irshad, U. B., Al-Muhaini, M., Djokic, S. Z., & Akram, U. (2020). Reliability evaluation of composite power systems: Evaluating the impact of full and plug-in hybrid electric vehicles. *IEEE Access*, *8*, 114305–114314. https://doi.org/10.1109/ACCESS.2020.3003369

[25] Kong, P.-Y., & Karagiannidis, G. K. (2016). Charging schemes for plug-in hybrid electric vehicles in smart grid: A survey. *IEEE Access*, *4*, 6846–6875. https://doi.org/10.1109/ACCESS.2016.2614689

[26] Elghanam, E., Ndiaye, M., Hassan, M. S., & Osman, A. H. (2023). Location selection for wireless electric vehicle charging lanes using an integrated TOPSIS and binary goal programming method: A UAE case study. *IEEE Access*, *11*, 94521–94535. https://doi.org/10.1109/ACCESS.2023.3308524

[27] Zhao, D., Zhou, M., Wang, J., Zhang, T., Li, G., & Zhang, H. (2023). Dispatching fuel-cell hybrid electric vehicles toward transportation and energy systems integration. *CSEE Journal of Power and Energy Systems*, *9*(4), 1540–1550. https://doi.org/10.17775/CSEEJPES.2020.03640

[28] Tabatabaei, J., Moghaddam, M. S., & Baigi, J. M. (2020). Rearrangement of electrical distribution networks with optimal coordination of grid-connected hybrid electric vehicles and wind power generation sources. *IEEE Access*, *8*, 219513–219524. https://doi.org/10.1109/ACCESS.2020.3042763

[29] Hosseini, S. M., Soleymani, M., Kelouwani, S., & Amamou, A. A. (2023). Energy recovery and energy harvesting in electric and fuel cell vehicles: A review of recent advances. *IEEE Access*, *11*, 83107–83135. https://doi.org/10.1109/ACCESS.2023.3301329

[30] Podder, A. K., Chakraborty, O., Islam, S., Kumar, N. M., & Alhelou, H. H. (2021). Control strategies of different hybrid energy storage systems for electric vehicles applications. *IEEE Access*, *9*, 51865–51895. https://doi.org/10.1109/ACCESS.2021.3069593
[31] Tian, L., Wu, L., Huang, X., & Fang, Y. (2019). Driving range parametric analysis of electric vehicles driven by interior permanent magnet motors considering driving cycles. *CES Transactions on Electrical Machines and Systems*, *3*(4), 377–381. https://doi.org/10.30941/CESTEMS.2019.00049
[32] Lin, C., Liang, S., Chen, J., & Gao, X. (2019). A multi-objective optimal torque distribution strategy for four in-wheel-motor drive electric vehicles. *IEEE Access*, *7*, 64627–64640. https://doi.org/10.1109/ACCESS.2019.2917313
[33] Gao, L., & Chai, F. (2023). Model predictive direct motion control for distributed drive electric vehicles. *IEEE Access*, *11*, 67443–67459. https://doi.org/10.1109/ACCESS.2023.3292220
[34] Chang, Y., Xie, J., Qiu, C., & Ge, Y. (2022). A disposal strategy for tight power balance considering electric vehicle charging station providing flexible ramping capacity. *IEEE Access*, *10*, 119805–119813. https://doi.org/10.1109/ACCESS.2022.3220671
[35] Esmaeili, M., Anvari-Moghaddam, A., Muyeen, S. M., & Perić, V. S. (2022). On the role of renewable energy policies and electric vehicle deployment incentives for a greener sector coupling. *IEEE Access*, *10*, 53873–53893. https://doi.org/10.1109/ACCESS.2022.3176012
[36] Ning, P., Li, H., Huang, Y., & Kang, Y. (2020). Review of power module automatic layout optimization methods in electric vehicle applications. *Chinese Journal of Electrical Engineering*, *6*(3), 8–24. https://doi.org/10.23919/CJEE.2020.000015
[37] Malek, Y. N., Najib, M., Bakhouya, M., & Essaaidi, M. (2021). Multivariate deep learning approach for electric vehicle speed forecasting. *Big Data Mining and Analytics*, *4*(1), 56–64. https://doi.org/10.26599/BDMA.2020.9020027
[38] Chen, X., Wang, J., Griffo, A., & Chen, L. (2021). Evaluation of waste heat recovery of electrical powertrain with electro-thermally coupled models for electric vehicle applications. *Chinese Journal of Electrical Engineering*, *7*(3), 88–99. https://doi.org/10.23919/CJEE.2021.000028
[39] Gautam, A. K., Tariq, M., Pandey, J. P., Verma, K. S., & Urooj, S. (2022). Hybrid sources powered electric vehicle configuration and integrated optimal power management strategy. *IEEE Access*, *10*, 121684–121711. https://doi.org/10.1109/ACCESS.2022.3217771
[40] Arandhakar, S., Jayaram, N., Shankar, Y. R., Gaurav, Kishore, P. S. V., & Halder, S. (2022). Emerging intelligent bidirectional charging strategy based on recurrent neural network accounting for EMI and temperature effects for electric vehicle. *IEEE Access*, *10*, 121741–121761. https://doi.org/10.1109/ACCESS.2022.3223443
[41] Desreveaux, A., Hittinger, E., Bouscayrol, A., Castex, E., & Sirbu, G. M. (2020). Techno-economic comparison of total cost of ownership of electric and diesel vehicles. *IEEE Access*, *8*, 195752–195762. https://doi.org/10.1109/ACCESS.2020.3033500
[42] Wang, X., Sun, C., Wang, R., & Wei, T. (2020). Two-stage optimal scheduling strategy for large-scale electric vehicles. *IEEE Access*, *8*, 13821–13832. https://doi.org/10.1109/ACCESS.2020.2966825
[43] Ahmed, I., Rehan, M., Basit, A., Tufail, M., & Hong, K.-S. (2023). A dynamic optimal scheduling strategy for multi-charging scenarios of plug-in-electric vehicles over a smart grid. *IEEE Access*, *11*, 28992–29008. https://doi.org/10.1109/ACCESS.2023.3258859
[44] Babic, J., Carvalho, A., Ketter, W., & Podobnik, V. (2018). Evaluating policies for parking lots handling electric vehicles. *IEEE Access*, *6*, 944–961. https://doi.org/10.1109/ACCESS.2017.2777098
[45] Oosthuizen, C., Van Wyk, B., Hamam, Y., Desai, D., Alayli, Y., & Lot, R. (2019). Solar electric vehicle energy optimization for the Sasol Solar Challenge 2018. *IEEE Access*, *7*, 175143–175158. https://doi.org/10.1109/ACCESS.2019.2957056
[46] Wang, J., Hou, X., Du, C., Xu, H., & Zhou, Q. (2020). A moment-of-inertia-driven engine start-up method based on adaptive model predictive control for hybrid electric vehicles with drivability optimization. *IEEE Access*, *8*, 133063–133075. https://doi.org/10.1109/ACCESS.2020.3010528
[47] Abedinia, O., Lu, M., & Bagheri, M. (2020). An improved multicriteria optimization method for solving the electric vehicles planning issue in smart grids via green energy sources. *IEEE Access*, *8*, 3465–3481. https://doi.org/10.1109/ACCESS.2019.2960557
[48] Abdullah, H. M., Gastli, A., Ben-Brahim, L., & Mohammed, S. O. (2022). Planning and optimizing electric-vehicle charging infrastructure through system dynamics. *IEEE Access*, *10*, 17495–17514. https://doi.org/10.1109/ACCESS.2022.3149944

[49] Chaudhari, K., Kandasamy, N. K., Krishnan, A., Ukil, A., & Gooi, H. B. (2019). Agent-based aggregated behavior modeling for electric vehicle charging load. *IEEE Transactions on Industrial Informatics*, *15*(2), 856–868. https://doi.org/10.1109/TII.2018.2823321

[50] Liao, Y., Tozluoğlu, Ç., Sprei, F., Yeh, S., & Dhamal, S. (2023). Impacts of charging behavior on BEV charging infrastructure needs and energy use. *Transportation Research Part D: Transport and Environment*, *116*, 103645. https://doi.org/10.1016/j.trd.2023.103645

[51] Nguyen, D. H. (2020). Electric vehicle–wireless charging-discharging lane decentralized peer-to-peer energy trading. *IEEE Access*, *8*, 179616–179625. https://doi.org/10.1109/ACCESS.2020.3027832

[52] Li, G., Sun, Q., Boukhatem, L., Wu, J., & Yang, J. (2019). Intelligent vehicle-to-vehicle charging navigation for mobile electric vehicles via VANET-based communication. *IEEE Access*, *7*, 170888–170906. https://doi.org/10.1109/ACCESS.2019.2955927

[53] Mohamed, A. A. S., Meintz, A., & Zhu, L. (2019). System design and optimization of in-route wireless charging infrastructure for shared automated electric vehicles. *IEEE Access*, *7*, 79968–79979. https://doi.org/10.1109/ACCESS.2019.2920232

[54] Tajmohammadi, M., Mazinani, S. M., Nikooghadam, M., & Al-Hamdawee, Z. (2019). LSPP: Lightweight and secure payment protocol for dynamic wireless charging of electric vehicles in vehicular cloud. *IEEE Access*, 7, 148424–148438. https://doi.org/10.1109/ACCESS.2019.2946241

[55] Kong, P.-Y., & Karagiannidis, G. K. (2016). Charging schemes for plug-in hybrid electric vehicles in smart grid: A survey. *IEEE Access*, *4*, 6846–6875. https://doi.org/10.1109/ACCESS.2016.2614689

[56] Straka, M., et al. (2020). Predicting popularity of electric vehicle charging infrastructure in urban context. *IEEE Access*, *8*, 11315–11327. https://doi.org/10.1109/ACCESS.2020.2965621

[57] Fathollahi, A., Derakhshandeh, S. Y., Ghiasian, A., & Masoum, M. A. S. (2022). Optimal siting and sizing of wireless EV charging infrastructures considering traffic network and power distribution system. *IEEE Access*, *10*, 117105–117117. https://doi.org/10.1109/ACCESS.2022.3219055

[58] Xiang, L., Li, X., Tian, J., & Tian, Y. (2018). A crossed DD geometry and its double-coil excitation method for electric vehicle dynamic wireless charging systems. *IEEE Access*, *6*, 45120–45128. https://doi.org/10.1109/ACCESS.2018.2864999

[59] Amini, M. H., McNamara, P., Weng, P., Karabasoglu, O., & Xu, Y. (2018). Hierarchical electric vehicle charging aggregator strategy using Dantzig-Wolfe decomposition. *IEEE Design & Test*, *35*(6), 25–36. https://doi.org/10.1109/MDAT.2017.2759505

[60] Singh, B., & Dubey, P. K. (2022). Distributed power generation planning for distribution networks using electric vehicles: Systematic attention to challenges and opportunities. *Journal of Energy Storage*, *48*, 104030. https://doi.org/10.1016/j.est.2022.104030

[61] Dubey, P. K., Singh, B., Singh, D., & Patel, D. K. (2023). Distributed generation current scenario in the world. *International Journal for Multidisciplinary Research*, *5*(4). https://doi.org/10.36948/ijfmr.2023.v05i04.4625

[62] Dubey, P. K., Singh, B., Kumar, V., Singh, D. et al. (2023). A novel approach for comparative analysis of distributed generations and electric vehicles in distribution systems. *Electrical Engineering*. https://doi.org/10.1007/s00202-023-02072-2

[63] Dubey, P. K., Singh, B., & Singh, D. (2024). Integration of distributed generation and electric vehicles in the distribution system. *Engineering Applications of Artificial Intelligence*. https://doi.org/10.1016/j.engappai.2024.109036

[64] Dubey, P. K., Singh, B., Singh, D., Shalini, S., Kushwaha, S., Singh, M. K., & Jahnavi. (2024). A review on electric vehicles and distributed generation planning. *Organizational Development Journal*, *12*(12), 313–355.

[65] Singh, B., Dubey, P. K., & Singh, S. N. (2022). Recent optimization techniques for coordinated control of electric vehicles in super smart power grids network: A state of the art. *2022 IEEE 9th Uttar Pradesh Section International Conference on Electrical, Electronics and Computer Engineering (UPCON)*, 1–7. https://doi.org/10.1109/UPCON56432.2022.9986471

[66] Yadav, S., Sudman, M. S. I., Dubey, P. K., Srinivas, R. V., Srisainath, R., & Devi, V. C. (2023). Development of a GA-RBF based model for penetration of electric vehicles and its projections. *2023 International Conference on Self Sustainable Artificial Intelligence Systems (ICSSAS)*, 1–6. https://doi.org/10.1109/ICSSAS57918.2023.10331883

[67] Dubey, P. K., Singh, B., Singh, D., Verma, P. N., Verma, K. S., & Jahnavi. (2024). Energy storage technologies: Types, recent trends, and development. *PARC 2024, GLA Mathura, India (IEEE)*. https://doi.org/10.1109/PARC59193.2024.10486412

[68] Singh, B., Dubey, P. K., Kumar, V., & Kumar, M. (2023). Grid integration of EV. In *Electric vehicles components charging technologies: Design and modeling* (Chap. 13). IET Digital Library. https://doi.org/10.1049/PBTR039E_ch12

[69] Dubey, P. K., Singh, B., Gautam, A. K., Singh, D., & Tripathi, M. N. (2024). A state of the art of recent trends in electric vehicles planning. In *Smart electric and hybrid vehicles: Advancements in materials, design, technology, and modelling*. Wiley. https://doi.org/10.1002/9781394225040.ch6

[70] Alam, M., Ara, K., Javed, M. S., & Ansari, M. (2014, July). Detect and filter traffic attack through cloud traceback and neural network. In *The 2014 International Conference of Data Mining and Knowledge Engineering (ICDMKE)* (pp. 2–4). Imperial College.

[71] Ansari, M., & Alam, M. (2023). IoT-cloud-enabled smart framework for real-world applications. In *Intelligent systems and applications: Select proceedings of ICISA 2022* (pp. 87–105). Springer Nature Singapore. https://doi.org/10.1007/978-981-19-6581-4_8

[72] Ansari, M., & Alam, M. (2024). A novel hybrid time-series approach for IoT-cloud-enabled environment monitoring. *Journal of Supercomputing*, *80*(7). https://doi.org/10.1007/s11227-023-05782-3

[73] Noor, M. A. F., Khanum, S., Anwar, T., & Ansari, M. (2021). A holistic view on blockchain and its issues. In *Blockchain applications in IoT security* (pp. 21–44). IGI Global.

[74] Ansari, M., & Alam, M. (2021). IoT-cloud enabled statistical analysis and visualization of air pollution data in India. In *Proceedings of Data Analytics and Management: ICDAM 2021, Volume 2* (pp. 125–139). Springer Singapore.

[75] Ansari, M., Ali, S. A., Alam, M., Chaudhary, K., & Rakshit, S. (2023). Unlocking the power of explainable AI to improve customer experiences in e-commerce. In *AI-based data analytics* (pp. 31–48). Auerbach Publications. https://doi.org/10.1201/9781032614083

[76] Ansari, M., Ali, S. A., & Alam, M. (2022). Internet of things (IoT) fusion with cloud computing: Current research and future direction. *International Journal of Advanced Technology and Engineering Exploration*, *9*(97), 1812–1825. https://doi.org/10.19101/IJATEE.2021.876002

[77] Ali, S. A., Ansari, M., & Alam, M. (2020). Resource management techniques for cloud-based IoT environment. In *Internet of Things (IoT) Concepts and Applications* (pp. 63–87). Springer International Publishing.

[78] Ansari, M., & Alam, M. (2024). An intelligent IoT-cloud-based air pollution forecasting model using univariate time-series analysis. *Arabian Journal for Science and Engineering*, *49*(3), 3135–3162.

[79] Ali, S. A., Ansari, M., Alam, M., & Rakshit, S. (2023). Federated learning for business intelligence: Predictive maintenance in Industry 4.0. In *AI-Based Data Analytics* (pp. 125–140). Auerbach Publications.

[80] Khan, S., Liu, X., Ali, S. A., & Alam, M. (2019). Storage solutions for big data systems: A qualitative study and comparison. *arXiv*. https://arxiv.org/abs/1904.11498

[81] Ansari, M., Chhabra, A., Hasan, N., & Alam, M. (2025). Quantum finance: Revolutionizing financial markets with quantum computing. In *AI-Driven Finance in the VUCA World* (pp. 96–105). Auerbach Publications. https://doi.org/10.1201/9781003482154

[82] Chhabra, A., Hasan, N., Ansari, M., & Alam, M. (2025). Algorithmic trading and AI: A synergy for risk management. In *AI-Driven Finance in the VUCA World* (pp. 27–45). Auerbach Publications. https://doi.org/10.1201/9781003482154

[83] Mughal, M. F., Ansari, M., & Alam, M. (2025). Revolutionizing financial forecasting: The rise of transformer models in long-term time series analysis. In *AI-Driven Finance in the VUCA World* (p. 142). Auerbach Publications. https://doi.org/10.1201/9781003482154

[84] Hasan, N., Ansari, M., Chhabra, A., & Alam, M. (2025). Revolutionizing finance: The transformative impact of artificial intelligence on the finance sector. In *AI-Driven Finance in the VUCA World* (pp. 172–186). Auerbach Publications. https://doi.org/10.1201/9781003482154

[85] Dubey, P. K., Singh, B., Singh, M. K., Patel, D. K., Singh, D., & Tripathi, M. N. (2024). Decision tree approaches for state of charge evaluation in hybrid electric vehicles: A comprehensive study. *2024 IEEE 11th Uttar Pradesh Section International Conference on Electrical, Electronics and Computer Engineering (UPCON)*, Lucknow, India, 1–7. https://doi.org/10.1109/UPCON62832.2024.10982717

Chapter 13

AI-driven energy optimization in edge data centers

The 4-Gear Predictive Threshold (4-GPT) framework

Mohit Singh Bisht and Vaneet Kumar

13.1 INTRODUCTION

Digital technologies are advancing rapidly, using cloud computing and edge-based services. This advancement has transformed IT operations (Gupta et al., 2021). Data has increased exponentially, and industries require improved computational power, enhanced storage facilities, and effective data management techniques. Cloud data centers (CDCs) are facing stress due to the continuous rise in demand. These CDCs play a crucial role in cloud infrastructure. They must fulfill the requirements of various applications (Singh & Kumar, 2020). Data centers are tasked with powering everything from simple storage to more sophisticated artificial intelligence (AI), machine learning (ML), and real-time data analytics. The rising energy demand in operating such data centers, however, poses a serious problem for organizations seeking to be sustainable (Zhang & Li, 2024).

Widely prevalent over the past few years, cloud services have a host of advantages such as the ability to scale storage and computing services on demand (Gupta et al., 2021). Nonetheless, the expansion of cloud-based applications has expanded exponentially the number of data centers that were needed to meet the enhanced computational demands of both companies and consumers (Sharma & Gupta, 2023). Estimates provided by the International Energy Agency (IEA) show that by 2020, data centers were the cause of the consumption of almost 1% of total global electricity. That amount is expected to continue increasing over the next few years due to the emergence of next-generation technologies like AI, Internet of Things (IoT), and big data analytics, which place a greater strain on computational headroom (Zhang & Li, 2024). There has been a rising demand in energy usage, yet as energy demand rises, there has emerged a situation where operational costs and environmental concerns of data center energy usage become vital issues (Gupta & Soni, 2020).

As an example, one often thinks of the development of AI and ML algorithms that may need large amounts of processing power (potentially massive scale) to build advanced models and undertake real-time data analysis (Sharma & Kapoor, 2022). The increase in data-processing requirements in more applications and devices with reduced latency has also contributed to overloading data centers due to the process of edge computing, which involves locating computation closer to a source of data generation. Such technologies require higher computing capability, which translates to more servers, hence rising use of energy (Yadav & Sharma, 2021).

With the rising pressure on energy resources, it is essential now more than ever to employ energy-efficient methodologies in ensuring that the operations of data centers are handled and optimized. Simultaneously, such plans should be able to strike the thin line between energy efficiency and performance. Minimization of energy consumption and maximized performance have emerged to be of major concern within research and industry (Zhang & Li, 2024) (Figure 13.1).

DOI: 10.1201/9781003685890-13

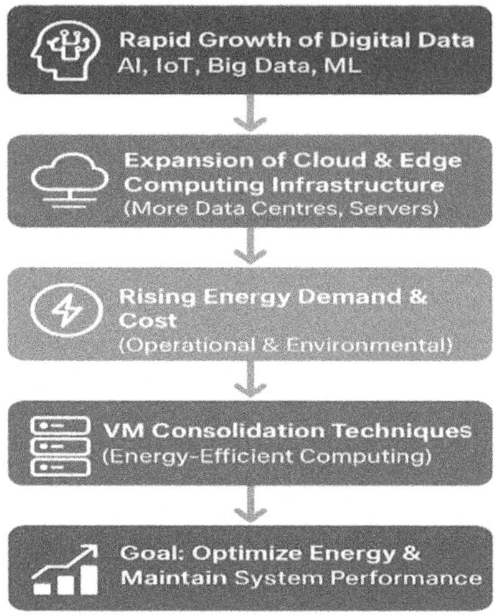

Figure 13.1 Overview of edge and cloud data center infrastructure with growing data demands.

Considering these issues, energy-efficient computing methods, such as virtual machine (VM) consolidation, have emerged as potential solutions. VM consolidation involves placing multiple VMs on a reduced number of physical servers. This approach reduces server usage, leading to energy savings. This technique may significantly reduce energy consumption. It requires fewer physical servers to perform the same workload. The essential factor is achieving an optimal balance. This balance must consider both system performance and energy efficiency.

Aggressive VM consolidation presents significant performance issues. The system may become less efficient. Aggressive VM consolidation may result in insufficient resources. This can lead to excessive resource provisioning, service level agreement (SLA) breaches, and degraded performance. Conversely, insufficient consolidation of VMs can result in inefficient resource utilization. This may lead to underutilized Physical Machines (PMs), consequently wasting energy. The proposed VM consolidation approach balances energy consumption and SLA compliance. This enhances the performance of CDCs.

The 4-Gear Predictive Threshold (4-GPT) algorithm is a novel approach. It enhances predictive capabilities. The algorithm operates on a four-gear mechanism. Each gear represents a distinct phase. This structure allows for improved accuracy in predictions.

13.2 THE 4-GEAR PREDICTIVE THRESHOLD ALGORITHM

We propose the 4-GPT method to address these issues. This algorithm offers a new method for energy conservation. It maintains system performance (Gupta & Soni, 2020). The 4-GPT technique employs predictive workload forecasting. This analysis examines historical data to determine potential workload requirements. This differs from conventional VM consolidation techniques. These algorithms employ a static approach. Such algorithms cannot distinguish between active load variations. On the other hand, 4-GPT allows predictions regarding future resource requirements according to PM utilization (Sharma & Gupta, 2023).

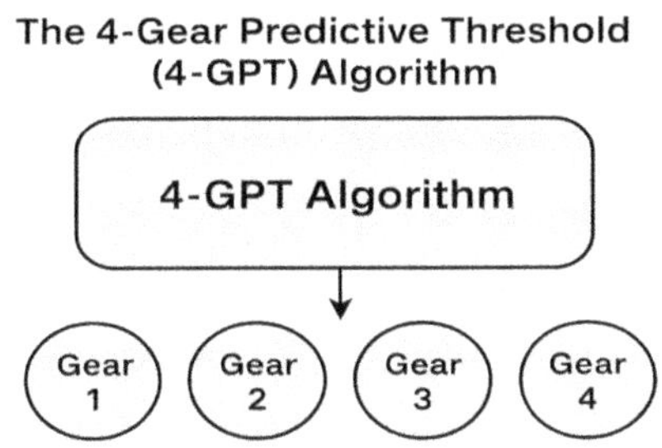

Figure 13.2 4-gear states of the 4-GPT algorithm.

The main principle of the 4-GPT algorithm is the opportunity to predict future demand by applying the concept of predictive modeling and dynamically change the operational condition of servers. The algorithm categorizes the working state of the server into four different gears, with each distinct gear representing a different performance and energy usage state. These gears allow servers to change gears corresponding to changing demands and reduce wastage of energy. An example is when the traffic is high; the system is able to log into a high-performance transmission gear in order to reach peak throughput. In low-demand periods, however, the system may shift to an energy-efficient gear where unnecessary energy use is minimized without ruining performance (Jain et al., 2023) (Figure 13.2).

The four gears in the 4-GPT algorithm mean the states of the different workings of the servers:

- **Gear 1**: High stress and high energy demand under peak loads. This device comes into play when the system is operating under heavy load and maximum output is necessary by the servers (Gupta et al., 2021).
- **Gear 2**: Balanced mode in which the best performance and energy efficiency is achieved. The mode is encountered when moderate performance is required by the workload (Singh & Kumar, 2020).
- **Gear 3**: Strong enough, but a decent power-saving mode. During low work times, this equipment is used since workloads are minimal and the use of resources is minimal (Jain et al., 2023).
- **Gear 4**: Completely low-energy setting with low performance. This gear is supported when there are no current loads or minimal loads, giving maximum energy savings with no performance impact (Zhang & Li, 2024).

Dynamic changing of the gears according to the guess at the workload is one of the main characteristics that separate the 4-GPT algorithm from other VM consolidation algorithms. The predictive component of the algorithm makes the system adequately ready ahead of workload requirements, and this limits the chances of resource contention and unnecessary migrations (Sharma & Gupta, 2023).

13.3 RELATED WORK

Energy consumption of data centers has always been the subject of research interest, particularly with the increase in computing power due to cloud, edge computing, AI, and big data applications. Data centers are well known to require tremendous energy usage, both to operate computing equipment and to cool it. The complexity of the task is in the optimization of energy consumption without compromising the high performance rate, especially in the environment of cloud computing and edge computing where the workload may be highly unpredictable and dynamic.

These issues have been proposed to be addressed by various energy-efficient approaches like VM consolidation, power-sensitive scheduling, and dynamically allocating resources. In this section, the major avenues of the field will be examined, along with the strengths and limitations and how the 4-GPT algorithm extends the prevailing limits (Table 13.1).

Table 13.1 Comparison of approaches for energy efficiency in data centers

Author	*Approach*	*Key Focus*	*Strengths*	*Limitations*
Gupta, Patel, & Verma (2021)	VM Consolidation	Reduces energy consumption by consolidating VMs onto fewer servers.	Minimizes power usage by reducing the number of active physical servers.	Performance degradation due to over-consolidation or resource contention.
Singh & Kumar (2020)	Predictive VM Consolidation	Uses predictive modeling for resource allocation.	Improves accuracy by forecasting workload demand.	Limited by the accuracy of the predictive models and dynamic workloads.
Jain, Singh, & Kaur (2023)	Power-Aware Scheduling	Dynamically allocates resources based on workload prediction.	Optimizes resource utilization and minimizes idle server time.	Inadequate for highly dynamic workloads (e.g., edge computing).
Zhang & Li (2024)	Dynamic Resource Allocation	Adjusts server performance based on predicted demand.	Can balance performance and energy efficiency in real time.	Complexity in handling highly variable and heterogeneous workloads.
Sharma & Kapoor (2022)	Hybrid AI Scheduling	Incorporates ML for workload prediction and scheduling.	Continuously improves scheduling by learning from past data.	May struggle with extreme workloads or under extreme conditions.
Mehta & Bansal (2020)	IoT-based Resource Optimization	Uses IoT sensors for real-time optimization of data center resources.	Enhances real-time decision-making for energy-efficient operations.	Can be limited by the number of IoT devices and infrastructure.
Srivastava & Garg (2021)	Energy-Efficient Resource Allocation	Utilizes algorithms to optimize the allocation of resources based on energy efficiency.	Minimizes operational cost by effectively managing power usage.	Limited by complexity in diverse workloads.
Tiwari & Patil (2020)	Green IT Practices	Implements eco-friendly practices in cloud and edge computing.	Reduces environmental impact and energy consumption.	May not be applicable for all types of data center configurations.
Liu & Chen (2022)	Cloud Data Center Scheduling	Optimizes VM and resource scheduling specifically in cloud environments.	Ensures resources are utilized based on demand, minimizing energy waste.	Lack of flexibility in highly dynamic workloads.
Gupta & Soni (2020)	AI-Driven VM Placement	Leverages AI to optimize the placement of VMs for energy savings.	Achieves higher efficiency by predicting workload spikes and adjusting accordingly.	Requires large-scale data for accuracy.

13.4 VM TECHNIQUE OF CONSOLIDATION

One of the most common ways that data centers have used to increase energy efficiency has been VM consolidation. The concept is straightforward: numerous VMs are put on a minority of physical machines, and in effect, this translates to fewer active physical servers and thus the power consumption will be minimized. The technique proves to be quite handy in a virtualized environment where the workload is dynamic and in an environment where the server infrastructure tends to be underutilized.

Many studies have aimed at enhancement of VM consolidation, involving the consideration of different parameters, including workload profile, resource loads, and power consumption. Gupta et al. (2021) considered a VM consolidation algorithm that would minimize power draw by detecting potentials to consolidate VMs depending on resource utilization patterns. Its concept is that VMs should be condensed onto fewer physical servers where the capacity of the server is not fully exploited. This method is effective in workloads that can be moderately foreseeable, and this method has its limitations. The method fails to consider the dynamic characteristic of the workloads where workloads may vary over time and this may be unpredictable, thereby causing performance to diminish due to failure in timely adjusting resource allocation.

Likewise, Singh and Kumar (2020) introduced a VM consolidation strategy that involved predictive modeling of future demands of resources using the past. Their contribution was an increment in that they incorporated workload prediction, but they quantified the workload forecasting as being static and did not adjust the servers in their approach dynamically in response to changing demands. Predictive modeling can be helpful, but their model lacks the dynamic gear shifting (server state optimizations) which is critical to strike a balance between energy efficiency and performance within highly dynamic workload environments such as edge computing.

Neither of the two studies discussed the necessity of dynamic, adaptive strategies, which should consider the changing nature of the workload in the optimization of energy usage. The 4-GPT algorithm fills this gap by using predictive workload forecasting and dynamic variations to server operations brought about by its gear-based mechanism. That is, in addition to VM consolidation being a significant factor, the 4-GPT algorithm can also turn on the fly to follow the changes in the workload and alter the state of the servers accordingly to minimize energy consumption at the cost of performance.

13.5 POWER-AWARE SCHEDULING AND RESOURCE ALLOCATION

Along with VM consolidation, another strategy necessary to optimally manage energy in data centers is power-aware scheduling. Power-aware scheduling algorithms seek to maintain dynamically allocated resources against estimated workloads to reduce wastage of energy at idle times or stress-free times. The algorithms assign resources to servers depending on actual and predicted workload and ensure that servers operate at their most ideal power state that may change depending on the workload demands at any time.

In a given sample, Jain et al. (2023) suggested a scheme of power-aware scheduling, which aims to optimize the conditions of server use, taking into account the predicted workload to achieve a higher energy efficiency rate through minimal idle time. The algorithm used by them is based on advanced prediction of workload and increasing or decreasing usage of servers respectively. Although such an approach resembles the prediction part of the 4-GPT algorithm, it does not allow dynamically varying the server operational states based on the dynamically changing real-time workload. This is especially important in settings with volatile workloads, e.g. edge computing settings where workloads tend to fluctuate significantly within seconds or even minutes.

In line with this, the energy-sensitive VM placement policy investigated and proposed by Zhang and Li (2024) is an attempt to strike an equilibrium between power and performance. This policy is based on a threshold-driven approach, that is, causing migrations only when a need arises and therefore eliminating overprovisioning of resources. Nonetheless, the technique is inadequate since it does not allow the dynamic alterations of server performance based on alternating workloads. It pays special attention to minimizing migration, and it is the right thing to do; however, it also fails to note the significance of putting the server in an operation state that would result in maximum energy savings. It is through this dynamic adjustment that the 4-GPT algorithm stands out because it can limit migrations as well as change server states to optimally utilize energy dynamically.

The 4-GPT algorithm therefore is an extension over such power-sensitive scheduling systems in that it dynamically adjusts server performance depending upon anticipated changes in the workload instead of merely minimizing migrations or idle times. While hopping servers across gears of operations, the algorithm will guarantee that power usage is efficient in all demand times such as in the high- and low-demand states (Gupta et al., 2021).

13.6 DYNAMIC RESOURCE PROVISIONING AND SLAS

Service level agreements in cloud computing and edge data centers describe the levels of performance that should be provisioned as response time, availability, and throughput. SLAs are a paramount factor to take into consideration when undertaking the strategy of optimizing energy because minimization of performance may lead to customer dissatisfaction and violation of SLAs (Sharma & Kapoor, 2022).

The goal of many energy-efficient scheduling and resource provisioning techniques is the energy-SLA trade-off. Most solutions fail to account for the changing nature of workloads over time. In edge computing, workloads can change rapidly. This variability depends on user demands, device activities, and environmental conditions. Standard energy optimization plans based on specific criteria may not meet SLA requirements. This is particularly true during abrupt energy spikes or periods of high workload.

The 4-GPT method addresses this issue by considering SLA limitations directly. It modifies server statuses in real time. The technique effectively predicts workloads and state changes, such as gear switching. It is beneficial for energy optimization and potential future workloads. Dynamic reactivity to varying workloads is essential. It sustains constant performance levels. This is important for fulfilling commitments in SLA agreements. This is particularly true during times when energy concerns are absent. The algorithm can predict future demand spikes. This reduces the likelihood of breaking SLAs (Sharma & Gupta, 2023).

Current energy-efficient resource provisioning systems often provide excessive resources. Alternatively, they may fail to utilize resources effectively, leading to an imbalance between energy consumption and the fulfillment of SLAs. The 4-GPT algorithm provides a dynamic and proactive methodology of maintaining performance and energy utilization within limits in addition to existing within the limits of the SLA commitments.

13.7 SUMMARY OF GAPS AND CONTRIBUTION OF THE 4-GPT ALGORITHM

Most energy optimization strategies, even though diverse, still use static rules or forecasts which do not take into consideration the dynamic and variable character of workloads in contemporary data centers. VM consolidation, power-aware scheduling, and other techniques provide a certain improvement, but they do not dynamically scale server performance in anticipation of workload demand. Also, most of the available algorithms lack a comprehensive solution that incorporates predictive workload forecasting with dynamically changing resources.

The 4-GPT algorithm fills this gap as VM consolidation and dynamic changes of the server states based on estimated workloads are achieved. This dynamic adjustment is largely applicable in edge computing environments owing to the unpredictable nature of workloads, which may become extreme. The fact that the algorithm predicts future demand, shifts gears on servers, and minimizes migrations makes it different from existing methods. Incorporating the 4-GPT algorithm in servers and relying on a predictive workload model and iterative servers available, it can reconcile energy efficiency and performance and provides the perfect solution to modern data centers, especially where the variability of the workloads is the highest in edge computing contexts (Zhang & Zhang, 2023).

13.8 OPPORTUNITIES AND CHALLENGES

13.8.1 Opportunities

13.8.1.1 The smart entry edge computing

With the continuing growth in demand for low-latency services, edge computing is getting more important. Processing of data is done closest to the source in edge computing data centers, which reduces latency and optimizes bandwidth utilization. These environments are particularly suitable for using the 4-GPT algorithm because it can adapt in real time to such settings to optimize resource utilization and energy efficiency. The predictive performance of the 4-GPT algorithm is suitable because the workload of edge computing is quite moving and localized. Without varying the server performance according to these changing demands, server energy consumption would not be efficient enough to support the best performance. However, 4-GPT would help optimize server performance and energy consumption to reach a satisfactory point.

13.8.1.2 Hybrid AI: hybrid computation AI and machine learning

The application of ML and AI to the 4-GPT method could enhance the precision of workload prediction significantly. AI-driven models enable continuous training using currently available data. This increases the likelihood of accurately estimating future resource demands and shortfalls. AI can adapt to emerging trends and patterns. This adaptation enhances the predictive capabilities of the 4-GPT algorithm. This ensures optimal energy utilization and enhances system efficiency. ML models are improving. The algorithm can adapt to changing workloads. This adaptation occurs more quickly and smoothly. As a result, energy savings increase. Performance also improves (Gupta & Soni, 2020).

13.8.1.3 Environmental sustainability

The increasing focus on environmental sustainability offers considerable opportunities for algorithms such as 4-GPT. These algorithms consume less energy and can have a meaningful impact. Data center operators face pressure to enhance environmental sustainability. They must work to lower their carbon footprint. The 4-GPT algorithm is crucial to this study. It enables data centers to minimize energy consumption, reduce waste, and engage in environmentally beneficial practices. Sustainability holds significance across all disciplines globally. Energy-efficient algorithms, such as 4-GPT, assist data centers in adhering to international environmental regulations. They contribute to a cleaner and more sustainable future.

13.8.2 Challenges

13.8.2.1 Predictive accuracy

The 4-GPT algorithm must address the accuracy of its workload predictions. This aspect is crucial for its effectiveness. Dynamic workloads in edge computing environments can change rapidly and unexpectedly as demand fluctuates. Predicting future resource needs can be difficult. This is particularly true in areas where changes occur unexpectedly. Inaccurate forecasts can result in poor resource utilization. This may occur through excessive or insufficient use of available resources. This may lead to fixed consumption of energy units or a failure that undermines the advantages of the algorithm. Improved predictive accuracy enhances system efficiency. It optimizes energy use and maximizes performance (Sharma & Kapoor, 2022).

13.8.2.2 The overhead VM relocation

The 4-GPT technique aims to minimize VM migrations. Transferring a VM may result in increased energy consumption. It may reduce effectiveness. Transferring VMs necessitates network and computational resources. This operation may consume additional energy. It could also reduce the performance of both the migrating VM and the host servers. Frequent reasons for migration may lead to increased energy consumption. This may consume significantly more energy than the system conserves in total. Finding a balanced approach to migration is crucial. Limiting migration has benefits. These must exceed the related costs (Gupta & Soni, 2020).

13.8.2.3 Heterogeneous workloads

Typical concrete examples of use cases found in edge computing environments involve sensor processing, real-time analytics, and multimedia processing and have considerably heterogeneous workloads. Such varying work patterns each have varying resource requirements, thus complicating the prediction of their behavior. It is a major task to design a predictive model capable of predicting what is needed by such diverse workloads. This heterogeneity has to be addressed effectively by the 4-GPT algorithm so that the energy optimization strategies become accurate and adaptive. Developing a model that can generalize over such a wide range of tasks with sufficient performance and energy efficiency remains one of the most important areas that still need improvement (Yadav & Sharma, 2021).

13.9 METHODOLOGY

The 4-Gear Predictive Threshold algorithm uses predictive modeling and dynamic resource adjustment to optimize energy consumption in edge computing data centers. The main idea of the algorithm is the ability to forecast the resource requirements of VMs based on past data so that the system can foresee future demand and scale down or up the operational state of servers. This forward planning guarantees the effective utilization of resources and because it operates under low energy consumption, its productivity is high (Gupta et al., 2021).

13.9.1 STEP 1: data collection/pseudo protein preprocessing

The procedure starts by obtaining a history of resource consumption on the VMs in the data center. Such information includes vital performance data such as central processing unit (CPU) usage, memory usage, disk Input and Output, and network traffic. The data are also preprocessed so that any noise and outliers causing distortion to trends are removed in order to make accurate predictions. Only cleaned data with substance are kept intact, and this provides an excellent foundation toward accurate workload modeling (Singh & Kumar, 2020).

13.9.2 Step 2: prediction of workload

Once the data have been preprocessed, an algorithm based on a predictive model is used to predict the future resource requirements per VM, e.g. time-series analysis, regression models, or ML algorithms. The modes forecasts the extent of CPUs and memory that will be utilized in the next hour or day. The related insights from these predictions enable proper setting of the servers to be in the best state of operation in order to ensure that resources are consumed optimally in relation to the expected demands (Jain et al., 2023).

13.9.3 Step 3: gearing

The 4-GPT algorithm modulates the operational status of servers using four gears, each of which enables the server to consume minimal energy without compromising the targeted levels of performance:

- **Gear 1**: High-performance, high-energy mode of execution in order to deal with maximum workloads where maximum resources are necessary.
- **Gear 2**: Balanced mode, in which performance and energy efficiency are traded off to yield a reasonable compromise between the two.
- **Gear 3**: Energy-saving mode that reduces performance in order to conserve energy during nonpeak hours when there is no need to exhaust the resources fully.
- **Gear 4**: Ultra-low-energy mode which is entered when idle and requires very minimal resources (Zhang & Li, 2024).

13.9.4 Step 4: VM reduction-moving

The next important aspect of the 4-GPT algorithm is that it reduces the number of unnecessary migrations of VMs. The system can avoid over-migrating by correctly predicting the most desirable VM placement according to the workloads that are expected. Migrations are set up only when performance degradation or underutilization of the resources cannot be eliminated by setting the server into another status with or without any addition of resources. The plan saves overhead costs on VM migrations to make sure that there is effective use of resources and less consumption of energy (Gupta & Soni, 2020).

13.9.5 Step 5: dynamic adaptation

The system is an innovation made to adapt continuously. Since the workloads evolve in the course of time, the algorithm adjusts the workload predictions and changes the operational conditions of servers accordingly. Such adaptive flexibility is cognizant of maintaining the system as energy-efficient and high-performing notwithstanding the varying demand. Real-time adjustment capability enables the data center to be proactive in adapting to fluctuations in the condition will be able to balance energy saving and performance maximization at all times (Sharma & Gupta, 2023).

13.10 RESULTS

13.10.1 Reduction in the breach of SLA

With the introduction of the 4-GPT algorithm, the number of SLA violations considerably reduced, as it showed a 55% decrease as opposed to the MAD (Median Absolute Deviation) method. This decrease is a testament to the better capacity of the algorithm to conform to the performance standards in order to achieve higher standards of meeting the service expectations. As compared to the IQR (Interquartile Range) method, the 4-GPT algorithm outperformed the latter one by 26%, which shows greater potential of the former to keep the service levels. As well, compared to the THR (threshold) method, the 4-GPT algorithm proved itself significantly in the minimization of SLA violations (39%), which underlines the presence of efficiency in reaching a balance between requirements for higher performance and operational efficiency. Such findings emphasize the effectiveness of the 4-GPT algorithm in achieving a stable performance with a low potential of violating SLAs, which can be introduced in the environment where service target achievement is paramount (Sharma & Gupta, 2023) (Figure 13.3).

13.10.2 Energy consumption

Another positive effect is the energy efficiency rates that were exhibited by the 4-GPT algorithm. It saved 17% energy as compared to the MAD method, which forms another strong point of it and demonstrates that it is able to maximize energy use without decreasing the performance of the rest of the system. Comparing the IQR approach, a 12% energy saving could be seen, which again proves that the algorithm is effective in maintaining power efficiencies and workload control. Even compared to the THR technique, the 4-GPT algorithm saved 6% in terms of energy consumption, which is evidence of its future potential to drive sustainability in high-performance computing facilities such as data centers. These results indicate the potential of the algorithm to increase energy efficiency, thus resulting in cost savings, operational and environmental effects for organizations that are looking to optimize their operational performance as well as energy footprint (Zhang & Li, 2024) (Figure 13.4).

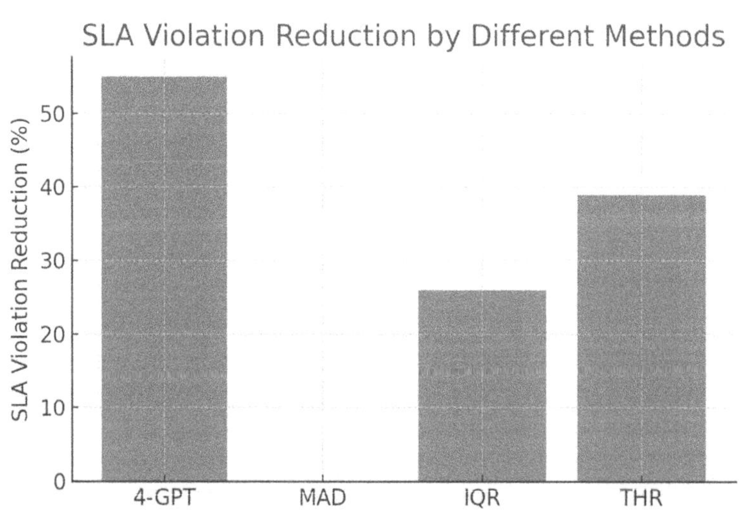

Figure 13.3 SLA violation reduction by different methods.

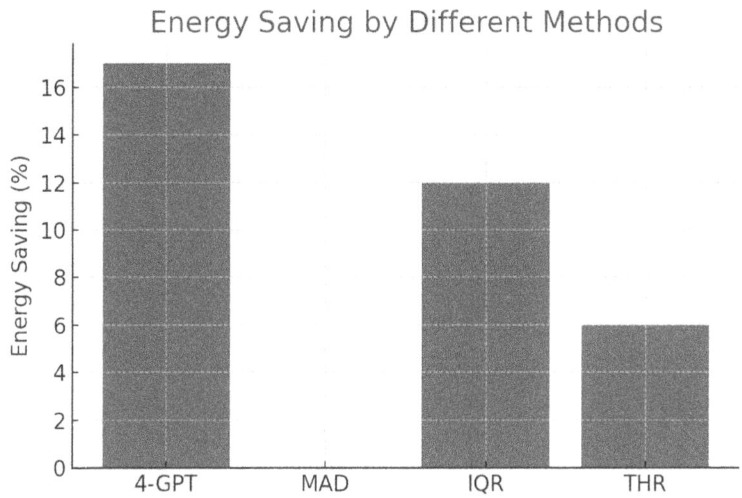

Figure 13.4 Energy saving by different methods.

13.10.3 Missing reduction of VM migration

The 4-GPT algorithm was also an important success due to the minimization of the migrations of VMs. It was able to minimize the rate of migrating VMs by as much as 40% compared with conventional techniques where it normally migrates frequently in order to manage load and maximize resources. This decrease is imperative because keeping excessive VM migrations can cause degradation in performance, an increase in overhead, and larger energy consumption because of the increase in resources needed to complete the migration process. By limiting the requirement for VM migrations, the 4-GPT algorithm does not only drive down the operational overheads, but it also creates optimization of the energy cost, as the less the migration, the less the system operation and use of resources. This qualifies the 4-GPT algorithm to be a more efficient and sustainable dynamic data center operation algorithm that is effective in the management of workload to avoid wastage of resources (Gupta & Soni, 2020) (Figure 13.5 and Table 13.2).

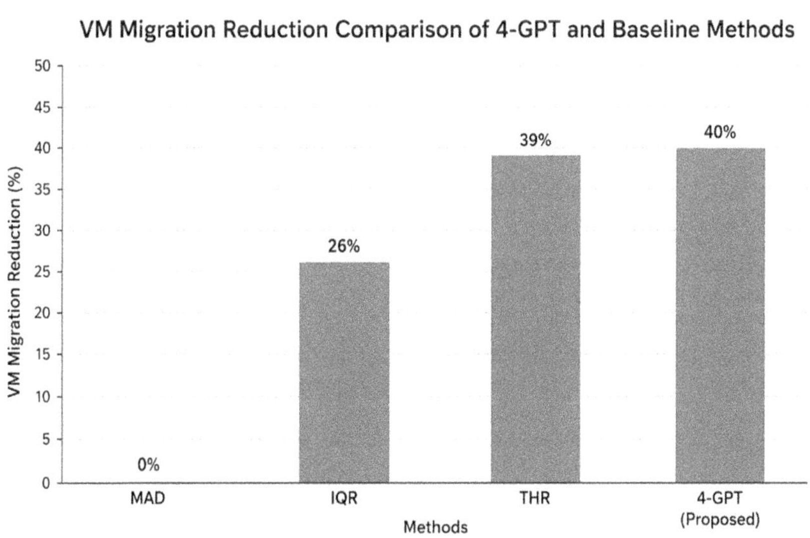

Figure 13.5 VM migration reduction achieved by 4-GPT.

Table 13.2 Performance Comparison of 4-GPT and baseline methods in edge data centers

Method	*SLA Violation Reduction (%)*	*Energy Savings (%)*	*VM Migration Reduction (%)*
4-GPT	**55**	**17**	**40**
MAD	0	0	N/A
IQR	26	12	N/A
THR	39	6	N/A

13.11 CONCLUSION

An innovation in optimizing energy use and performance in edge computing data centers is the 4-GPT algorithm. The incorporation of predictive workload forecasting and dynamic scaling of server activity in four different gears, 4-GPT makes sure that energy consumption is minimal and the performance remains high on a consistent basis. Contrary to the more commonly used techniques, with this algorithm, workload changes can be dynamically adapted with efficiently used resources, which in itself contributes to the overall lesser amount of SLA violations, energy use, and frequent VM migrations. Such attributes make 4-GPT an ideal match for modern edge computing environments with highly variable and rapidly changing resource demand. The superiority of the algorithm is confirmed through experimental trials, with the results indicating that the algorithm could be significantly better than traditional energy optimization approaches in a real-world setting. With data centers increasingly starting to get larger and more complex, intelligent energy solutions such as the 4-GPT are all the more essential. In the future, the conducted study can be improved based on the needs of predictability, accuracy, and reliability of the algorithm. It has also considered extending its use to heterogeneous data center situations, where more of the different hardware and software combinations are optimized. It will also boost energy efficiency, sustainability, and performance, further generating the viability of large-scale dynamic data centers in the future.

REFERENCES

Gupta, D., & Soni, M. (2020). A survey of energy-efficient computing in cloud and edge data centres. *IEEE Access*, 8, 135097–135111.

Gupta, R., Patel, S., & Verma, A. (2021). Energy-efficient virtual machine consolidation strategies for cloud data centres. *Journal of Cloud Computing*, 10(2), 118–130.

Jain, S., Singh, R., & Kaur, M. (2023). Power-aware scheduling in virtualized data centres: An adaptive approach. *IEEE Transactions on Cloud Computing*, 11(4), 451–463.

Liu, Y., & Chen, X. (2022). Cloud data centres and energy efficiency: A comprehensive review. *Energy Reports*, 8, 788–799.

Mehta, S., & Bansal, A. (2020). VM migration in virtualized cloud environments: A comprehensive review of algorithms and solutions. *Cloud Computing and Distributed Systems*, 21(1), 31–44.

Sharma, R., & Kapoor, S. (2022). Intelligent workload prediction in edge data centres: Challenges and solutions. *IEEE Access*, 10, 50329–50341.

Sharma, V., & Gupta, N. (2023). Efficient energy-aware dynamic resource allocation for virtualized data centres. *Energy Reports*, 9, 1921–1931.

Singh, M., & Kumar, V. (2020). Predictive modeling for VM consolidation in energy-efficient data centres. *International Journal of Computer Science and Engineering*, 8(5), 251–264.

Singh, R., & Kaur, M. (2023). Predictive VM resource scheduling for energy optimization in cloud data centres. *Journal of Cloud Computing: Advances, Systems and Applications*, 12(1), 1–16.

Srivastava, P., & Garg, V. (2021). Energy-efficient task scheduling in cloud computing environments. *International Journal of Cloud Computing and Services Science*, 10(3), 13–25.
Tiwari, S., & Patil, R. (2020). Sustainable cloud computing through green IT solutions. *Journal of Computing and Sustainable Development*, 11(1), 51–61.
Yadav, M., & Sharma, S. (2021). Minimizing energy consumption in edge computing environments: A data-driven approach. *Journal of Energy*, 22(4), 452–467.
Zhang, H., & Li, S. (2024). Energy-aware VM allocation policy for cloud data centres. *Sustainable Computing: Informatics and Systems*, 27, 100453.
Zhang, X., & Zhang, Z. (2023). Machine learning-based energy optimization for virtualized cloud environments. *Journal of Computational Science*, 68, 105601.

Chapter 14

Transforming agriculture and environmental monitoring with artificial intelligence

Advancing sustainability and precision

M. Robinson Joel

14.1 INTRODUCTION

Over the last few decades, the agricultural industry has seen substantial technological developments, from the usage of automated technology to precision farming practices. But even with these advancements, environmental monitoring is still a crucial field that has to be developed further to meet escalating climate issues, sustainable agriculture, resource management, and transformation. Traditional agricultural methods are failing to maintain environmental sustainability and production due to the growing frequency of adverse weather patterns, changing climatic conditions, along with the need for more effective use of fertilizers and water [1]. Advanced technologies that support farm management techniques through real-time monitoring, information-driven decision-making, and predictive analysis are becoming more and more necessary in response to these issues [2].

By fusing artificial intelligence (AI) with the Internet of Things (IoT), this system seeks to close this gap and provide a complete environmental monitoring system. Real-time environmental data, such as soil moisture levels, temperatures, humidity levels, pollution levels, and weather, are gathered by the suggested system using IoT sensors [3]. After that, the data is sent to a centralized system for processing, enabling farmers to remotely monitor environmental and health indicators on their farm and take prompt action as needed. By giving farmers information about the factors influencing crop growth and yield potential, real-time data gathering helps them make better decisions by maximizing resource utilization and eliminating waste [4]. Figure 14.1 shows the overall view of transforming agriculture and environmental bonitoring with AI.

The use of AI technology, namely in the shape of an automated weather prediction model, is one of the project's major advances. Farmers may arrange for and handle resources proactively by using this predictive power to foresee future weather conditions [5]. The method enables farmers to implement preventive measures, like modifying irrigation schedules or shielding crops from unfavorable weather, by predicting variables like rainfall, temperature swings, or the possibility of frost. This improves agricultural systems' resistance to climatic unpredictability as well as enhancing the effectiveness of farming operations [6].

With the use of cutting-edge technology, the AI Environmental Monitoring Equipment is a complete, integrated solution intended to improve farm automation. It effortlessly integrates a number of technology elements, including ThingSpeak, a platform that uses the cloud for data storage and visualization, Espressif Systems Platform (ESP) modules for data transfer over wireless networks, and an Arduino for data gathering [7]. With the help of this networked architecture, farmers can keep an eye on their farming environment in real time and make data-driven decisions that increase sustainability and production.

The Arduino, which serves as the system's central data collection device, collects important environmental data including temperature, humidity, soil moisture, and air quality.

 DOI: 10.1201/9781003685890-14

Figure 14.1 Overall view of transforming agriculture and environmental monitoring with artificial intelligence.

The system uses ESP modules to provide smooth connectivity, sending the gathered data to ThingSpeak for cloud processing and storage [8].

The main platform is ThingSpeak, which provides a user-friendly web-based interface for real-time data visualization. Farmers may quickly understand the environmental conditions of their farm using simple-to-read infographics and dashboards, allowing for prompt actions when needed [7]. The incorporation of AI-powered weather forecast capabilities is what distinguishes this system. The AI module can estimate future weather conditions, including rainfall, temperature changes, and the probability of severe occurrences, by analyzing previous meteorological data [5]. This enables farmers to take preventive measures. The same online interface presents these forecasts, providing a comprehensive picture of the state of affairs both now and in the future [3].

Whether it's modifying irrigation schedules, controlling crop protection, or arranging planting and harvesting periods to coincide with predicted weather patterns, this predictive feature helps farmers manage their operations more efficiently [2]. This approach is an important first step toward closing the gap between conventional agricultural methods and the quickly changing, technologically advanced agriculture of the decades to come. It provides a dependable, continuous tracking solution with the added benefit of predictive capabilities by combining IoT, cloud computing, and AI [1]. When these technologies are combined, farm operations become more efficient, less waste is generated, and crops are better equipped to withstand changing environmental conditions [6].

14.2 TRANSFORMATION IN AGRICULTURE PRECISION FARMING

When compared to conventional cultivation methods, precision farming increases average yields by using inputs in precisely the right proportions [9]. In order to boost production efficiency, enhance product quality, raise crop chemical use efficiency, preserve energy, and protect the environment, it is a complete system that uses essential information, technology, and leadership components to optimize production. Thus, precision agriculture is a compelling idea, and its tenets logically raise the hope that farming inputs may be used more efficiently, leading to higher profitability and less harmful output for the environment. Technology for tomorrow's environmentally friendly agriculture can be provided by today's advancements in precision farming [2].

Today's global food system confronts enormous obstacles that will only become worse over the next four decades. With enough desire and money, a lot may be accomplished right away with the knowledge and technology available today. However, more significant adjustments to the food industry and funding for research to develop fresh approaches to cutting-edge issues will be needed to meet future difficulties. Concerns about agricultural growth and development have grown to include the following: declining total productivity, depleting and deteriorating natural resources, advancing farm incomes, lack of an eco-regional approach, declining and fragmented land holdings, trade liberalization in agriculture, inadequate employment opportunities in industries other than agriculture, and global climate variation [10]. As a result, using recently developed technologies is seen to be one way to boost agricultural output going forward. Precision farming acknowledges site-specific variations within fields and modifies management measures appropriately, as opposed to managing a whole field based on some fictitious average state that might not exist throughout the field. Most farmers are aware that the yields in their farms vary depending on the terrain [4].

14.2.1 Crop health monitoring and disease detection

In the current scenario, when pest intervention and diseases, especially post-harvest losses, cause around 20% to 25% of crop production to be lost, crop health management is absolutely essential [11]. It results in a significant loss of income and has a variety of effects on the volume of output. Additionally, maintaining the crops requires time, work, fertilizer, water, and other resources. Figure 14.2 explains the crop health monitoring and overview of disease detection.

Thus, IoT-based crop health tracking offers a data-driven method for real-time crop monitoring through the use of sensor gadgets, gateway connectivity, and an intuitive dashboard [3]. In order to make better decisions on the quality and health of crops, it involves monitoring a variety of characteristics, including temperature, humidity, precipitation, insect intrusion, and seed and soil quality. It entails the prompt participation of farm managers, even from distant places when they are unable to be on-site, in case of emergency scenarios such as nighttime freezing or insect incursions. The farming industry may gain from IoT crop health surveillance systems in several ways, including lower expenses and labor costs. It also

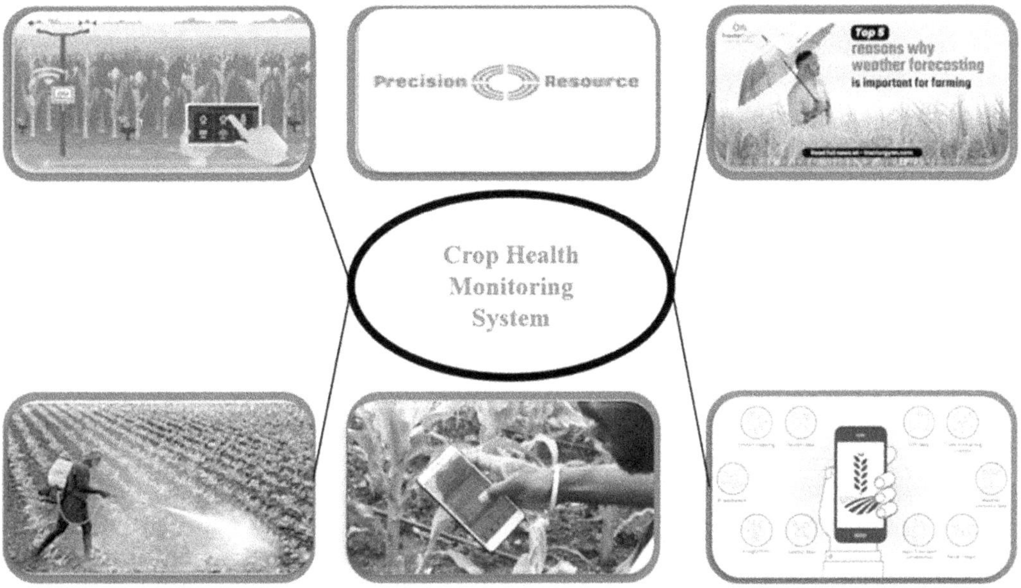

Figure 14.2 Crop health monitoring and disease detection overview.

addresses the large volume of data and recommends appropriate actions that should be performed to increase crop output [2].

A crucial component of contemporary agriculture is crop health monitoring. It enables agronomists and farmers to monitor plant health, identify illnesses early, and maximize available resources. Utilizing cutting-edge technology such as drones, IoT sensors, and satellite imaging, producers may minimize their environmental effect, save expenses, and increase output [6]. Plant health was manually evaluated in traditional farming, which took a lot of time and was prone to human error. Farmers may now make better decisions thanks to data-driven monitoring technologies that offer precise, real-time information. By boosting productivity and encouraging more environmentally friendly methods, these technologies are completely changing the agricultural industry.

Preventing major crop loss and minimizing the need to feed excessive chemical usage are two benefits of identifying diseases and insect infestations early on. Farmers may minimize waste and lessen their influence on the environment by applying insecticides, fertilizer, and water exactly where they are required [12]. Monitoring makes sure that crops get the proper nutrients and attention when they need it, which increases yield and improves the quality of the products. More lucrative agricultural operations are made possible by automated monitoring, which lowers labor costs and guards against financial losses due to undiagnosed crop stress. Farmers may adapt their operations to climate change by monitoring weather and soil moisture, which lowers the chance of damage from drought or flooding [7]. Monitoring promotes long-term mineral health and biodiversity by lowering chemical abuse and enhancing soil management.

14.2.2 Smart irrigation systems

Smart agriculture is the use of modern technology in agricultural management to facilitate resource optimization, remote monitoring, and system automation. This idea is related to the idea of an intelligent irrigation system that is used in this paper. This idea is the addition of intelligence to automatic irrigation systems, either through the use of intelligent control techniques like fuzzy logic or neural networks to improve irrigation decision-making, or through the use of deep learning or machine learning (ML) techniques to improve irrigation modeling or control by incorporating large amounts of data into the analysis. Intelligent irrigation systems administer water to crops at the correct time, in the right amounts, and in the right locations in an effort to boost agricultural output while lessening the impact of crops on the environment [13].

This article presents the findings of a comprehensive study of the literature that was prompted by interest in irrigation methods. Its initial design took into account earlier research on similar subjects. This led to an assessment of disruptive and emerging technologies in agriculture in urban areas, which included papers from 2015 through 2021. IoT, automation, as well as AI are the most often mentioned technologies in the documents, and those technologies are also the most widely used in irrigation, according to that review [14].

The bar chart Figure 14.3 and Table 14.1 illustrate the distribution of various publication types, highlighting a significant disparity in the number of records across categories. Conferences lead with a substantial count of 1,733 publications, indicating that the majority of research outputs or academic contributions in this dataset were presented at conferences.

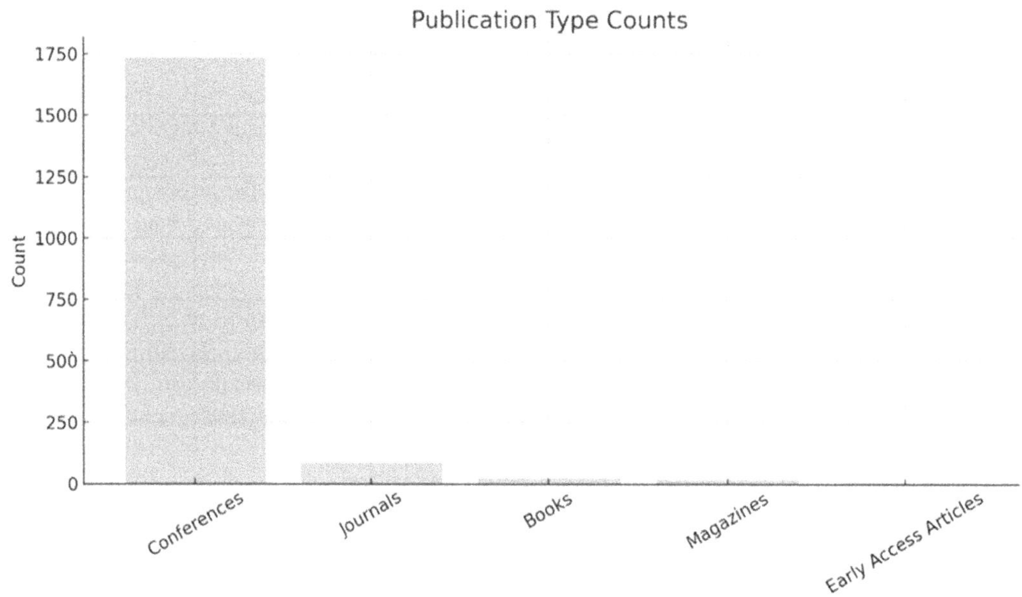

Figure 14.3 No. of articles published in IEEE forum based on smart irrigation systems.

Table 14.1 Number of articles published in IEEE forum based on smart irrigation systems

Publication type	*Count*
Conferences	1,733
Journals	85
Books	22
Magazines	16
Early Access Articles	4

This trend suggests a strong inclination among researchers toward disseminating their work in conference settings, likely due to faster publication timelines and opportunities for immediate peer interaction and feedback.

Journals follow distantly with 85 publications, representing more formal and peer-reviewed contributions that often require longer preparation and review processes. Books (22 publications) and Magazines (16 publications) contribute modestly, suggesting these formats are less commonly used for academic or technical dissemination within this context. The smallest category, Early Access Articles with only four records, reflects the emerging but still limited use of prepublication or early access dissemination strategies.

Overall, the data demonstrates the dominant role of conferences in the current research publishing landscape, while also highlighting the supplementary roles of journals, books, and other publication formats. This distribution can be valuable for researchers, institutions, and policymakers to understand the preferred communication channels in specific fields and optimize their dissemination strategies accordingly.

14.2.3 Yield prediction and crop forecasting

Many industries employ ML techniques, from retailers to forecast how customers will use their phones to assessing customer behavior. Agriculture has also been using ML for a number of years. One of the difficult issues in precision agriculture is predicting crop production, and several models have been put forth and proven to work thus far. Since agricultural production depends on a wide range of parameters, including soil, fertilizer use, climate, weather, and seed type, this challenge necessitates the use of many different datasets. This suggests that predicting agricultural production involves a number of intricate procedures and is not a simple operation. Although agricultural yield prediction programs can now fairly forecast the actual yield, improved yield prediction performance is still preferred [15]. Figure 14.4 shows and explains the crucial subjects, including crop forecasting and yield prediction.

Based on a number of characteristics, ML—a subfield of AI that focuses on learning—is a useful strategy that can improve yield prediction. From datasets, ML may extract knowledge by identifying patterns and connections. Datasets that depict the results based on prior experience must be used to train the models. During the training phase, historical data is used to define the parameters that govern the models since the predictive model is constructed utilizing many characteristics. A portion of the past data that hasn't been utilized during training is utilized for performance evaluation during the testing phase [16].

As shown in Figure 14.4, the crucial subjects, including crop forecasting and yield prediction, are explained in detail. The fundamental techniques for estimating agricultural productivity include both contemporary ML algorithms and conventional statistical methods. Weather data, satellite imaging, and soil sensors are examples of data sources for forecasting that offer vital inputs for creating accurate prediction models. AI and ML in yield prediction highlight the use of sophisticated algorithms such as Random Forests, SVMs, and deep learning (CNN, LSTM) in order to improve prediction accuracy. Drones and satellite data are used by geospatial and remote sensing technologies to anticipate yields and track the health of plants at a regional level. Impact of global warming investigates how crop yield is impacted by changing weather patterns along with extreme occurrences, as well as how models adjust to these developments.

By accounting for soil fertility, water content, and nutrient amounts, soil and nutrient management guarantees accurate forecasting; accuracy and model validation entail assessing prediction models employing statistical metrics (e.g., RMSE, R2) to ensure dependability and direct decision-making; and these elements collectively constitute a comprehensive strategy for crop performance prediction, agricultural planning optimization, and global food security initiatives.

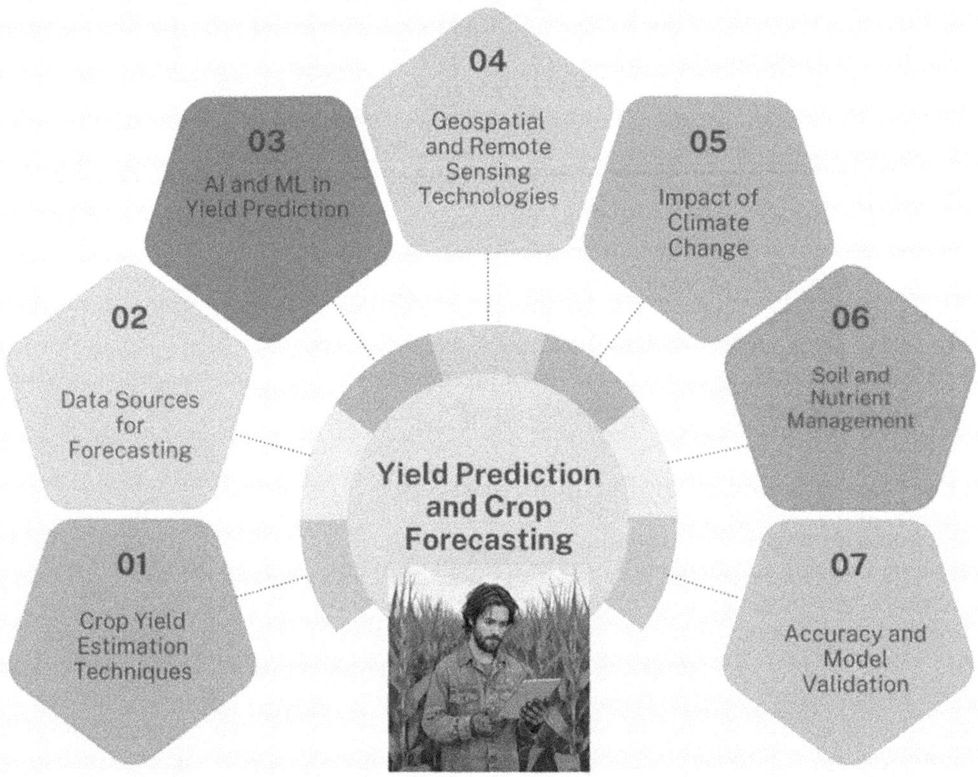

Figure 14.4 Crucial subjects include crop forecasting and yield prediction.

14.2.4 Weed and pest detection

Fast identification of weed and insect infestations is necessary to prevent a widespread infestation and to apply targeted pesticides, such as through precision farming methods, in order to ensure crop output and agricultural land quality. An automated monitoring system can also assist farmers in using effective and environmentally friendly management techniques in relation to climate change and its associated consequences, such as shifting soil conditions, flora, and wildlife. An ML model that analyzes satellite and unmanned aerial vehicle (UAV) imagery—which is used for automated and real-time monitoring of agricultural areas—can be used to monitor agricultural land and detect pest and weed infestations automatically [17]. This model shown in Figure 14.5 provides data in the visible as well as radar spectrum. Such a model may identify trends in the picture data, such as weed and pest infestation-induced changes in the color and form of the acres, and categorize the various items according to their kind of vegetation or infestation.

The biggest obstacle to automated weed and pest infestation identification is still the need for a sizable data collection that covers all significant infestation kinds in various stages in order to produce accurate and trustworthy findings. In the upcoming years, automated field management and drone precision farming will not be feasible due to the high cost of acquiring UAV or drone imagery. However, satellite imaging, at least freely available satellite imagery, lacks the necessary geographic resolution to identify infestations early on and may only be practical later in the infestation's life cycle [18].

An ML method for image segmentation and classification is needed to identify and categorize various objects and patterns in the picture data in order to detect pest and weed

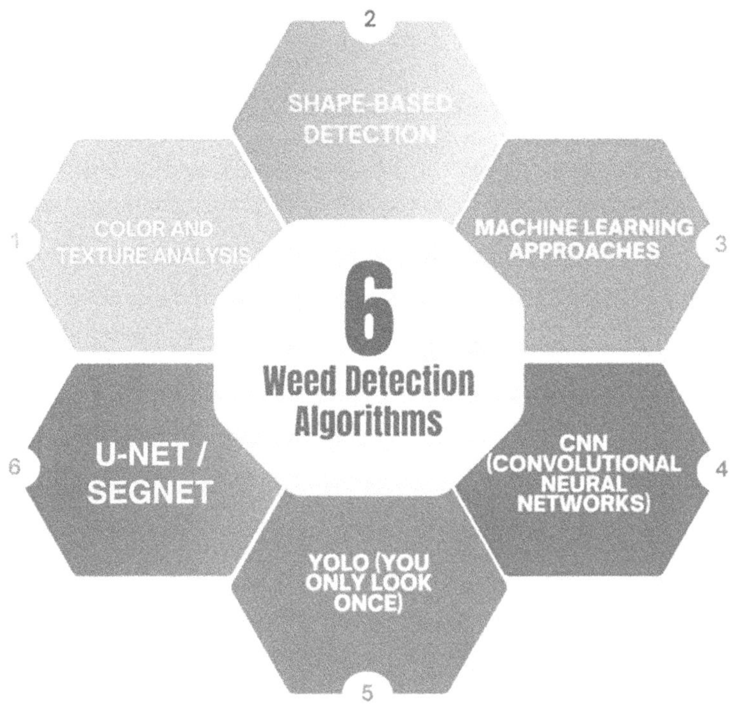

Figure 14.5 Some of the weed detection algorithms.

infestation. Certain convolutional neural networks (CNNs), such as the U-Net or Mask R-CNN architectures, are frequently used to handle such computer vision problems. These can identify infected regions using labeled training data, as well as the types of infestations based on their impact on crops and plants. They can also detect variations in the color and form of agricultural landscapes [19].

14.3 TRANSFORMATION IN ENVIRONMENTAL MONITORING

The use of AI in monitoring the environment provides thorough air and water quality monitoring, precise disaster forecasts, and the identification of pollutant sources. A brief summary of the benefits of environmental tracking, the drawbacks of traditional techniques, and possible AI-based solutions is given in this article. Highlighted are a number of noteworthy AI applications in tracking the environment that demonstrate how they support efficient environmental management. By making it possible to better comprehend, anticipate, and mitigate environmental dangers, AI technologies improve environmental monitoring. However, obstacles including a lack of specialized AI specialists in the environmental field and issues with data access, management, and privacy stand in the way of achieving AI's full potential.

14.3.1 Changes in climate using AI

Artificial intelligence has been programmed to assess iceberg changes 10,000 times quicker than a person could, and it is aware of where and how quickly icebergs are melting. Scientists will be better able to comprehend the amount of meltwater icebergs release into the ocean as a result of climate change warming the atmosphere. According to the European Space Agency, researchers at the University of Leeds in the United Kingdom claim that their AI can trace massive Antarctic icebergs over satellite photos in a tenth of a second [20]. This process takes a long time for people, and it is challenging to spot icebergs in the sea ice and cloud cover.

The influence of deforestation on environmental issues is also being mapped using AI, satellite imagery, and ecological knowledge. Headquartered in Edinburgh, Scotland, Space Intelligence claims to be active in over 30 nations and has used satellite data to map more than one million hectares of surface area from space. Metrics including rates of deforestation and the amount of carbon stored in forests are measured remotely by the company's technology [21].

A United Nations initiative in Africa is using AI to assist populations in Burundi, Chad, and Sudan that are at risk from climate change. In order to help communities and authorities better plan strategies to adapt to and minimize the effects of climate change, the IKI Project employs AI technology to help anticipate weather patterns. This entails expanding the availability of renewable energy, putting in place suitable waste management systems, and promoting reforestation [22].

By improving trash management, another AI system is assisting in the fight against climate change. According to the US Environmental Protection Agency, waste is a major source of methane and accounts for 16% of greenhouse gas emissions worldwide. An AI system created by London, UK-based software business Greyparrot examines trash processing and recycling facilities in order to assist them in recovering and reusing more garbage. According to the business, it identified an average of 86 tonnes of material that might be recovered but was disposed of in landfills in 2022, out of 32 billion trash pieces across 67 waste categories [23].

AI along with other technologies are being used by The Ocean Cleanup, an environmental organization in the Netherlands, to assist in removing plastic debris from the ocean. The group is using object detection AI to produce comprehensive maps of ocean debris in far-flung areas. After that, the ocean garbage may be collected and taken out, which is more effective than earlier cleanup techniques that used trawlers and airplanes. Plastic pollution damages the environment and releases greenhouse gases, which fuel climate change.

A startup named Sipremo in São Paulo, Brazil, is utilizing AI to forecast the types of climatic disasters that will occur as well as their locations and timing. The goal is to assist governments and companies in properly preparing for the effects of climate change and the escalating problems it presents to communities. The company's research of crisis circumstances and variables like air quality may help determine whether to postpone or cancel events in a variety of businesses, including insurance, energy, logistics, and sports.

The California-based company Eugenie.ai has created an emissions-tracking technology that integrates data from equipment and processes with satellite images. After that, AI examines this data to assist businesses in tracking, tracing, and cutting emissions by 20–30%. Approximately 30% of greenhouse gas emissions worldwide are produced by the industrial sector [24].

14.3.2 Air and water quality monitoring

Economic growth, environmental sustainability, as well as human health all depend on the quality of water. Drinking, farming, sanitation, and industrial operations all depend on clean water. Waterborne illness transmission, aquatic environment contamination, and food chain disruption are all consequences of poor water quality. Water that is contaminated, frequently by chemicals, pathogens, and especially heavy metals, can cause long-term environmental harm and pose major health hazards to people, animals, and plants. These health hazards include neurological diseases, reproductive issues, and gastrointestinal illnesses [25]. By supplying safe, dependable water for ecological and human requirements, maintaining high water quality promotes sustainable development, preserves biodiversity, and keeps ecosystems healthy. The preservation of life and the welfare of future generations ultimately depend on the quality of water.

Through predictive modeling, ML algorithms, including sophisticated data analysis, AI is revolutionizing air quality monitoring. Real-time insights, early notifications of pollution surges, and more effective regulatory actions are made possible by these technologies [26]. AI-powered systems use vast networks of IoT monitors that continually gather data in real time, in contrast to traditional monitoring systems that depend on fixed stations that deliver data at predetermined intervals. Consequently, this offers a more thorough and instantaneous comprehension of trends in air quality [27].

Predictive modeling is one of AI's most useful features. AI algorithms are able to predict pollution levels with a high degree of accuracy by examining both historical data and the current state of the environment. These forecasts assist researchers and decision-makers in anticipating patterns in air pollution before they materialize. AI also forecasts pollution hotspots, finds connections and trends that could otherwise go unnoticed, and helps communities take preventive action to lessen poor air quality. When early warnings are available, people with respiratory disorders can adjust their schedules to minimize exposure to dangerous pollutants [28].

In addition to monitoring and forecasting, AI helps organizations and governments put more efficient pollution management plans into action. Deep learning algorithms optimize traffic flow and lower car emissions by assisting authorities in making data-driven judgments. Better industrial zoning regulations based on air quality effect evaluations are also supported by AI-driven insights. AI is now being used in many smart city projects to dynamically modify environmental laws in reaction to changing pollution levels, therefore enhancing the quality of the air in cities [29].

14.3.3 Disaster prediction and management

We have benefited from AI in a number of areas, including health care, commerce, and customer service. Additionally, scientists have shown that AI may be used to forecast natural disasters [30]. Using enormous amounts of high-quality data, AI can predict the onset of a number of natural disasters, which might literally mean the distinction between existence and demise for hundreds of thousands of people. To help lessen the harm caused by natural catastrophes, local governments can employ automated responses, forecast and early detection methods, and more. With today's capabilities, residents are being alerted, mitigation is being initiated, and emergency professionals are being assisted in responding more effectively. Because AI develops, sensors multiply, and data collectors grow, prediction and detection systems are expected to become increasingly accurate and effective [31]. Figure 14.6 explains the disaster prediction and management model and its insights.

Researchers are using deep learning algorithms to analyze enormous amounts of seismic data. AI can evaluate earthquake patterns and magnitudes using seismic data. Such data could be helpful in forecasting when earthquakes would occur. For instance, Google and Harvard are developing an AI system that can predict earthquake aftershocks. Scientists studied more than 131,000 earthquakes including aftershocks to build a neural network. When tested on 30,000 events, the neural network fared better than conventional techniques in forecasting the locations of aftershocks.

Through Google Maps and Google Search, Google is creating an AI framework to predict floods in India and notify users. Data for the AI system's training is gathered using flood simulations and rainfall records. Similar to this, scientists are developing AI systems that can derive information from climate and rainfall data. These systems are being tested using flood

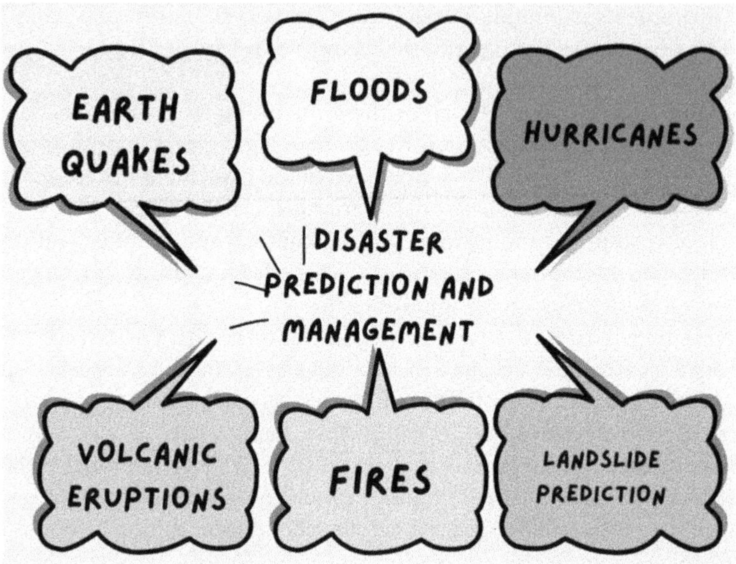

Figure 14.6 Disaster prediction and management.

scenarios to see whether or not they can predict floods more accurately than conventional techniques. Monitoring urban floods can be done with AI. Twitter and other mobile applications are being used by study participants at the University of Dundee within the United Kingdom in order to gather crowdsourced data on urban floods. The AI recognizes the data, which includes images and details about the place and circumstances in a locality.

14.3.4 Biodiversity and wildlife conservation

At a time when we confront more complicated ecological concerns and the pressing need to preserve our natural environment and distinct biocultural legacy, AI is becoming a potent ally for communities involved in conservation initiatives. Our capacity to monitor and protect ecosystems, reduce conflicts between humans and animals, maximize resource management, and promote sustainable cohabitation between individuals and animals might all be improved by the combination of AI and conservation.

Figure 14.7 explains the biodiversity and wildlife conservation in details in the upcoming content. By evaluating ecological data and suggesting suitable restoration methods, AI aids in ecosystem restoration initiatives. The efficiency and efficacy of conservation initiatives can be raised by using AI algorithms to optimize interventions like invasive species control and replanting [32]. In order to estimate the effects of climate change on ecosystems, AI algorithms can examine ecological characteristics, species distributions, and climatic data [33]. Finding climatic refugia, improving ecosystem resilience to shifting environmental circumstances, and creating adaptation strategies for climate change can all be facilitated by this information. AI improves conservation resource management, including intelligent project monitoring, predictive repair for conservation equipment, and the effective deployment of park rangers. AI-powered solutions increase the overall efficacy of conservation efforts by automating repetitive procedures and increasing the efficiency of data collecting [34].

AI has many applications in conservation, but its incorporation into citizen science and involvement in conservation is still relatively young. More experts with local expertise who can connect semantics, social behavior, and conservation patterns are desperately needed in

Figure 14.7 Biodiversity and wildlife conservation.

the domains of conservation culturomics along with computational sustainability. Significant breakthroughs in a number of pressing conservation areas, especially those driven by technology, are also anticipated. By taking into account a number of variables, including habitat fragmentation, ecological interconnectedness, and the efficacy of protected areas, AI will help to optimize conservation strategy [35]. Taking into account a variety of biological and socioeconomic factors, AI algorithms currently recommend the best sites for wildlife corridors, protected areas, and habitat restoration projects.

14.4 CROSS-CUTTING THEMES IN AGRICULTURE

Agriculture has a significant role in addressing a number of contemporary societal concerns, such as food security, animal and human infectious and noncommunicable illnesses, and climate change adaptation and mitigation. Many government agencies and international organizations now prioritize the transformation of agricultural and food systems. Researchers studying food and agriculture must successfully use the global convergence of interests in order to break through research "silos" and implement long-term, profound systemic change. In this article, we highlight the linkages between federal and international efforts related to food safety and security, equality and inclusion, animal wellness and general well-being health for humans and nutrition, as well as climate mitigation as well as adaptation.

Figure 14.8 explains the elements involved in the cross-cutting themes in agriculture. Many worldwide issues pertaining to the health of humans, animals, and the environment might all be impacted by agriculture at the same time. Researchers, legislators, nonprofits, farmers, and other interested parties have pushed for changes in the agriculture industry in recent years. "Deep and sustained, unpredictable change in systems, generally involving interpersonal, technological in nature, economic, social, as well as environmental processes" is what defines transformation.

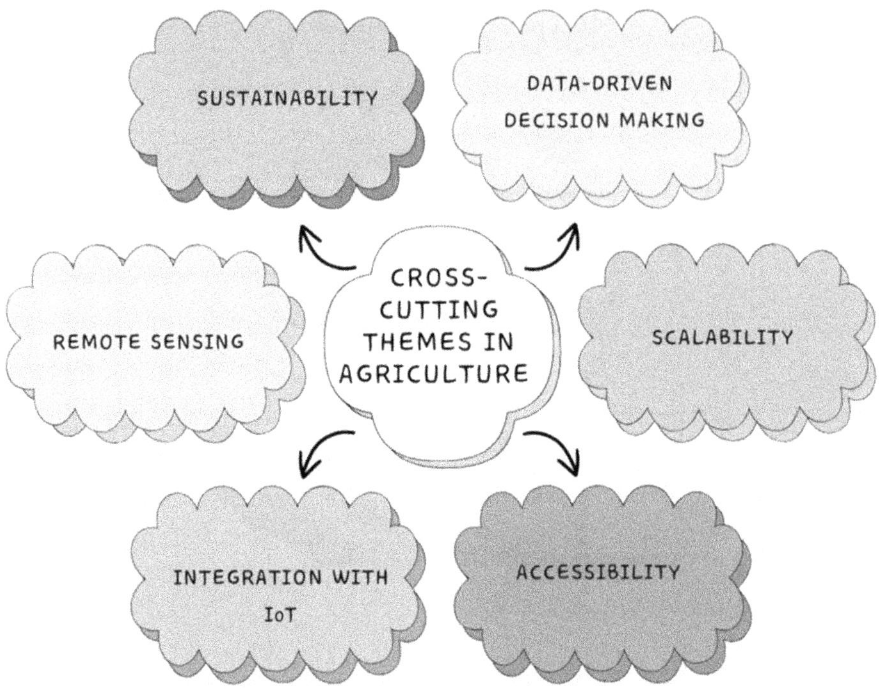

Figure 14.8 Elements in cross-cutting themes in agriculture.

14.4.1 Scalability and accessibility for agriculture and environmental monitoring

Monitoring of the farming environment is necessary for optimal crop development and resource efficiency. Big data, cloud computing, and AI have revolutionized traditional agriculture by promising higher-quality goods and more production [36]. The popularity of these technologies contributes to their extensive usage in safety monitoring. This technology gets around the problems with traditional systems, such maintenance, costs, and investment, by making it easier to gather and move data between devices. This paper analyzed the shortcomings of traditional distance vector hopping (DV hop) positioning algorithms and proposed the use of cloud computing-based AI optimization technology and big data networks to monitor the secure operation of the agricultural production environment [37].

A scalable and adaptable infrastructure for effective data processing, analysis, and storage is offered by cloud computing. Agricultural stakeholders may use a robust computational platform that can easily integrate different components and services needed for controlling and monitoring the secure functioning of the agricultural manufacturing ecosystem by utilizing cloud computing [38]. Real-time decision-making and optimization are made possible by this technology's centralized data storage, which makes it easier to integrate data from many sources, including weather, soil, and pest monitoring data. The environment safety surveillance system's capabilities are further improved by combining cloud computing alongside AI optimization approaches. Large datasets may be analyzed by AI algorithms, which can then spot trends, connections, and abnormalities that human observers would miss at first glance [39].

14.4.2 Integration with IoT and remote sensing

The picture "Integration with IoT and Remote Sensing" shows how the use of smart technology is changing contemporary farming methods. IoT gadgets and remote sensing instruments are at the core of this change, enabling data-driven farming decision-making and real-time monitoring [38]. Traditional farmers, who stand in for traditional agricultural practices, are seen on the left planting and caring for crops. The right side, on the other hand, emphasizes technical developments: a farmer-controlled drone is seen gathering data from the fields. Precision agriculture is made possible by this drone's use of remote sensing to examine crop health, soil moisture, and insect infestation from above [41].

Connected by signs and symbols like as sensors and mobile devices, the "IoT" moniker represents the way data is gathered from several sources, including satellite imaging, drones, and soil sensors, and sent to farmers or agricultural scientists. These data streams are utilized to improve yields, forecast illnesses, automate watering, and track the weather [40]. The gap between intelligent farming systems and manual labor is closed by this connection. It provides farmers, particularly those in rural regions, with easily available technology to boost output, save resources, and guarantee sustainable farming methods [41]. In general, the picture illustrates how current innovation and old knowledge may work together to create farming that is more intelligent and effective.

By improving crop monitoring and management, the combination of IoT and AI technologies is revolutionizing precision agriculture. Figure 14.9 explains the integration with IoT

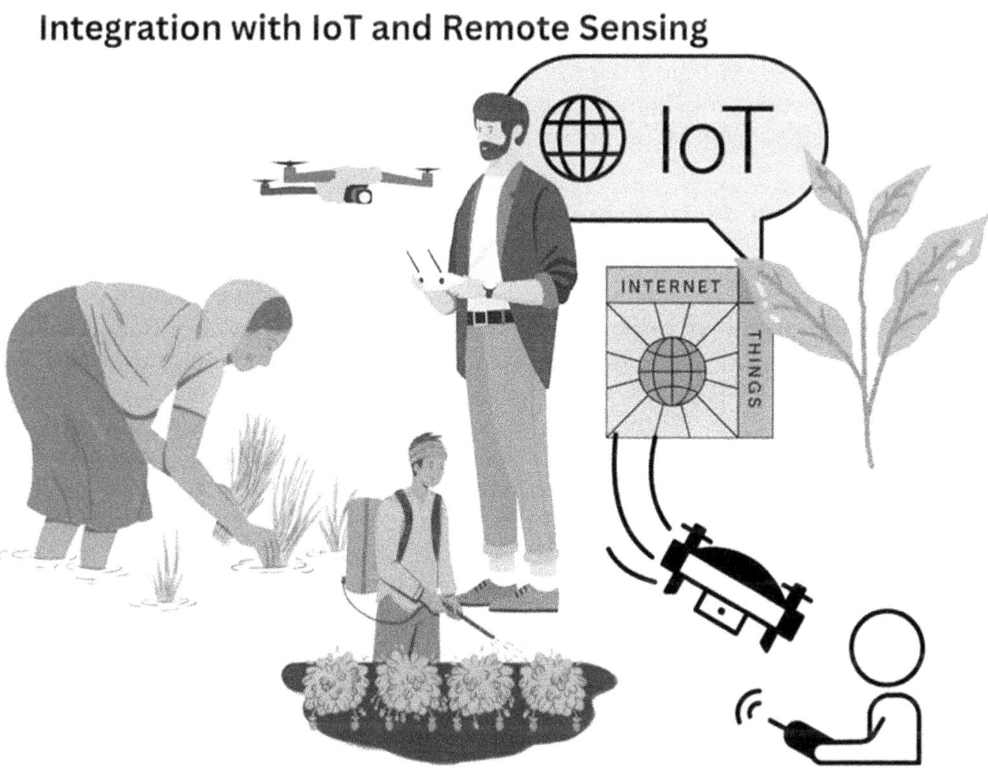

Figure 14.9 Integration with IoT and remote sensing in agriculture.

and remote sensing in agriculture. This study examines state-of-the-art techniques and developments in contemporary agriculture, such as autonomous agricultural robots (AgroBots), remote sensing, and high-throughput phenotyping. By automating processes like harvesting, organizing, as well as weed identification, these technologies drastically lower labor costs and their negative effects on the environment. High-throughput phenotyping uses robotics, spectrum imaging, and remote sensing to gather information on plant characteristics so that decisions about irrigation, fertilization, and pest control may be made with knowledge. For the purpose of assessing soil conditions and tracking crop health, DGPS as well as remote sensing provide accurate, real-time data.

Using information and communication technology (ICT) to control the temporal and geographical variability of fields is known as precision agriculture. Areas with comparable attributes, referred to as management regions, can be found by evaluating these differences in field along with crop parameters. Precision farming is typically credited to its founder, Professor Pierre C. Robert, who defined it as a knowledge revolution made possible by modern technology. He emphasized that precision agriculture creates a more accurate and advanced farm management system, going beyond just implementing new technology. Food production must increase significantly in the coming years to meet the needs of a world population that is expected to exceed 9.7 billion people by the middle of the century.

A variety of approaches, including ground-based sensor systems, have been developed to handle field variability. These systems include tractors and mobile platforms that have cameras, light detection and ranging (LiDAR) sensors—also referred to as laser sensors—and ultrasonic sensors (sonar). The twentieth century saw the widespread mechanization of agriculture, which increased land productivity and achieved economies of scale by substituting machines for human labor. Precision farming has great potential to improve agricultural products' intrinsic and extrinsic quality, boost farmer earnings, and lessen farming's negative environmental consequences. Precision farming may make a substantial contribution to more sustainable agriculture methods, even though it is not a panacea.

14.4.3 Data-driven decision-making

Global population growth and climate change demand a paradigm change toward more intelligent farming methods. There is a great deal of promise for increasing production, sustainability, and efficiency through smart agriculture, which combines ICTs alongside agricultural methods [39]. Sensing including monitoring, data collecting, communication technology, and decision support systems are some of its essential parts. Since it enables data-driven decision-making in a range of situations, data is crucial to smart agriculture. Knowledge on the weather, the earth, crops, animals, development phases, and yields is crucial for making educated decisions. However, analyzing agricultural data may be challenging due to its volumetric, diverse, and real-time nature. AI is a powerful tool for optimizing data in smart agriculture [38].

Thanks to developments in ML, deep learning, as well as computer vision, AI can now accurately anticipate harvests, monitor crops, and detect pests and diseases [36]. By combining AI with data-driven decision-making, farmers are able to optimize yields, reduce waste, and better allocate resources. For broad use, nevertheless, issues including data security, privacy issues, and talent shortages must be resolved. AI and smart technology enable precision agriculture, which optimizes inputs and reduces environmental impact by enabling focused field-level treatments. Automation and intelligent agricultural equipment also increase the effectiveness of farm management [42].

Platforms for data integration are essential for merging data from many sources, and strong data management techniques tackle issues like security, scalability, and interoperability. In the future, cutting-edge technologies like blockchain as well as the IoT have enormous potential to further transform smart agriculture. AI will pave the path for a safe, effective, as well as resource-conscious agricultural future if it is continuously innovated and ethical and sustainable issues are prioritized.

14.5 PROPOSED SYSTEM IN AGRICULTURE

In the realm of agriculture, drones have been receiving a lot of interest. They may be used to monitor pests, apply pesticides, and measure crop development, among other things. In agriculture, sensors are also frequently employed to track environmental factors like temperature and soil moisture. The restricted use of high-density detection systems in agriculture is caused by the high cost of radio-wave modules and communication infrastructure. We suggest an agricultural sensor network system that uses drones combined optical camera communication (OCC) to solve this problem. The plan is to use a drone-mounted camera to receive sensor data that is sent via LED panels placed on sensor nodes. This makes it possible to deploy high-density sensing in places with radio silence and inadequate infrastructure at a reasonable cost. To effectively gather the sensor data, we provide a trajectory control technique for the receiving drone. We verified through computer simulations that the suggested technique, in comparison to a shortest-path algorithm, cuts the overall flight duration by 30%. To train YOLOv7 for object identification, we used a drone-mounted camcorder to take 5178 pictures of LED panels.

The 3D bar chart presents the mean Average Precision (mAP) at Intersection over Union (IoU) thresholds 0.5 to 0.95 for the YOLOv5l model trained under different configurations. The mAP metric is a widely used performance measure in object detection tasks, reflecting both precision and recall across various IoU thresholds. Figure 14.10 and Table 14.2 explain the models and their feature values. Each of the four bars represents a specific experiment where the YOLOv5l model was trained for 300 epochs, but with variations in batch size and image resolution. Batch Size 16, Image Size 640. This configuration achieved the highest mAP

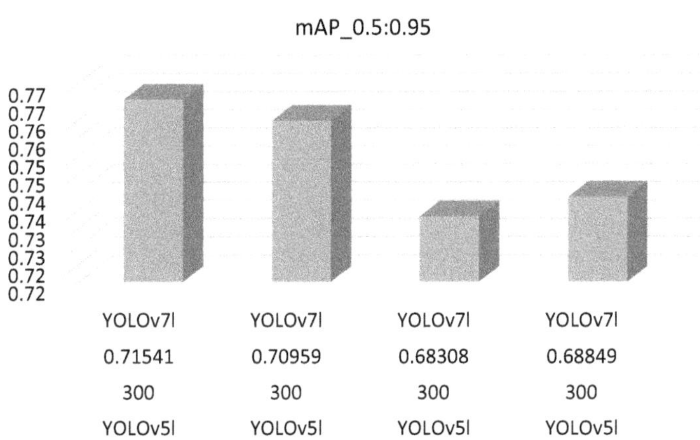

Figure 14.10 Models and their feature values.

Table 14.2 Models and their feature values

Batch	*Img-Size*	*Weight*	*Epochs*	*mAP_0.5:0.95*	*Proposed Model*	*mAP_0.5:0.95*
16	640	YOLOv5l	300	0.71541	YOLOv7l	0.77
16	320	YOLOv5l	300	0.70959	YOLOv7l	0.76
8	640	YOLOv5l	300	0.68308	YOLOv7l	0.73
8	320	YOLOv5l	300	0.68849	YOLOv7l	0.74

of 0.71541, indicating superior detection performance. The larger batch size allows better generalization, and the high-resolution input provides finer spatial details, improving model accuracy.

Batch Size 16, Image Size 320. This setup yielded an mAP of 0.70959, which is slightly lower than the previous one. While the batch size remains optimal, reducing the image resolution slightly decreases the model's ability to detect fine-grained features. Batch Size 8, Image Size 640, With a smaller batch size, the model scored an mAP of 0.68308. Although the image resolution is high, the smaller batch size might have reduced the effectiveness of batch normalization and generalization during training. Batch Size 8, Image Size 320. This configuration achieved an mAP of 0.68849, marginally better than the previous one. This suggests that image resolution has a smaller impact when batch size is already low, but neither setting is optimal in this case.

In conclusion, the model performs best with a larger batch size and higher resolution input, validating the importance of both factors in optimizing object detection using YOLOv5l.

14.6 CONCLUSION

This piece of literature examined the revolutionary potential of AI in agriculture, emphasizing how it may improve agricultural yields, maximize resource use, and tackle pressing issues like environmental sustainability and global food security. We illustrated how AI may enhance crop results, boost operational effectiveness, and lessen environmental impact through a thorough examination of AI applications, ranging from robots and ML to agricultural precision. Participants also talked about the technological, sociological, and regulatory obstacles that need to be removed before AI can be fully implemented in agriculture. To tackle these issues, we underlined the necessity of cooperation between legislators, IT developers, stakeholders in agriculture, and academic institutes. This kind of cooperation is necessary to manage privacy and ethical concerns while fostering an atmosphere that supports AI. The potential application of AI in agriculture was also discussed in our article, with a focus on the significance of ongoing innovation, education, and technological adaptability. By creating a farming environment that is more efficient and sustainable, AI has the potential to completely transform the agriculture sector. In order to fully achieve this potential, stakeholders must take advantage of the chance to promote an equitable and sustainable farming environment, guaranteeing that agriculture can meet future challenges while promoting environmental health and global food security.

REFERENCES

[1] A. A. Ayele, A. Y. Asmare and A. B. Alehegn, "Smart Agriculture and Sustainable Farming Using IoT and Machine Learning Technologies," *IEEE Access*, vol. 12, pp. 29023–29035, 2024.
[2] H. A. Rahman, M. A. Mahmud and M. H. Kabir, "A Review of AI-Based Decision Support Systems for Precision Agriculture," *IEEE Internet of Things Journal*, vol. 11, no. 5, pp. 4012–4025, Mar. 2024, doi: 10.1109/JIOT.2023.3339887
[3] M. T. Rahman et al., "IoT-Enabled Smart Agriculture: Toward a Data-Driven Decision Support Framework," *IEEE Sensors Journal*, vol. 24, no. 3, pp. 1221–1233, Feb. 2024.
[4] P. Zhang and J. Wu, "Data Analytics in Smart Farming: Enhancing Crop Management through IoT and AI," *IEEE Transactions on Industrial Informatics*, vol. 20, no. 1, pp. 310–319, Jan. 2024.
[5] A. Kumar, S. Shukla, and M. Patel, "AI-Based Weather Forecasting System for Smart Farming Applications," *IEEE Transactions on Artificial Intelligence*, vol. 4, no. 2, pp. 186–197, Apr. 2024.
[6] L. Fan, Y. Chen, and H. Tang, "A Deep Learning Framework for Crop Health Monitoring and Climate Adaptation," *IEEE Transactions on Geoscience and Remote Sensing*, vol. 62, pp. 1–12, 2024, doi: 10.1109/TGRS.2024.3369021
[7] R. U. Rehman, M. A. Khan and Z. A. Khan, "IoT-Driven Environmental Monitoring System Using Arduino and ThingSpeak," *IEEE Internet of Things Magazine*, vol. 7, no. 1, pp. 34–39.
[8] S. N. Patel and A. D. Joshi, "Design and Implementation of IoT-Based Smart Farm Monitoring System," *IEEE Access*, vol. 12, pp. 14720–14730, 2024.
[9] K. P. Sriram, P. Kola Sujatha, S. Athinarayanan, G. Kanimozhi and M. R. Joel, "Transforming Agriculture: A Synergistic Approach Integrating Topology with Artificial Intelligence and Machine Learning for Sustainable and Data-Driven Practice," *2024 2nd International Conference on Sustainable Computing and Smart Systems (ICSCSS)*, Coimbatore, India, 2024, pp. 1350–1354, doi: 10.1109/ICSCSS60660.2024.10625446
[10] A. M. Sharif et al., "Challenges and Future Directions in Sustainable Agriculture: Technological and Socioeconomic Perspectives," *IEEE Access*, vol. 11, pp. 117540–117553, 2023.
[11] S. Athinarayanan, R. Selvi, K. Rajakumari, R. Arunadevi, and M. Robinson Joel, "An Advanced Strategy for Efficient Agricultural Production using Machine Learning," *2023 2nd International Conference on Automation, Computing and Renewable Systems (ICACRS)*, Pudukkottai, India, 2023, pp. 680–686, doi: 10.1109/ICACRS58579.2023.10405182
[12] A. Singh, R. Roy, and P. Ghosh, "Early Detection of Crop Diseases Using IoT and ML for Sustainable Farming," *IEEE Access*, vol. 11, pp. 104221–104231, 2023.
[13] R. J. Almeida, J. C. D. Santos, and R. A. R. Oliveira, "Smart Irrigation System Using Deep Learning and IoT for Precision Agriculture," *IEEE Internet of Things Journal*, vol. 11, no. 2, pp. 1443–1455, Jan. 2024.
[14] S. J. Watson, L. F. Ramos, and M. A. Molina, "A Systematic Review on the Use of AI and IoT in Smart Irrigation Systems: Trends and Gaps," *IEEE Access*, vol. 11, pp. 94567–94582, 2023, doi: 10.1109/ACCESS.2023.3299850
[15] N. S. Reddy, P. S. Kumar, and S. Chatterjee, "Crop Yield Prediction Using Machine Learning: A Comparative Study," *IEEE Access*, vol. 11, pp. 121345–121357, 2023.
[16] A. Verma, R. K. Sharma, and D. Singh, "Machine Learning Techniques for Agricultural Yield Prediction: Recent Trends and Future Directions," *IEEE Internet of Things Journal*, vol. 11, no. 1, pp. 987–997, Jan. 2024.
[17] A. Mehmood, H. M. Asif, and S. S. Gill, "Deep Learning-Enabled UAV and Satellite Imagery Analysis for Smart Pest Detection in Agriculture," *IEEE Transactions on Industrial Informatics*, vol. 20, no. 2, pp. 1892–1902, Feb. 2024.
[18] R. Jain, P. V. Krishna, and M. S. Reddy, "Cost-Effective Crop Monitoring Using Freely Available Satellite Data for Weed Detection," *IEEE Access*, vol. 11, pp. 117420–117432, 2023.
[19] S. Kumar, A. Sharma, and R. K. Saini, "CNN-Based Image Segmentation for Weed and Pest Detection Using U-Net and Mask R-CNN Models," *IEEE Access*, vol. 11, pp. 101982–101995, 2023, doi: 10.1109/ACCESS.2023.3302173

[20] European Space Agency, "AI maps Antarctic ICEBERGS 10,000 Times Faster Than Humans," *ESA News*, 2024. https://www.esa.int
[21] Space Intelligence, "Using AI to Tackle Deforestation and Climate Change," *Company Report*, 2024. https://www.space-intelligence.com
[22] United Nations, "AI supports vulnerable African Communities in Adapting to Climate Change," *UN Climate Initiative Report*, 2024. https://www.un.org
[23] Greyparrot, "AI-Powered Waste Recognition for Smart Recycling," *Greyparrot Annual Impact Report*, 2023. https://www.greyparrot.ai
[24] A. Jain, M. Padmanaban, J. Hazra, R. Guruprasad, H. Syam, "AI enabled Decarbonization Framework," In *ACM SIGKDD International Conference on Knowledge Discovery and Data Mining*, 2022 Aug. 14.
[25] World Health Organization, *Water Quality and Health Review of Turbidity: Information for Regulators and Wa ter Suppliers*, WHO, Geneva, 2017. https://www.who.int
[26] B. Zhang et al., "Air Pollution Prediction Using Machine Learning Algorithms: A Case Study in China," *IEEE Access*, vol. 7, pp. 150781–150796, 2019. https://ieeexplore.ieee.org
[27] M. A. Hossain et al., "A Smart City IoT-Based Air Quality Monitoring System Using Machine Learning," *IEEE Sensors Journal*, vol. 21, no. 5, pp. 5562–5573, Mar. 2021. https://ieeexplore.ieee.org
[28] S. Devarakonda, P. Sevusu, H. Liu, R. Liu, and A. B. Srivastava, "Real-Time Air Quality Monitoring through Mobile Sensing in Metropolitan Areas," *IEEE Internet of Things Journal*, vol. 2, no. 5, pp. 414–423, Oct. 2015. https://ieeexplore.ieee.org
[29] S. Kumar, R. Shrestha, D. Niyato, and Y. L. Guan, "Smart Urban Air Pollution Monitoring Using UAVs and Edge Computing," *IEEE Internet of Things Journal*, vol. 6, no. 3, pp. 4941–4951, June 2019.
[30] A. Hossain, M. A. Rahman, and Y. B. Mahdy, "A Comprehensive Survey on Artificial Intelligence for Disaster Management," *IEEE Access*, vol. 9, pp. 123578–123611, 2021. https://ieeexplore.ieee.org
[31] S. V. Ukkusuri and L. Lu, "Emerging AI Technologies in Emergency Management: An Overview," *IEEE Intelligent Systems*, vol. 36, no. 4, pp. 72–79, July–Aug. 2021. https://ieeexplore.ieee.org
[32] A. A. Besson, M. N. Peterson, and S. A. Moore, "Artificial Intelligence and Conservation Planning: Optimizing Replanting and Invasive Species Control," *IEEE Transactions on Computational Social Systems*, vol. 8, no. 6, pp. 1443–1453, Dec. 2021. https://ieeexplore.ieee.org
[33] M. D. Potts et al., "Climate Change Modeling for Biodiversity Conservation Using AI," *IEEE Access*, vol. 10, pp. 98534–98547, 2022. https://ieeexplore.ieee.org
[34] R. Kumar, L. Singh, and K. Agarwal, "AI-Powered Monitoring and Predictive Maintenance in Ecological Conservation," *IEEE Internet of Things Journal*, vol. 9, no. 3, pp. 2345–2354, Feb. 2022. https://ieeexplore.ieee.org
[35] J. C. Dale, P. T. Johnson, and H. Li, "Optimizing Conservation Strategies Using Machine Learning: Socio-Ecological Data Integration," *in IEEE Transactions on Emerging Topics in Computing*, vol. 11, no. 2, pp. 288–298, April–June 2023. https://ieeexplore.ieee.org
[36] A. Kamilaris, A. Kartakoullis, and F. X. Prenafeta-Boldú, "A Review on the Practice of Big Data Analysis in Agriculture," *Computers and Electronics in Agriculture*, vol. 143, pp. 23–37, Dec. 2017. https://ieeexplore.ieee.org
[37] X. Liu, Y. Wang, and J. Zhang, "An Improved DV-Hop Localization Algorithm Based on Cloud Computing for Smart Agriculture," *IEEE Access*, vol. 8, pp. 76543–76552, 2020. https://ieeexplore.ieee.org
[38] T. Wolfert, L. Ge, C. Verdouw, and M. J. Bogaardt, "Big Data in Smart Farming—A Review," *Agricultural Systems*, vol. 153, pp. 69–80, May 2017. https://ieeexplore.ieee.org
[39] S. Li, L. Da Xu, and S. Zhao, "The Internet of Things: A Survey," *Information Systems Frontiers*, vol. 17, no. 2, pp. 243–259, Apr. 2015. https://ieeexplore.ieee.org
[40] A. Kamilaris and F. X. Prenafeta-Boldú, "Deep Learning in Agriculture: A Survey," *Computers and Electronics in Agriculture*, vol. 147, pp. 70–90, Apr. 2018. https://ieeexplore.ieee.org

[41] M. K. Jat, A. Garg, M. Sharma, and S. S. Sidhu, "Smart Farming Using IoT and Drone-Based Remote Sensing in Agriculture," *IEEE Internet of Things Journal*, vol. 8, no. 10, pp. 8223–8230, May 2021. https://ieeexplore.ieee.org
[42] M. Tzounis, N. Katsoulas, T. Bartzanas, and C. Kittas, "Internet of Things in Agriculture, Recent Advances and Future Challenges," *Biosystems Engineering*, vol. 164, pp. 31–48, Aug. 2017. https://ieeexplore.ieee.org

Chapter 15

A comprehensive review on AI technologies for detecting defects from fruits and vegetables

Anjali Nambiyar and Rajkumar Kannan

15.1 INTRODUCTION

India is a developing country that cultivates a broad range of farming products, such as fruits and vegetables. The objective of object detection is to accurately identify and localize multiple objects of interest in an image, providing both their class labels and spatial information in the form of bounding boxes. These products are available at various places, such as markets, local stores, and from handcart farmers. While storing these products, they face a lot of internal and external damage. These damages are identified using computer vision, which is also known as defect detection. The identification of damage to fruits and vegetables using computer vision technology is especially important.

External defect detection is defined as identifying defects on the outer layer of the fruits and vegetables being processed. A qualitative review of image classification was done using recent deep learning and machine learning algorithms. The most important aspect of quality measurement, quality analysis, quality control, and determining the product's lifespan is identifying defects in manufactured products. Examining these defect detection methods supports the manufacturer in removing damaged goods from the final product, hence increasing efficiency. Advanced technologies are used in existing approaches for defect detection systems in products to ensure quality standards and reliability are maintained. Previous research has reviewed techniques and methods for disease detection, object detection, quality and quantity analysis, and defect detection in products over the past few years. This research review focuses on defect detection and types of defects in fruits and vegetables using recent techniques.

Defects include identifying cracks, dents, scratches, and other imperfections in any product. It is a tedious process. This review looks at one such issue: defects in fruits and vegetables. Fruits and vegetables are used in various places, such as markets and local stores. This study provides an image processing-based system for detecting external defects in fruits and vegetables, namely color, size, and shape. The summarization of the article and the number of algorithms are discussed for quality assessment. This review focuses on the similarity of different deep learning and machine learning approaches, recent preprocessing, feature extraction, and classification with regard to evaluating real-time data on fruits and vegetables.

This article defines a trend review related to disease detection and flaw detection as examined in the most recent state-of-the-art procedures. In this article, it is important to bring out new methods to improve the inspection of fruits and vegetables compared with machine learning and deep learning techniques [1]. An insufficiently complex technique for extensive contrast enhancement was used. By improving the bi-linear evaluation method, the mean and variance of the contrast enhancement using a Gaussian curve gain redistribution function are decreased completely, and the procedure is fully automatic [2].

DOI: 10.1201/9781003685890-15

Efforts have been made to develop analytical methods for the early detection of bacterial plant diseases affecting vegetable production systems, particularly in India, such as Ralstonia solanacearum [3]. Multiple omics processes have explored a deep overview of physiological and chemical pathways during the postharvest stages of fruits and vegetables. The multiple omics approach aims to improve postharvest reviews and enhance the image quality and shelf life of fruits and vegetables [4]. This review compared the different phases between methods for many applications of computer vision based on 86 articles using recent techniques.

The objective of the article is based on recent artificial techniques. The image processing techniques consist of various steps. The reviewed methods can be depicted in the form of a generic architecture, as shown in Figure 15.1. Reviews of the major methodology processes are defined based on these architecture techniques. Input image: Input images are collected through mobile phones, and the raw image data is processed and classified.

- **Dataset:** This research reviews the results of the artificial intelligence (AI) techniques for preprocessing, segmentation, and canny edge detection defined in this article using real-time datasets such as muskmelon, chickoo, chayote squash, and banana. The applications are quantity, recognition, object detection, defect detection, disease detection, quality measurement, quality analysis, quality control, and so on.
- **Data Preprocessing:** This step involves various preprocessing operations, represented by a parallelogram: Resize: a process to adjust the image size if needed. Normalization, Noise Reduction: removing noise or artifacts from the image, and Contrast Enhancement: improving the image's contrast. Color Conversion: converting the image to grayscale or other color spaces if necessary.
- **Feature Extraction:** Shown as a rectangle, this step involves extracting the relevant features from the preprocessed image. This step might include: Histograms: capturing

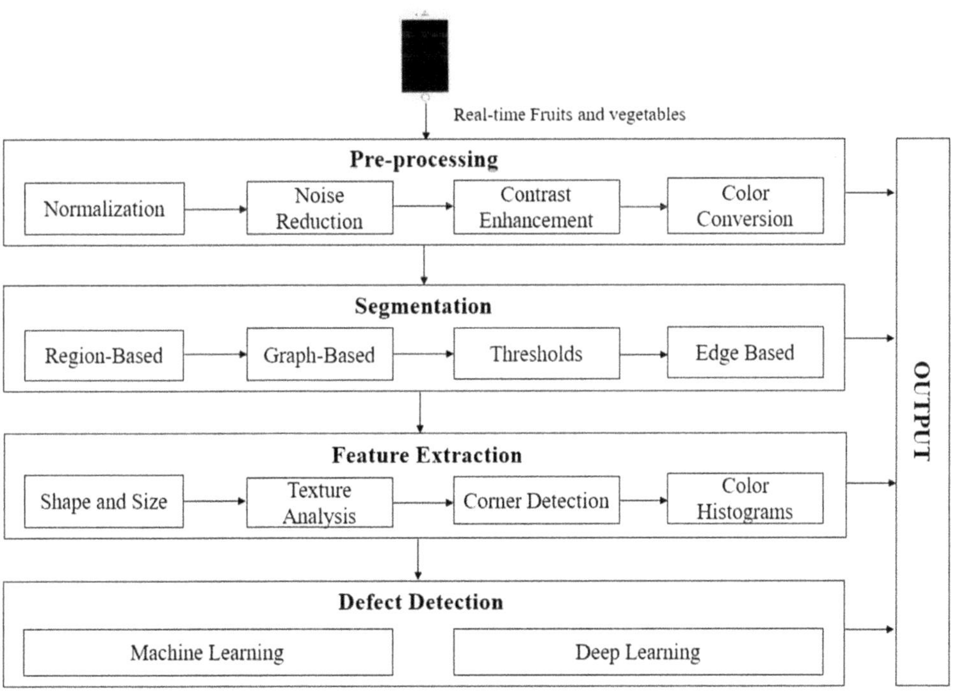

Figure 15.1 Review methodology.

pixel intensity distribution. Texture Analysis: identifying patterns and textures. Edge Detection: detecting edges and boundaries. Deep Learning Features, using pretrained neural networks for feature extraction.

- **Classification:** Represented by types of defect detection, internal and external defects. This step might involve various classification machine learning and deep learning algorithms, such as support vector machine (SVM), Random Forest, K-Nearest Neighbor (KNN), convolutional neural network (CNN), and transfer learning. Output (Result): displaying the classification result, including the predicted class label or category.

The rest of the document is categorized as follows: Based on our corpus of articles, 2. Categorization of defect types in fruits and vegetables, 3. Deals with the feature extraction methods and techniques, 4. Defines AI techniques for defect detection and identification bringing the recent methods and techniques. 5. Discussion and Prospect. Finally, Conclusion and Future Scope.

15.2 CATEGORIZATION OF DEFECT TYPES IN FRUITS AND VEGETABLES

The classification of defects from fruits and vegetables is a crucial step in the data analysis pipeline, and it is important for the quality and reliability of research outcomes. In the context of a summary of preprocessing techniques, it's important to cover various aspects such as data cleaning, transformation, and noise reduction. A real-time database of over 200 fruit and vegetable photos of various sizes and shapes has been compiled. There are two categorizations of defect detection: one is external defect and internal fruit defects. The defects in different types of healthy and damaged samples collected from our local fruits and vegetables are shown in Figure 15.2. These are a good example for preprocessing.

Figure 15.2 Few samples for fruits and vegetables with different sizes and shapes. (1) Healthy i.e. (a) Muskmelon (b) Chickoo (c) Chayote squash, and (2) Damaged i.e. 2(a) Muskmelon 2(b) Chickoo 2(c) Chayote squash.

15.2.1 External defect detection

The review typically refers to a research article or publication providing a comprehensive study and analysis of various external defect detection techniques and performance evaluation. This paper aims to summarize multiple approaches and critically assess the existing work on external defects contrast enhancement in image processing. One technique for improving disparity in pictures with inferior contrast detection is caused by varying radiance. The Single Scale Retinex method, determined by the best contrast parameters, can yield optimal results [5].

Weak contrasts and discoloration are generated to brighten the images. Image deconvolution techniques include SIDD, DND, and RENOIR. This article examines algorithms such as Peak Signal Noise Ratio (PSNR) and Structural Similarity Index Measure (SSIM) compared to current vision techniques [6]. Another approach is an underwater contrast enhancement system using a medium transmission-guided multicolor space encoder comparable to Ucolor. It improves the visual quality of underwater photos and achieves superior performance when using bi-interval contrast and color correction [7].

A modern adaptive heuristic equalization (HE) method is fused with gamma transformation advancements. The sub-histograms' probability distribution functions (PDFs) are computed, preserving the original characteristics of the input photos while improving image contrast [8]. A sparse gradient reduction sub-network is established to extract low-amplitude networks from the images. The edge detection method brings out paired radiance maps of short- or normal-context images [9]. This work describes a simple and successful picture contrast enhancement technique to attain a high dynamic range. Pixel estimation, pre-gamma correction, impulsive Contrast Limited Adaptive Histogram Equalization, and post-gamma correction are applied to improve qualitative and quantitative evaluations [10].

The fusion of Discrete Wavelet Transform (DWT) and corner information strengthens virtual images by estimating Gaussian noise on a standard LIVE image database. These findings show that the algorithm performs far better than traditional methods across a wide range of noise levels [11]. The morphological traits are hard to see and interpret due to the low-contrast imagery. Consequently, several contrast-improving techniques, including HE, AHE, CLAHE, and GE, are investigated to improve the contrast of malaria images [12].

A versatile framework for algorithmic parametric contrast enhancement optimization includes a deep neural feature-based regression module for picture quality assessment that serves as the objective of a Bayesian optimization process to determine the ideal parameters in contrast enhancement approaches [13]. To enhance the contrast of dark areas in low-contrast images, this article presents a fresh method for increasing contrast. It can improve both areas of the image. The results of the three metrics used are Feature Similarity Index (FSIM), Structural Similarity Index (SSIM), and Absolute Mean Brightness Error (AMBE), compared to PSNR, MSE, NIQE, and BRISQUE [14].

For the purpose of enhancing underwater images, this article suggests a method for adaptive contrast augmentation and color correction. The ratio of the variations between the highest and lowest color channels is used to determine these values. The GA algorithm produces higher-quality photos than any other method [15]. The adaptive beamforming by deep learning (ABLE) technique is improved for image contrast enhancement. It compares the image quality of contrast-enhanced ultrasound images obtained with a sparse spiral ensemble to delay-and-sum and spatial coherent techniques [16]. The contrast enhancement using the nine-image fusion method and eight blind image assessments creates a contrast map from the luminance channel intensity. A grayscale map of the Degree of Linear Polarization (DoLP) is used for Color Polarization high-frequency image enhancement [17].

This proposed article suggests an artificial neural network (ANN)-based low-contrast picture enhancement algorithm. The performance parameters PSNR, MSE, Entropy, QI, QRCM,

CQE, SSIM, and computational time are compared between the new ANN-AHE method and the existing techniques [18]. In order to achieve two enhancements, the image in this work is divided into two frequency bands, with the low-frequency component undergoing contrast enhancement and the high-frequency component undergoing edge enhancement using Gaussian filters and simultaneous histogram equalization techniques [19]. Here are a few sample images of fruits and vegetables showing defects due to a small dataset. After collecting a large dataset, data augmentation was used. The segmentation process uses various image segmentation methods. These methods detect the image parts where the fruits and vegetables are damaged or unhealthy, as shown in gray. Traditional methods are used to segment these images specifically to depict the defective and damaged parts.

This approach to infrared image contrast enhancement is superior to previous methods in terms of maintaining overall image brightness, enhancing contrast, suppressing artifacts, and accurately representing detail in dark areas [20, 21]. There are significant disparities in the quality control of tomato suppliers with respect to health-related events. The compilation of high-quality fruit and vegetable images is one of the work's most important contributions [22]. The conveyor, which operated at a rate of five fruits per second, required the samples to be manually placed. Each camera captured three different sections of the fruit surface at three random positions while the fruit rotated during movement [23].

An automated technique for locating welding defects is suggested, utilizing the enhanced U-net network from digital X-rays with data augmentation. It is recommended to use the upgraded U-net network to identify welding defects with high precision automatically [24]. Following the proposal of contrast enhancement techniques to improve object contrast with the background, a global-to-local thresholding strategy based on entropy maximization and Otsu's techniques is put forth for SAR object segmentation [25]. A unique algorithm for segmenting brain tumors with missing modalities is presented in this research. When a modality is absent, the segmentation becomes more robust based on the correlation representation that is produced [26]. From the standpoint of four chamber views, the conventional semantic segmentation process of the fetal heart is both defective and normal. The validation evaluation outcomes of segmentation predictions made with the U-NET model's feature extraction performance in segmentation instances [27].

The suggested article technique has several drawbacks, including increased computing time, incorrect segmentation, and complexity. Three levels are utilized to overcome these problems: multidimensional scaling in an image, quick mapping, and principal component analysis. These techniques offer substantial and rapid benefits. Additionally, the visual quality has been enhanced [28]. The data is segmented using an improved fuzzy-based k-means clustering technique, detecting diseased parts with high efficiency and precision [29]. External fruit and vegetable faults, as well as different image processing techniques, are used in this study for quality inspection. The techniques include FCM and FCM-NPGA algorithms to detect segmented apple defect images [30]. Figure 15.3 shows the steps for contrast enhancement of a sample muskmelon fruit image, with the third step illustrating contrast enhancement using a sample fruit image collected in real-time from mobile phones.

15.2.2 Internal defect detection

Defect detection is crucial for identifying damages to objects, including major defects in the agricultural industry affecting fruits and vegetables. Recent trend reviews show that defects are a significant factor in skin damage classification. The main goal of this article is to present an image-based categorization algorithm for detecting defective portions of fruits and vegetables.

Figure 15.3 Samples of mobile camera images from real-life scenes for the preprocessing of fruit. 1. A sample of muskmelon i.e. (a) Test input image, (b) Gray image, (c) Contrast enhanced, (d) Conversion to HIS, and 2. A sample of muskmelon i.e. (e) Hue images, (f) Saturation images, (g) Intensity images.

Postharvest pear fruit internal browning defects were assessed nondestructively using a capacitive instrumental-chain in the super low–low radio-frequency region combined with image analysis. The FRED pear fruit was classified into two classes, "unhealthy" and "healthy," using unsupervised fuzzy C-means clustering analysis [31]. Blackheart is one of the main diseases widely affecting pomegranate fruit, caused by the fungus Alternaria. The presence of the infection in three levels of severity was analyzed by opening the samples in half. The highest accuracy models were obtained using all texture features and histograms with bins. Compared to manual inspection, X-rays showed better detection of incipient infected fruits [32].

Identification and classification of healthy fruit and fruit with Mediterranean fly infection were conducted. This work obtained accuracy values of 98.33%, 99.33%, 98.36%, and 99.34%, correspondingly. The hyperspectral detection technique is used for detecting internal defects from fruits, and compared to an RGB camera, a hyperspectral camera is more costly but has been used to detect pineapple damages [33]. The ResNet50 CNN architecture is used in a transfer learning-based model to classify fresh and rotting fruits in a binary manner. A dropout layer and a Gauss filter improve performance, and it can surpass earlier attempts and reach high accuracies of over 99% on test data that hasn't been seen yet. There are theoretical and practical ramifications to this study [34].

The proposed technique involves apple fruit, constructing the neural network model, training it, and testing it to predict the freshness of apples. The suggested CNN transfer learning-based technique's performance experiment results show that the proposed approach, with 98% accuracy, outperforms other deep learning models [35]. MATLAB was used to develop the damage detection system, and the experimental results showed that, even with poor image resolution and negligible skin color changes, the accuracy was 94.33% when compared to the transfer learning approach. With this method, detection speed is increased by more than 30 times compared to typical manual detection [36].

The fruits of 17 cultivars from a single dataset were used, and prediction models were developed based on a multilayer perceptron ANN combined with the Levenberg–Marquardt

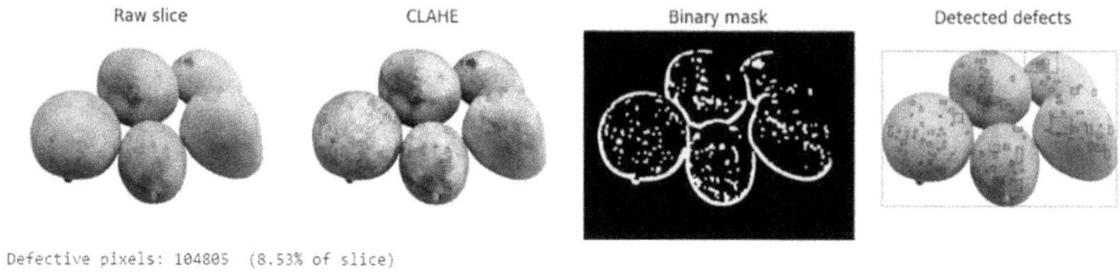

Figure 15.4 X-ray-based diagnosis of internal fruit defects using deep learning.

learning algorithm. Regarding the Vis/NIR spectrophotometer dataset, good predictive performances with high accuracy were achieved [37]. A method is proposed for classifying mango spectroscopy data into two classes like internal defects and healthy mangoes. Advanced feature selection techniques such as recursive feature elimination (RFE) and sequential feature selection (SFS) methods were used for selecting optimal features and classification. The empirical results show that the proposed technique has superior accuracy over the existing contemporary state-of-the-art model [38].

The creation of digital horticulture systems aims to provide customers with high-quality fruit. With X-ray Computed Tomography (CT) producing both 2D and 3D pictures, X-ray imaging has lately shown itself to be a reliable method for assessing the internal quality of fruits and vegetables. The 3D and 2D BraeNet achieved 96 ± 1% accuracy for both CT and radiography data after being trained on both healthy and damaged apple fruit from storage under N_2 or high CO_2 atmospheres [39]. Figure 15.4 indicates internal defects differently using X-rays than damaged or decayed areas. Defective regions appear darker or lighter on grayscale X-ray images depending on their composition. This makes X-ray an ideal modality for internal defect detection.

15.3 FEATURE EXTRACTION METHODS AND TECHNIQUES

In feature extraction, images are evaluated after segmentation for subsequent analysis. This method focuses on extracting image attributes based on color, texture, form, and size descriptors. Scale invariant feature transform (SIFT), SURF, ORB, BRISK, KAZE, AKAZE, edge detection, texture analysis, and color-based methods and techniques are reviewed. These techniques are used to classify the proposed methods better than the others. In order to detect scratches, this study presents a design that uses smart edge detection to lower image noise. CNNs are used to categorize the vehicle's surface area into scratch and non-scratch zones. The proposed several feature extraction methods, techniques, and edge or corner detection points are discussed in this concept. Corner detection refers to recognizing discontinuities in picture features such as color, shape, or object edges [40].

The Adaptive Information Entropy Feature (AIEF) method is offered as a way to extract feature points. In order to prevent high-density extracted feature points from becoming stuck, the image blocks are homogenized. Finally, validation of the AIEF technique is performed to differentiate it from recent algorithms, SIFT, SURF, and ORB-SLAM2 [41]. A saliency model called Edge-aware Multi-Level Interactive Network is used to locate flaws on the surface of the strip steel. This concept specifically uses the U-shape diagram. It produces a superior saliency map capable of precisely identifying and classifying the flaws [42].

The eye tracking simulation is an intelligent edge detection method used in the electric wheelchair (EWC) that distinguishes it from other current systems and yields superior outcomes compared to other edge detection techniques [43]. Identifying and categorizing hand movements is one of the various methods explored. Various methods of preprocessing are employed, such as principal component analysis, local binary patterns, and a gradient histogram. Following preprocessing, a novel model for ORB and canny edge detection is presented in this study. It has been found that the models' accuracy exceeds that of the current model [44]. This article aims to evaluate the performance of four well-known state-of-the-art algorithms for feature extraction (SIFT, SURF, BRISK, and ORB) in terms of their objectives. In the context of multi-view object tracking in VSNs, bitrate savings and multi-view matching accuracy are discussed [45].

The histogram of orientated gradient (HOG) function is used to detect the eye corner regions, while the KAZE technique extracts features from the upper eye fold region. Additionally, the current method suggests extracting the neighboring gradient features using a sub-image-based neighbor gradient (SING) feature extraction method [46]. A novel feature extraction technique is proposed from the perspective of electrochemistry, applying recent transformer-based neural network techniques for State of Health (SoH) assessment. This method is compared to the original data before feature extraction [47].

The article describes a fully automated technique for arterial segmentation, vessel labeling, and asymmetry feature extraction applied to the supra-aortic area. Thirty-three geometrical and morphological characteristics of the supra-aortic region's vascular anatomy were measured [48]. This paper discusses various feature extraction techniques and the classification of point and line edge detection methods. The merits and demerits of the chosen algorithms are discussed. SIFT, SURF, ORB, BRISK, and KAZE are a few prominent examples of scale and rotation invariant local features that were initially developed [49].

An extensive analysis of the transformation of common SIFT, SURF, ORB, BRISK, KAZE, and AKAZE algorithms is presented. In perspective transformation, the BRISK and AKAZE algorithms outperform the SURF method. The BRISK and SURF algorithms perform better against jitter, according to the second set of experimental data [50]. These days, canny edge detection algorithms have seen several advancements. In this study, median filtering is used in place of Gaussian filtering to improve the noise robustness of the algorithm. The suggested approach produces outcomes that are more noise-resistant and capable of preserving more significant edge information [51]. After image segmentation, defected edges are detected. If canny edge detection does not perfectly detect the defective edges, an improved canny edge detector is used to detect the edges properly. A sample edge detection result is shown in Figure 15.5.

A global context-aware module uses contextual data to identify pedestrians. Feature representations are improved by merging global context with data from the viewable portion of obscured pedestrians. The implemented results challenge the effectiveness of existing

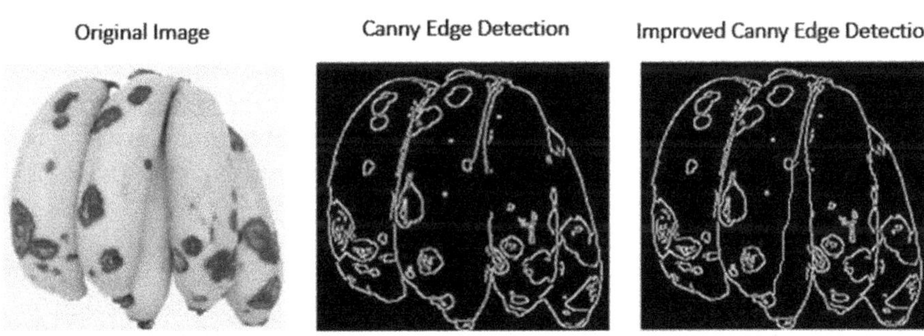

Figure 15.5 Improved canny edge detection.

techniques [52]. Hybrid feature extraction techniques are differentiated from many prior state-of-the-art techniques. Parameter estimation experiments show these methods can reduce incorrect image evaluations and parameter predictions [53]. The proposed MMFCNN consists of three blocks: a feature extraction block, a feature multimodal block, and an internal defect detection system block. The results demonstrate that MMFCNN is effective for the internal flaw detection system of decorative tiles [54].

This paper proposes a novel method to enhance the face recognition accuracy rate using a CNN in conjunction with feature extraction, HOG, SIFT, Gabor, and canny contour detector approaches to detect the presence of variation or occlusion [55]. This research focuses on the latter, where contrast-enhanced canny edge detected (CEED-Canny) X-ray pictures and deep learning are used to diagnose tuberculosis. The lung X-ray edge detected pictures were created using the CEED-Canny [56, 49]. Intelligent categorization of fruits and vegetables requires an efficient fault detection system. The goal of this proposed system is to use morphology, color, and texture to detect exterior abnormalities in fruits and vegetables.

15.4 AI TECHNIQUES FOR DEFECT DETECTION AND IDENTIFICATION

The proposed framework uses a CNN model to identify defective fruit in one of six images taken by two cameras; conventional methods also classify the fruits as defective [57, 58]. To develop a system based on the Internet of Things (IoT) for healthy grape plant growth, prompt recognition, and finding diseases on grapes, additional measures are necessary to stop the spread of illness [59, 60]. A novel method has been proposed for utilizing deep learning models and IoT devices for clearly and efficiently identifying undamaged fruits and vegetables [61].

15.4.1 Machine learning techniques for defect detection

The fruit and vegetable fault detection training and testing methodologies currently used for classification are discussed. This paper's contribution is the combination of preexisting structures with the suggestion of new adversarial generative networks for images of fruits and vegetables. Conditional Wasserstein GAN with Gradient Penalty (CWGAN-GP) and Auxiliary Classifier Wasserstein GAN with Gradient Penalty (ACWGAN-GP) are utilized, and additionally, the Progressive Spinal fully connected layer is used for freshness classification [62]. This article provides a state-of-the-art description of several well-known problem domains, including quality grading of fruits, vegetables, and produce; detecting diseases of fruits, vegetables, and crops; and recognizing unique fruits, vegetables, and crops [63].

The importance of quality inspection is highlighted by categorizing products in two steps: good illumination and classification based on detected defects. Various machine learning classification methods employ a combination of MV and IP to classify products depending on the kind and quality of their flaws [64]. An MSVM classifies the faulty samples according to defect vectors. Identifying early stage defects in numerous fruits and vegetables is a major concern. Various features of images are recognized, analyzed, and understood using preprocessing methods with the best accuracy [65].

Traditional algorithms, such as SVMs and random forests, have been investigated for automated fruit and vegetable quality assessment. Additionally, acoustic and vibrational sensors have been explored for nondestructive evaluation of fruit and vegetable quality, providing valuable information for assessing produce ripeness and shelf life [66]. From the perspective of distinguishing qualities, this technique consists of an online machine vision-based agro-medical expert system procedure and a set of features. The disease-attacked region is separated from the input using the k-means clustering approach, and then the necessary features are extracted with the aid of an SVM to identify the diseases. Results with an accuracy rate of over 90% have been attained [67].

To convolve the input image, healthy and damaged fruits and vegetables are classified and compared to the proposed model using HOG, LBP, KNN, ANN, Fuzzy, SVM, and DT algorithms. The suggested CNN-based model outperforms comparable techniques in terms of overall performance and can classify sour lemons with 100% accuracy [68]. A hybrid approach is put forward in the current study to increase the precision and prediction of classification. To validate and forecast the model accuracy, the Pima Indian Diabetes Dataset (PIDD) from the Machine Learning Repository dataset was utilized [69].

This work presents a novel approach to hyperspectral image categorization utilizing SVMs, bidimensional empirical mode decomposition (BEMD), and principal component analysis (PCA) [70].

For damage identification in computer vision systems, a range of approaches such as KNN, SVMs, ANN, and M5 Rules has been developed for the goal of identifying flaws in fruits and vegetables. The most energy-intensive system in a building is the air conditioning system. Three types of supervised learning classifiers were created: multilayer perceptrons (MLP), SVMs, and deep learning. Six distinct types of defects were used to examine each classifier's performance [71].

15.4.2 Deep learning techniques for defect detection

The suggested system is made up of a surface defect detection algorithm and an edge-cloud architecture. Subsequently, a deep learning-based technique is designed to use multimodal data, including image and depth data, to identify surface defects. The suggested approach achieved good accuracy and recall in tiny defect detection, according to experiments [72]. The structure of a unique stream CNN is recommended for the identification of infections affecting citrus fruits. Four contrast enhancement operations are used in the data augmentation process. The next step involves selecting and adjusting a CNN, MobileNet-V2, for this approach. As a result, an improved Whale Optimization Algorithm (IWOA) is used [73].

The challenge with this method is estimating the object count as a classification task rather than focusing on object detection, which uses more resources. A new architecture has been unveiled, consisting of a decision tree for object counting based on parameter weight and prior classification results and an ensemble of EfficientNet for image classification [74]. This model is trained in its entirety with binary classification followed by multi-classification recognition. Neural "necrosis" affects the GELU activation function in ConvNeXt, whereas the effective activation function is ERelu. This was validated to be effective in the multi-classification task of fruit and vegetable diseases [75]. This proposed system is based on the receptive field Block Net (RFB Net), and it aims to increase the RFB module by integrating the lower and higher layers for defect detection. The result shows that the accuracy of this model is higher than RFB Net on both texture and simple background product images [76].

There are several types of surface imperfections on steel, and the damage that is currently present includes many small patches and fuzzy edges. The outcomes of the experiment indicate that the enhanced YOLOv4 model outperforms the original method in terms of accuracy [77]. An enhanced YOLOv4 algorithm for fabric flaw identification is more accurate and employs a new SPP structure that uses SoftPool rather than MaxPool. According to test findings, the improved YOLOv4 effectively raises the mAP in comparison to the standard YOLOv4. The enhanced YOLOv4 method can be used in other defect detection industries and is fast and accurate in pinpointing the location of flaws [78]. Feature extraction was used to train deep residual neural network (ResNet) classifiers to identify external faults, and CV was used to refine a benchmark for external defect detection.

The existing work classified the fruit maturity and determined the state of the fruits. Digital photographs of samples were collected and used to classify tomato defects. The regions are segmented using the Gabor wavelet transform to identify the principal faults in the fruits [79]. A simple contrast enhancement technique, including grayscale conversion, was employed. Canny edge detection is utilized to identify the input fruit boundaries during the segmentation stage. It was evident from the comparison study with other multi- or single-fruit systems that the suggested approach produced greater accuracy [80]. The approach classified the quality of the fruit and set pricing accordingly. The CNN models that the categorization system operated on were Xception, InceptionV3, VGG16, MobileNet, and NASNetMobile. With 97.34% accuracy, the InceptionV3 model is the most accurate [81].

Finding and categorizing road damage is one of the trickiest issues public and private road maintenance organizations deal with. With a global flaw detection rate of 91%, the YOLO algorithm operates effectively [82]. To address the bugs and introduce a new taxonomy for DDoS attacks, including a new classification based on flows network, the CICDDoS2019 dataset is used in this study. The long short-term memory (LSTM) algorithm and deep neural network (DNN) are applied for DDoS attack detection [83]. Effective fused modal biometric sensor systems based on similarity index sequence fusion of the left, right, and face handprints are proposed in this work. CNNs and KNNs are used in multimodal biometric identification systems, which are trained to identify and recognize people based on multimodal biometric scores [84].

This research work presents a survey of the suggested algorithm together with the findings of a preliminary experiment using denial of service (DoS) attacks. The evaluation's findings demonstrate that deep learning can increase heterogeneous IoT detection accuracy [85]. This research presents the proposal of a novel intrusion detection system (IDS) built on a deep learning CNN, known as BotIDS. Developing, implementing, and testing the IDS against several popular Botnet attacks utilizing a particular Bot-IoT dataset is the primary goal of this study, relative to other deep learning methods like GRU, LSTM, and simple RNN [86]. An overview of study articles using various preprocessing and feature extraction algorithms has been presented in Table 15.1, which indicates the classification of recent methods based on various categories with corresponding articles.

The conventional approaches rely on manually designed features or shallow learning methods that can identify flaws and are applied in certain detection scenarios or with prior knowledge. The overall category-based classification systems are shown in Figure 15.6.

The recent research conducted by various authors related to the detection of application-specific defects and quality analysis from recent techniques and methods in fruits and vegetables is summarized in Table 15.2.

Table 15.1 Overview of various categories with corresponding articles

Applications	*Classifiers*	*Model*		*Category*	*Ref*
Fruits, vegetables, plants, and other applications	External and Internal Defect Detection Preprocessing and Segmentation			C1-C13	[5–37] [38, 39]
	Feature Extraction			C14-C20	[40–56]
	AI Techniques for Defect Detection	Machine Learning		C21-C33	[57–72]
		USL	USL		
		Deep Learning		C34-C53	[73–86]
		SL	USL		

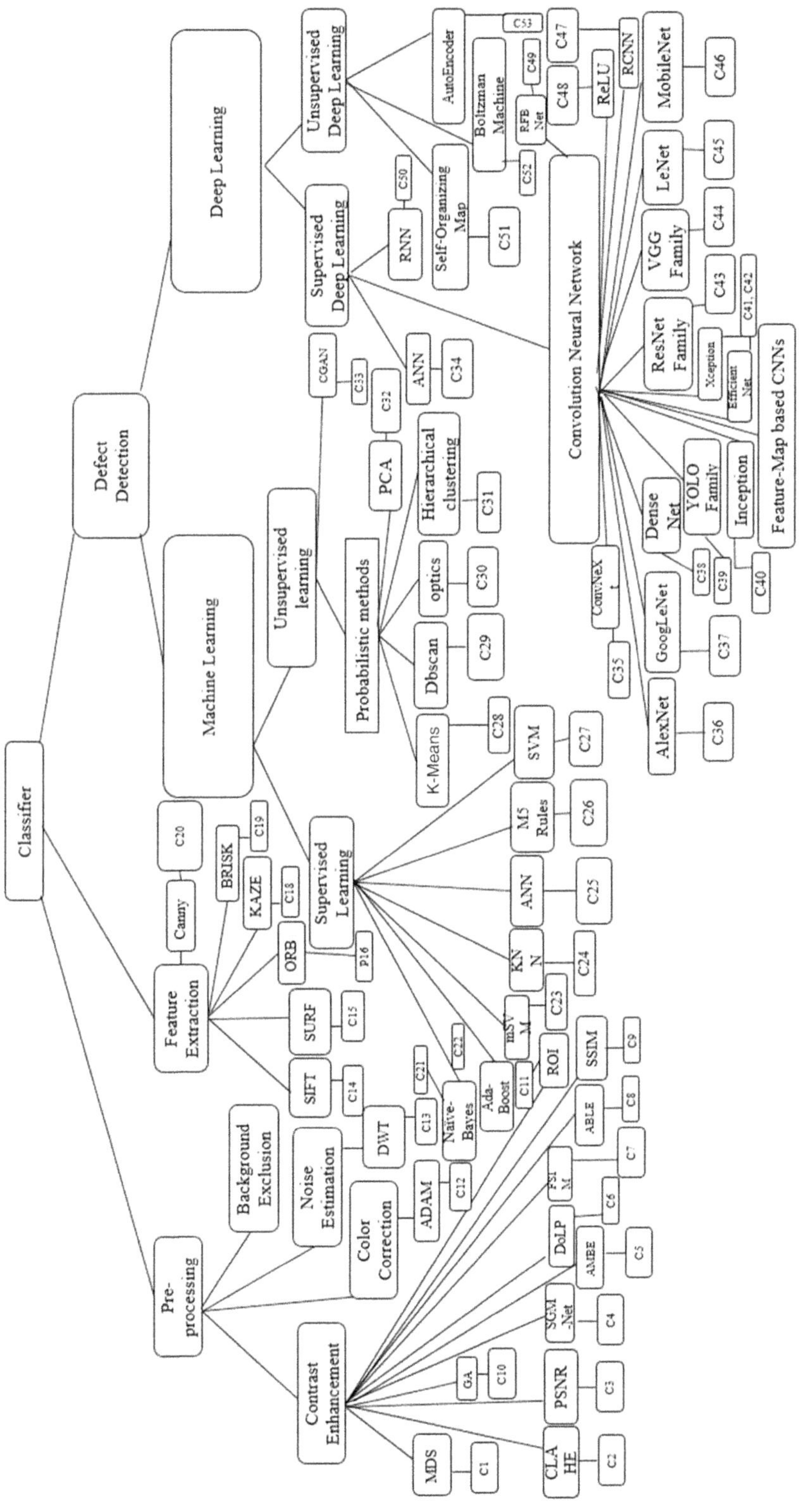

Figure 15.6 Classification of recent trend methods and techniques.

Table 15.2 The performance of recent methods and techniques

Ref	*Bibliography*	*Categorization*	*Preprocessing*	*Feature extraction*	*Methods*	*Accuracy*
[22]	Arthur Z et al. (2020)	External Defect Detection	Noise Reduction	Fine-tuning	ResNet50	94%
[23]	Shuxiang Fan et al., (2020)	Defect Detection	Noise Removal	Color Correction	SVM CNN	90%
[65]	Seyed Iman Saedi et al. (2020)	Image Detection	RGB	Size and Feature maps	VGG11, ResNet50, ResNet152 and YOLOv3	99.36%
[70]	Ade Iriani Sapitri (2021)	Defect Detection	RGB Contrast Enhancement	Co-occurrence and Statistical	SVM	90.15%
[72]	Yirui Wu et al. (2022)	Identification of Healthy and Damaged fruit	Resize	Features Shape and Texture	HOG, LBP, KNN, (ANN), Fuzzy, SVM, DT, CNN	99%
[74]	Kirill Bogomasov, et al. (2021)	Recognizing Citrus Fruit Diseases	Contrast Enhancement	Size, Shape, Color	MobileNet-V2 IWOA	99.7%
[75]	Yumin Zhang, et al. (2022)	Early Identification	RGB	Size and color based	ConvNeXt	89.20%
[84]	S.Dhakshin et al. (2020)	Defect Detection	Image Enhancement	Feature Prediction	SVM	0.97
[81]	Md Sohel Miah et al. (2021)	Identification of Fresh and Rotten Fruits	Noise Removal	Color Based	Xception, InceptionV3, VGG16, MobileNet, and NASNetMobile	97.34%

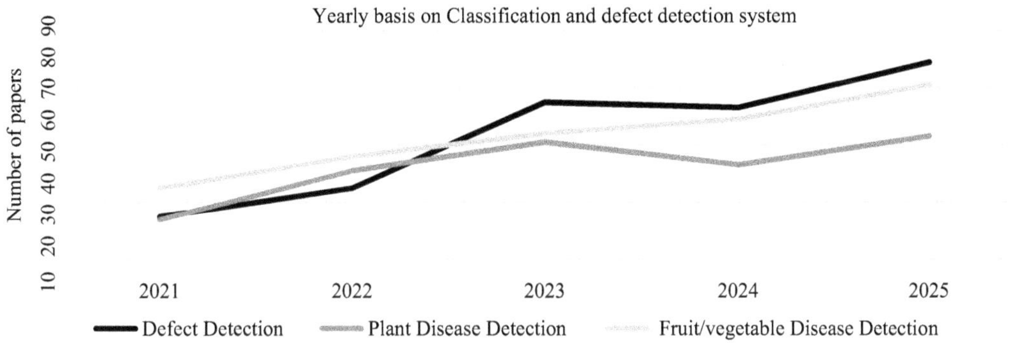

Figure 15.7 Distribution of the article corpus per year and per type of detection.

Table 15.3 Performance metrics for defect classification

Defect type	*Accuracy (%)*	*Precision (%)*	*Recall (%)*	*F1 score (%)*
External_defect	94.5	95.2	93.8	94.5
Internal_ defect	88.7	87.0	90.2	88.6
No_ defect	91.2	89.5	92.0	90.7

Table 15.4 Dataset categories

Dataset type	*Fruit type*	*Nondefect*	*Mild defect*	*Large defect*	*Total*
Existing	Banana	155	75	125	355
	Mango	270	250	200	720
Real_Time	Musk melon	200	190	150	540
	Chayote squash	180	150	125	455
	Chickoo	150	125	120	395
Total		955	790	720	2465

In Figure 15.7, the number of articles increases on a yearly basis in this domain. The X-axis denotes the yearly basis for classification systems and defect detection systems, and the Y-axis denotes the number of papers published. The targeted articles in this research work are shown using a graph method.

In Table 15.3, a performance table summarizes how well the classification model distinguishes between different defect classes using metrics like accuracy, precision, recall, and F1 score.

Therefore, improvement of the imbalanced dataset was carried out by adding collected real-time images for each class, 100 and above, captured using SAMSUNG mobile phones in 255 × 255 pixel image format to perform this experiment. The different dataset categories are mentioned in Table 15.4.

15.5 DISCUSSION AND PROSPECTS

The ultimate goal of applying computer vision techniques is to enhance the grade and amount assessment of fruits and vegetables, as well as defect identification in real-time agricultural fields. This vision-based system has the advantages of rapidity, persistence, and nondestructiveness. Fruit and vegetable detection is flexible under difficult conditions such as lighting, multiple angles, variability, cropping, and occlusion.

15.5.1 Challenges

- Using real-time images from markets and mini-shops as datasets.
- Reducing labor, effort, and cost is one of the advantages of computer vision for farmers.
- The traditional approach required significant time for defect processing, whereas computer vision for defect identification uses defined and automatic processing, lowering the required time.
- Identifying fruits and vegetables can be challenging due to their unique shapes, sizes, colors, and edges.
- Classifying defect detection using images from various parts of both fruits and vegetables. These classifications use a width multiplier and checking to build a training model.

15.5.2 Limitations

- One of the fundamental drawbacks of current computer vision methods is that they necessitate the preparation of a dataset, which is a time-consuming process.
- Another significant disadvantage is that fruit and vegetable defects are more dependent on the environment, as the same image processing approach can produce varying degrees of validity on different datasets.
- There are occasions when accuracy is slightly hampered by improper exploration of the interior portions of fruits and vegetables, which reduces the likelihood of identifying damages and quality analysis.
- To extend this method as a technical framework that could be applied to various industrial applications, such as fruit and vegetable quality defect detection [65].
- Furthermore, color and texture influence detection; for example, flaws are significantly more noticeable on yellow banana peels. As a result, further studies should improve this [69].
- Existing feature extraction approaches are currently unable to obtain spectral information from various regions of the fruit. The development of a multiple spectroscopic system that can be used to inspect interior flaws in apples and other fruits in real-time is necessary.

15.6 CONCLUSION

The study in this paper has covered and analyzed various themes, such as existing defect detection, identification, and defect classification, i.e., external and internal defect methods applied to fruits and vegetables using current techniques. A survey was conducted based on recent trends in fruits and vegetables among various markets and mini-shops using real-time images. The datasets were collected from local markets, mini-shops, and handcart farmers. In addition to conducting the study, this article has surveyed 86 articles published from 2020 to the present. It was concluded that the system developed for identifying fruits and vegetables, along with related sample approaches using visual sensing techniques, is one of the most successful and efficient approaches available.

It is strongly supported that researchers should apply the developed system for detecting defects in fruits and vegetables and compare the various methods or algorithms used by researchers at each phase. Feature extraction becomes feasible through the use of different descriptors, such as corner detection. Then, to determine the accuracy of the suggested system, several AI techniques and methods should be considered in different areas of real-time datasets in the agricultural field and markets. The same performance metrics can be applied to various datasets to evaluate accuracy. Many academics have recently focused on deep learning approaches due to their promise to provide high-quality and accurate results. These methods can lead to various other inventions in defect detection and quality analysis, raising the opportunity for future work in this area. This review will help farmers, stakeholders, and mini-shopkeepers.

REFERENCES

1. N. Anjali, Rajkumar Kannan, Frederic Andres, Georgeghita Ghinea. (2021). Trend review related to defects detection from fruits and vegetables. Preprints, Posted: November 2, 2021 by the author(s). Distributed under a Creative Commons CC BY license, DOI: https://doi.org/10.20944/preprints202111.0035.v1
2. Xiongwei Sun, Xiong Wei, Qinshan Xu, Cheng Zhang (2021). An effective contrast enhancement approach for wide dynamic range image. *2021 5th International Conference on Vision, Image and Signal Processing (ICVISP)*. IEEE, DOI: https://doi.org/10.1109/ICVISP54630.2021.00061
3. A.N. Tripathi, S.K. Tiwari, S.K. Sharma, P.K. Sharma, T.K. Behera. (2024). Current status of bacterial diseases of vegetable crops. *Vegetable Science*, Vol 51, pp. 106–117, DOI: https://doi.org/10.61180/vegsci.2024.v51.spl.10
4. Fariborz Habibi, Daniel A. Boakye, Yuru Chang et al. (2024). Molecular mechanisms underlying postharvest physiology and metabolism of fruit and vegetables through multi-omics technologies. *Scientia Horticulturae*, Vol. 324, p. 112562, DOI: https://doi.org/10.1016/j.scienta.2023.112562
5. H. Tanaka, A. Taguchi. (2021). Image contrast enhancement with high dynamic range using single scale retinex. *2021 International Symposium on Intelligent Signal Processing and Communication Systems (ISPACS)*. IEEE, https://doi.org/10.1109/ISPACS51563.2021.9651001
6. Jiao Wang, Yanzhu Hu. (2020). An improved enhancement algorithm based on CNN applicable for weak contrast images. *IEEE Access*, Vol. 8, pp. 8459–8476, DOI: http://creativecommons.org/licenses/by/4.0/
7. M. Benedict Tephila, B. Maruthi Shankar, S. Abhinandhan et al. (2022). An experimental approach in colour correction and contrast improvement for underwater images. *2022 Second International Conference on Artificial Intelligence and Smart Energy (ICAIS)*. IEEE, DOI: https://doi.org/10.1109/ICAIS53314.2022.9742723
8. B. Aswini, M.J. Josemartin. (2021). Adaptive heuristic histogram equalisation approach combined with gamma transformation for image contrast enhancement. *2021 Fourth International Conference on Microelectronics, Signals & Systems (ICMSS)*. IEEE, DOI: https://doi.org/10.1109/ICMSS53060.2021.9673652
9. Wenhan Yang, Wenjing Wang, Haofeng Huang, Shiqi Wang, Jiaying Liu. (2021). Sparse gradient regularized deep retinex network for robust low-light image enhancement. *IEEE Transactions on Image Processing*, Vol. 30, pp. 2072–2086, DOI: https://doi.org/10.1109/TIP.2021.3050850
10. Tzu-Chia Tung, Chiou-Shann Fuh, Chiou-Shann Fuh. (2021). ICEBIN: Image contrast enhancement based on induced norm and local patch approaches. *IEEE Access*, Vol. 9, pp. 23737–23750, DOI: https://doi.org/10.1109/ACCESS.2021.3056244
11. Varad A. Pimpalkhute, Rutvik Page, Ashwin Kothari, Kishor M. Bhurchandi, Vipin Milind Kamble. (2021). Digital image noise estimation using DWT coefficients. *IEEE Transactions on Image Processing*, Vol. 30, pp. 1962–1972, DOI: https://doi.org/10.1109/TIP.2021.3049961

12. Doni Setyawan, Retantyo Wardoyo, Moh Edi Wibowo, E. Elsa Herdiana Murhandarwati. (2022). Improvising low contrast malaria images using contrast enhancement techniques on various color models. *2022 Seventh International Conference on Informatics and Computing (ICIC)*. IEEE, DOI: https://doi.org/10.1109/ICIC56845.2022.10006911
13. Simone Zini, Marco Buzzelli, Simone Bianco, Raimondo Schettini. (2022). A framework for contrast enhancement algorithms optimization. *2022 IEEE International Conference on Image Processing (ICIP)*. IEEE, DOI: https://doi.org/10.1109/ICIP46576.2022.9897184
14. Meysam Alavi, Mehrdad Kargari (2022). A new contrast enhancement method for Color dark and low-light images. *2022 9th Iranian Joint Congress on Fuzzy and Intelligent Systems (CFIS)*. IEEE, DOI: https://doi.org/10.1109/CFIS54774.2022.9756426
15. Weidong Zhang, Xipeng Pan, Xiwang Xie, Lingqiao Li, Zimin Wang, Chu Han. (2021). Color correction and adaptive contrast enhancement for underwater image enhancement. *Computers & Electrical Engineering*, Vol. 91, DOI: https://doi.org/10.1016/j.compeleceng.2021.106981
16. Mohamadali Mahmoodpour, Abdolah Amirany, Mohammad Hossein Moaiyeri, Kian Jafari. (2022). A learning based contrast specific no reference image quality assessment algorithm. *2022 International Conference on Machine Vision and Image Processing (MVIP)*. IEEE, DOI: https://doi.org/10.1109/MVIP53647.2022.9738784
17. Jianwen Meng, Wenyi Ren, Ruoning Yu, Dan Wu, Rui Zhang, Yingge Xie, Jian Wang. (2023). Contrast enhanced color polarization image fusion. *Optik*, Vol. 284, pp. 1–9, DOI: https://doi.org/10.1016/j.ijleo.2023.170935
18. Versha Thakur, Harjinder Singh. (2021). An artificial neural network based adaptive histogram equalization algorithm for enhancement of low contrast images. IEEE, pp. 268–273, https://ieeexplore.ieee.org/document/9441356
19. Akira Taguchi, Wataru Mizuno. (2022). A method for integrated processing of edge enhancement and contrast enhancement. *2022 International Symposium on Intelligent Signal Processing and Communication Systems (ISPACS)*. IEEE, DOI: https://doi.org/10.1109/ISPACS57703.2022.10082831
20. Zeeshan Ali Haq, Mohd Ashhad, Zainul Abdin Jaffery, Shabana Mehfuz. (2023). Infrared imaging technique and multi-input neural network for potato defect detection. *2023 International Conference on Recent Advances in Electrical, Electronics & Digital Healthcare Technologies (REEDCON)*. IEEE, DOI: https://doi.org/10.1109/REEDCON57544.2023.10150501
21. Weitao Deng, Lei Liu, Huateng Chen, Xiaofeng Bai. (2022). Infrared image contrast enhancement using adaptive histogram correction framework. *Optik*, Vol. 271, pp. 1–16, DOI: https://doi.org/10.1016/j.ijleo.2022.170114
22. Arthur Z. da Costa, Hugo E.H. Figueroa, Juliana A. Fracarolli. (2020). Computer vision based detection of external defects on tomatoes using deep learning. *Biosystems Engineering*, Vol. 190, pp. 131–144, DOI: https://doi.org/10.1016/j.biosystemseng.2019.12.003
23. Shuxiang Fan, Jiangbo Li, Yunhe Zhang, Xi Tian, Qingyan Wang, Xin He, Chi Zhang, Wenqian Huang. (2020). Online detection of defective apples using computer vision system. *Journal of Food Engineering*, Vol. 286, pp. 1–10, https://doi.org/10.1016/j.jfoodeng.2020.110102
24. Lei Yang, Huaixin Wang, Benyan Huo, Fangyuan Li, Yanhong Liu. (2021). An automatic welding defect location algorithm based on deep learning. *NDT & E International*, Vol. 120, pp. 1–9, DOI: https://doi.org/10.1016/j.ndteint.2021.102435
25. Yusong Bai, Jian Kang. (2022). A global-to-local thresholding method for object enhancement in SAR images. *2022 3rd China International SAR Symposium (CISS)*. IEEE, DOI: https://doi.org/10.1109/CISS57580.2022.9971368
26. Tongxue Zhou, Stéphane Canu, Pierre Vera, Su Ruan. (2021). Latent correlation representation learning for brain tumor segmentation with missing MRI modalities. *IEEE Transactions on Image Processing*, Vol. 30, pp. 4263–4274, https://doi.org/10.1109/TIP.2021.3070752
27. Shouket Abdulrahman Ahmed, Hazry Desa, Abadal-Salam T. Hussain. (2023). Classification of semantic segmentation using fully convolutional networks based unmanned aerial vehicle application. *IAES International Journal of Artificial Intelligence (IJ-AI)*, Vol. 12, No. 2, pp. 641–647, DOI: http://doi.org/10.11591/ijai.v12.i2.pp641-647

28. R. Kabilan, G. Prince Devaraj, U. Muthuraman, N. Muthukumaran, J. Zahariya Gabriel, R. Swetha. (2021). Efficient color image segmentation using fastmap algorithm. *2021 Third International Conference on Intelligent Communication Technologies and Virtual Mobile Networks (ICICV)*. IEEE, DOI: https://doi.org/10.1109/ICICV50876.2021.9388470
29. Neeraj Kumari, Ashutosh Kr. Bhatt, Rakesh Kr. Dwivedi, Rajendra Belwal. (2020). Hybridized approach of image segmentation in classification of fruit mango using BPNN and discriminant analyser. *Multimedia Tools and Applications*, Vol. 80, pp. 4943–4973, DOI: https://doi.org/10.1007/s11042-020-09747-z
30. A. Sasikumar, M. Sathyanarayanan, A.N. Sriayapppan, R. Santhosh. (2023). A CNN-based canny edge detection approach for car scratch detection. *2023 International Conference on Inventive Computation Technologies (ICICT)*. IEEE, DOI: https://doi.org/10.1109/ICICT57646.2023.10134444
31. Eleonora Iaccheri, Annachiara Berardinelli, Gianni Ceredi, Luigi Ragni. (2025). Non-destructive assessment of the internal defects of FRED® pear by a low radio-frequency capacitive technique. *Computers and Electronics in Agriculture*, Vol. 229, p. 109709.
32. Sandra Munera, Alejandro Rodríguez-Ortega, Sergio Cubero, Nuria Aleixos, Jose Blasco. (2025). Automatic detection of pomegranate fruit affected by blackheart disease using X-ray imaging. *LWT – Food Science and Technology*, Vol. 215, p. 117248.
33. R. Hadipour-Rokni, E.A. Asli-Ardeh, A. Jahanbakhshi, S. Sabzi. (2023). Intelligent detection of citrus fruit pests using machine vision system and convolutional neural network through transfer learning technique. *Computers in Biology and Medicine*, Vol. 155, p. 106611.
34. Leopold Fischer-Brandies, Lucas Müller, Justus Johannes Riegger, Ricardo Buettner. Fresh or rotten? Enhancing rotten fruit detection with deep learning and gaussian filtering. *IEEE Access*, Vol. 13, pp. 31857–31869, DOI: https://doi.org/10.1109/ACCESS.2025.3542612
35. Mudasar Iqbal, Syed Tahseen Haider, Rana Saud Shoukat, Saif Ur Rehman, Khalid Mahmood, Santos Gracia Villar, Luis Alonso Dzul Lopez, Imran Ashraf. (2025). Canned apple fruit freshness detection using hybrid convolutional neural network and transfer learning. *Journal of Food Quality*, Vol. 2025, p. 8522400, DOI: https://doi.org/10.1155/jfq/8522400
36. Jiahao Li, Baofeng Mai, Tianhu Liu, Zicheng Liu, Zhaozheng Liang, Shuyang Liu. (2025). A machine vision method for detecting pineapple fruit mechanical damage. *Agriculture*, Vol. 15, p. 1063, DOI: https://doi.org/10.3390/agriculture15101063
37. Tiziana Amoriello, Roberto Ciorba, Gaia Ruggiero, Francesca Masciola, Daniela Scutaru et al. (2025). Vis/NIR spectroscopy and Vis/NIR hyperspectral imaging for non-destructive monitoring of apricot fruit internal quality with machine learning. *Foods*, Vol. 14, p. 196, DOI: https://doi.org/10.3390/foods14020196
38. Astrid Tempelaere, Leen Van Doorselaer, Jiaqi He, Pieter Verboven, Bart M. Nicolai. (2024). BraeNet: Internal disorder detection in 'Braeburn' apple using X-ray imaging data. *Food Control*, Vol. 155, p. 110092, DOI: https://doi.org/10.1016/j.foodcont.2023.110092
39. Chenjiao Tan, Changying Li, Penelope Perkins-Veazi, Heeduk Oh, Rui Xu, Massimo Iorizzo. (2025). High throughput assessment of blueberry fruit internal bruising using deep learning models. *Frontiers in Plant Science*, Vol. 16, DOI: https://doi.org/10.3389/fpls.2025.1575038
40. Touka Hafsia, Hatem Tlijani, Khaled Nouri. (2022). A comparative analysis of techniques for extracting features from the object in image processing. *2022 5th International Conference on Advanced Systems and Emergent Technologies (IC_ASET)*. IEEE, DOI: https://doi.org/10.1109/IC_ASET53395.2022.9765905
41. Dan Yin, Siwei Zhou, Pengcheng Wang. (2020). A feature points extraction algorithm based on adaptive information entropy. *IEEE Access*, Vol. 8, pp. 127134–127141, DOI: https://creativecommons.org/licenses/by/4.0/
42. Xiaofei Zhou, Hao Fang, Xiaobo Fei. (2021). Edge-aware multi-level interactive network for salient object detection of strip steel surface defects. *IEEE Access*, Vol. 9, pp. 149465–149476, DOI: https://doi.org/10.1109/ACCESS.2021.3124814
43. Ludymila R. Borges, Eduardo L.M. Naves, Angela A.R. Sa. (2022). Usability evaluation of an electric-powered wheelchair driven by eye tracking. *Universal Access in the Information Society*, Vol. 21, pp. 1013–1022, DOI: https://doi.org/10.1007/s10209-021-00809-z

44. Ashish Sharma, Anmol Mittal, Savitoj Singh, Vasudev Awatramani. (2020). Hand gesture recognition using image processing and feature extraction techniques. *Procedia Computer Science*, Vol. 173, pp. 181–190, DOI: https://doi.org/10.1016/j.procs.2020.06.022
45. Salma Eleuch, Nadia Khouja, Fethi Tlili, Tomaso Erseghe, Simone Milani. (2022). A comparative evaluation of well-known feature extractors for multi-view vehicle tracking in VSNs. *2022 IEEE Ninth International Conference on Communications and Networking (ComNet)*. IEEE, DOI: https://doi.org/10.1109/ComNet55492.2022.9998435
46. Sheela Ramachandra, Suchithra Ramachandran. (2022). Region specific and subimage based neighbour gradient feature extraction for robust periocular recognition. *Journal of King Saud University-Computer and Information Sciences*, Vol. 34, pp., 7961–7973, DOI: https://doi.org/10.1016/j.jksuci.2022.07.013
47. Kai Luo, Huiru Zheng, Zhicong Shi. (2023). A simple feature extraction method for estimating the whole life cycle state of health of lithium-ion batteries using transformer-based neural network. *Journal of Power Sources*, Vol. 576, DOI: https://doi.org/10.1016/j.jpowsour.2023.233139
48. P. Canals, S. Balocco, O. Díaz, J. Li, A. García-Tornel, A. Tomasello, M. Olive-Gadea, M. Ribo. (2023). A fully automatic method for vascular tortuosity feature extraction in the supra-aortic region: unraveling possibilities in stroke treatment planning. *Computerized Medical Imaging and Graphics*, Vol. 104, pp. 1–11, DOI: https://doi.org/10.1016/j.compmedimag.2022.102170
49. Ardiant Utomoa, Edwin Farrel Juniawana, Vincent Lioea, Diaz D. Santikaa. (2021). Local features based deep learning for mammographic image classification: In comparison to CNN models. *Procedia Computer Science*, Vol. 179, pp. 169–176, DOI: https://doi.org/10.1016/j.procs.2020.12.022
50. Yiwen Ou, Zhiming Cai, Jian Lu, Jian Dong, Jian Dong. (2020). Evaluation of image feature detection and matching algorithms. *ICCCS*, IEEE. DOI: https://doi.org/10.1109/ICCCS49078.2020.9118480
51. Chenhao Wu, Huijuan Ma, Hongqi Jiang et al. (2022). An improved canny edge detection algorithm with iteration gradient filter. *2022 6th International Conference on Imaging, Signal Processing and Communications (ICISPC)*. IEEE, DOI: https://doi.org/10.1109/ICISPC57208.2022.00011
52. Zhenxing Liu, Xiaoning Song, Zhenhua Feng, Tianyang Xu, Xiaojun Wu, Josef Kittler. (2023). Global context-aware feature extraction and visible feature enhancement for occlusion-invariant pedestrian detection in crowded scenes. *Neural Processing Letters*, Vol. 55, pp. 803–817, DOI: https://doi.org/10.1007/s11063-022-10910-w
53. Joydev Hazra, Aditi Roy Chowdhury, Kousik Dasgupta, Paramartha Dutta. (2022). A hybrid structural feature extraction-based intelligent predictive approach for image registration. *Innovations in Systems and Software Engineering*, Vol. 20, pp. 643–651, DOI: https://doi.org/10.1007/s11334-022-00436-8
54. Houhong Lu, Yangyang Zhu, Ming Yin, Guofu Yin, Luofeng Xie. (2022). Multimodal fusion convolutional neural network with cross-attention mechanism for internal defect detection of magnetic tile. *IEEE Access*, Vol. 10, pp. 60876–60886, https://creativecommons.org/licenses/by/4.0/
55. Hicham Benradi, Ahmed Chater, Abdelali Lasfar. (2023). A hybrid approach for face recognition using a convolutional neural network combined with feature extraction techniques. *IAES International Journal of Artificial Intelligence (IJ-AI)*, Vol. 12, No. 2, pp. 627–640, DOI: https://doi.org/10.11591/ijai.v12.i2.pp627-640
56. Lakshmi Shrinivasan, Reshma Verma, Mysore Dakshinamurthy Nandeesh. (2023). Early prediction of diabetes diagnosis using hybrid classification techniques. *IAES International Journal of Artificial Intelligence (IJ-AI)*, Vol. 12, No. 3, pp. 1139–1148, DOI: https://doi.org/10.11591/ijai.v12.i3.pp1139-1148
57. A. Wu, J. Zhu, T. Ren. (2020). Detection of apple defect using laser-induced light backscattering imaging and convolutional neural network. *Computers & Electrical Engineering*, Vol. 81, pp. 1–9, DOI: https://doi.org/10.1016/j.compeleceng.2019.106454
58. V.G. Narendra, Ancilla J. Pinto. (2020). Defects detection in fruits and vegetables using image processing and soft computing techniques. *International Conference on Harmony Search Algorithm*, Vol. 1275, pp. 325–337, DOI: https://doi.org/10.1007/978-981-15-8603-3_29

59. Wenzhuo Zhan, Juan Hu, Guoxiong Zhou, Mingfang He. (2020). Detection of apple defects based on the FCM-NPGA and a multivariate image analysis. *IEEE Access*, Vol. 8, pp. 38833–38845, DOI: https://doi.org/10.1109/ACCESS.2020.2974262
60. Kainjan Sanghavi, Mahesh Sanghavi, Archana M. Rajurkar. (2021). Early stage detection of Downey and Powdery Mildew grape disease using atmospheric parameters through sensor nodes. *Artificial Intelligence in Agriculture*, Vol. 5, pp. 223–232, DOI: https://doi.org/10.1016/j.aiia.2021.10.001
61. Yue Yuan, Xianlong Chen. (2024). Vegetable and fruit freshness detection based on deep features and principal component analysis. *Current Research in Food Science*, Vol. 8, p. 100656, https://doi.org/10.1016/j.crfs.2023.100656
62. Zhihong Zhao, Jieshun You, Yunong Cui. (2022). Freshness recognition of fruit and vegetable images using GANs series data augmentation. *Proceedings of the 2022 4th International Conference on Video, Signal and Image Processing*. ACM, pp. 29–36, DOI: https://doi.org/10.1145/3577164.3577169
63. Md. Tarek Habib, Dewan Mamun Raza, Md. Mohaiminul Islam et al. (2022). Applications of computer vision and machine learning in agriculture: A state-of-the-art glimpse. *2022 International Conference on Innovative Trends in Information Technology (ICITIIT)*. IEEE, DOI: https://doi.org/10.1109/ICITIIT54346.2022.9744150
64. Mahmoud Soltani Firouz, Hamed Sardari. (2022). Defect detection in fruit and vegetables by using machine vision systems and image processing. *Food Engineering Reviews*, Vol. 14, pp. 353–379, DOI: https://doi.org/10.1007/s12393-022-09307-1
65. Seyed Iman Saedi, Hossein Khosravi. (2020). A deep neural network approach towards real-time on-branch fruit recognition for precision horticulture. *Expert Systems with Applications*, Vol. 59, pp. 1–10, DOI: https://doi.org/10.1016/j.eswa.2020.113594
66. C. O'Brien, N. Falagán, S. Kourmpetli, S. Landahl, L.A. Terry, M.C. Alamar. (2024). Non-destructive methods for mango ripening prediction: Visible and near-infrared spectroscopy (visNIRS) and laser Doppler vibrometry (LDV). *Postharvest Biology and Technology*, Vol. 212, p. 112878, DOI: https://doi.org/10.1016/j.postharvbio.2024.112878
67. Md. Tarek Habib, Anup Majumder, A.Z.M. Jakaria, Morium Akter, Mohammad Shorif Uddin, Farruk Ahmed (2020). Machine vision based papaya disease recognition. *Journal of King Saud University-Computer and Information Sciences*, Vol. 32, pp. 1215–1225, DOI: https://doi.org/10.1016/j.jksuci.2018.06.006
68. Ahmad Jahanbakhshia.; Mohammad Momenyb.; Majid Mahmoudic.; Yu-Dong Zhangd. (2020). Classification of sour lemons based on apparent defects using stochastic pooling mechanism in deep convolutional neural networks. *Scientia Horticulturae*, Vol. 263, pp. 1–10, DOI: https://doi.org/10.1016/j.scienta.2019.109133
69. T. Subba Reddy, J. Harikiran. (2020). Hyperspectral image classification using support vector machines. *IAES International Journal of Artificial Intelligence (IJ-AI)*, Vol. 9, No. 4, pp. 684–690. DOI: https://doi.org/10.11591/ijai.v9.i4.pp684-690
70. Ade Iriani Sapitri, Siti Nurmaini, M. Sukemi, Naufal Rachmatullah, Annisa Darmawahyuni. (2021). Segmentation atrioventricular septal defect by using convolutional neural networks based on U-NET architecture. *IAES International Journal of Artificial Intelligence (IJ-AI)*, Vol. 10, No. 3, pp. 553–562, DOI: https://doi.org/10.11591/ijai.v10.i3.pp553-562
71. Mohamed Anis Benallal, Mustapha Si Tayeb. (2023). An image-based convolutional neural network system for road defects detection. *IAES International Journal of Artificial Intelligence (IJ-AI)*, Vol. 12, No. 2, pp. 577–584, DOI: https://doi.org/10.11591/ijai.v12.i2.pp577-584
72. Yirui Wu, Hao Cao, Guoqiang Yang, Tong Lu, Shaohua Wan. (2022). Digital twin of intelligent small surface defect detection with cyber-manufacturing systems. *ACM Transactions on Internet Technology*, DOI: https://doi.org/10.1145/3571734
73. Muhammad Hassam, Muhammad Attique Khan, Ammar Armghan, Sara A. Althubiti et al. (2022). A single stream modified mobilenet V2 and whale controlled entropy based optimization framework for citrus fruit diseases recognition. *IEEE Access*, Vol. 10, pp. 91828–91839, DOI: https://doi.org/10.1109/ACCESS.2022.3201338

74. Kirill Bogomasov, Stefan Conrad. (2021). Efficient fruit and vegetable classification and counting for retail applications using deep learning. *Proceedings of the 5th International Conference on Advances in Artificial Intelligence*. ACM, pp. 65–71, DOI: https://doi.org/10.1145/3505711.3505720
75. Yumin Zhang, Yiru Zhang, Yiheng Zhang. (2022). Fruit and vegetable disease identification based on updating the activation function for the ConvNeXt model. *2022 6th International Conference on Electronic Information Technology and Computer Engineering*. ACM, pp. 1045–1049, DOI: https://doi.org/10.1145/3573428.3573616
76. Yangsai Wang, Yulong Zhuang, Yongqi Han. (2021). A defect detection network based on an improved receptive field block. *2021 International Conference on Machine Learning and Intelligent Systems Engineering (MLISE)*. IEEE, DOI: https://doi.org/10.1109/MLISE54096.2021.00024
77. Yi Cheng, Siqing Wang. (2022). Improvements to YOLOv4 for steel surface defect detection. *2022 5th International Conference on Intelligent Autonomous Systems (ICoIAS)*. IEEE, DOI: https://doi.org/10.1109/ICoIAS56028.2022.9931299
78. Qiang Liu, Chuan Wang, Yusheng Li, Mingwang Gao. (2022). A fabric defect detection method based on deep learning. *IEEE Access*, Vol.10, pp. 4284–4296, DOI: https://doi.org/10.1109/ACCESS.2021.3140118
79. S. Dhakshina Kumar, S. Esakkirajan, S. Bama, B. Keerthiveena. (2020). A microcontroller based machine vision approach for tomato grading and sorting using SVM classifier. *Microprocessors and Microsystems*, Vol. 76, pp. 1–13, DOI: https://doi.org/10.1016/j.micpro.2020.103090
80. Yuping Huanga, Renfu Lu, Kunjie Chen. (2020). Detection of internal defect of apples by a multichannel Vis/NIR spectroscopic system. *Postharvest Biology and Technology*, Vol. 161, pp. 1–8, DOI: https://doi.org/10.1016/j.postharvbio.2019.111065
81. Md Sohel Miah, Tayeeba Tasnuva et al. (2021). An advanced method of identification fresh and rotten fruits using different convolutional neural networks. *ICCCNT*. IEEE, DOI: https://doi.org/10.1109/ICCCNT51525.2021.9580117
82. Mohamed Anis Benallal, Mustapha Si Tayeb. (2023). An image-based convolutional neural network system for road defects detection. *IAES International Journal of Artificial Intelligence (IJ-AI)*, Vol. 12, No. 2, pp. 577–584, DOI: https://doi.org/10.11591/ijai.v12.i2.pp577-584
83. Thapanarath Khempetch, Pongpisit Wuttidittachotti. (2021). DDoS attack detection using deep learning. *IAES International Journal of Artificial Intelligence (IJ-AI)*, Vol. 10, No. 2, pp. 382–388, DOI: https://doi.org/10.11591/ijai.v10.i2.pp382-388
84. Chahreddine Medjahed, Abdellatif Rahmoun, Christophe Charrier, Freha Mezzoudj. (2022). A deep learning-based multimodal biometric system using score fusion. *IAES International Journal of Artificial Intelligence (IJ-AI)*, Vol. 11, No. 1, pp. 65–80, DOI: https://doi.org/10.11591/ijai.v11.i1.pp65-80
85. Sharipuddin Sharipuddin, Benni Purnama, Kurniabudi Kurniabudi, Eko Arip Winanto, Deris Stiawan, Darmawijoyo Hanapi, Mohd Yazid Idris, Rahmat Budiarto. (2021). Intrusion detection with deep learning on internet of things heterogeneous network. *IAES International Journal of Artificial Intelligence (IJ-AI)*, Vol. 10, No. 3, pp. 735–742, DOI: https://doi.org/10.11591/ijai.v10.i3.pp735-742
86. Idriss Idrissi, Mohammed Boukabous, Mostafa Azizi, Omar Moussaoui, Hakim El Fadili. (2021). Toward a deep learning-based intrusion detection system for IoT against botnet attacks. *IAES International Journal of Artificial Intelligence (IJ-AI)*, Vol. 10, No. 1, pp. 110–120, DOI: https://doi.org/10.11591/ijai.v10.i1.pp110-120

Chapter 16

Generative AI for sustainable solutions

Prabhav Pushker and Saba Khanum

16.1 INTRODUCTION

16.1.1 The diversification of generative AI in sustainability settings

The story of artificial intelligence could be written as a science fiction novel, and it has over the last few years. We have moved from AI systems that perform specific tasks exactly as programmed to AI systems that can actually create, think, and generate entirely new forms of possibility. We are not simply referring to the technical progress of generative AI; it is a disruption in its very nature, and it opens up entirely new varied forms of thinking about sustainability problems more broadly [1]. Generative, or GenAI, provides a frontier also of its own, which includes entry points or openings inherently different from one another. To illustrate GenAI, let's use GANs or generative adversarial networks. These utilize a fanciful form of back and forth—generate realistic data while one seeks not to recognize certain data as being fake, respectively. The result—highly realistic synthetic data—fills gaps in environmental data [2]. Variational autoencoders (VAEs), on the other hand, include a probabilistic notion of generativity and as such can create aspects that are inherently different while being produced in a way that remains tightly aligned with patterns of the underlying data [7]. We have not even talked yet about a whole different invention with transformer architectures—developed for natural language processing features; they have arguably performed better for different time-series data typing and spatial modeling features [8].

The sustainability sector is where the fun starts to happen. Environmental data are notoriously messy—there is only sparse coverage of very remote areas; they are nonlinear (containing nonlinear relationships and not representing simple mathematical relationships); they contain complex temporal dependencies over the course of time; and so on. Traditional statistical or analytical techniques often struggle with this complexity; however, GenAI systems appear to manage this well. In addition to sufficiently covering the data by synthesizing realistic life-world scenarios, the systems may help researchers identify hidden patterns that humans may also not realize even exist. GenAI may also potentially help generate predictive models that accept uncertainty related to environmental systems as a consistent feature [9].

16.1.2 Research scope and quality procedures

Putting the analytical component of this inquiry together was not without complications. The analysis itself involved a thorough examination of more than 50 peer-reviewed articles published between 2020 and 2025, primarily from the following publications: IEEE, Elsevier, and Wiley. The research was based on a systematic review methodology inspired by the PRISMA-SCR protocols. This means that we were not just going to wing it, but we also did not want to miss something important, in addition to maintaining the kind of quality

DOI: 10.1201/9781003685890-16

methodology that research requires [10]. We adopted a multi-faceted approach that examined five areas where GenAI is making a noticeable impact, and these areas include renewable energy systems, green logistics and supply chains, smart city infrastructure, climate action initiatives, and Internet of Things (IoT) ecosystems. For all five areas, we examined the specifics—what the technologies look like when implemented, performance indicators, environmental impacts, and implementation challenges [11]. In our data extraction process, we focused on the factual—a percentage reduction in errors, an increase in efficiencies, and measures of resource consumption—all the quantification that allows us to distinguish between dogma and reality. We also considered some of the softer indicators—barriers to implementation, relevant policy considerations, and potential next steps—because at times the critical indicators come down to commentary rather than something captured in a spreadsheet [12].

16.1.3 The weaponization of AI: the need for sustainability

Let's be direct about it. Time is not on our side. Doing things in an academic time frame is irrelevant at this point, you know? Climate change is a problem now; it is not a future problem. Resource depletion is a reality; it is not an abstract idea. Environmental degradation is certain and irreversible. We need real-world solutions offered with real-world uncertainty and complexity, not real-world solutions offered with simplicity and clarity [13]. We are nearly out of options using existing environmental management frameworks. These frameworks are based solely on historical data sources, which are increasingly irrelevant in this challenging environment. They assume linear relationships based on the assumption that environmental systems can react in a linear fashion, when in fact environmental systems interact in ways that often produce nonlinear relationships. Trying to navigate a ship in a storm using only data from last year's weather map would be similar to the current state of affairs [14]. This is where GenAI enters the picture, not as a magical solution, but as a powerful new tool in our toolbox. These systems can operate on sparse data by generating synthetic data to fill in the gaps [2].

They can deal with the multi-faceted relationships that trip up traditional methods [7]. And, most importantly, they can generate probabilistic outputs that represent the uncertainty we actually experience in the world [9]. And perhaps most importantly of all, they can generate multiple scenarios to enable better multi-faceted planning and decision-making [15]. But there is a catch: using AI systems for sustainability is not as simple as turning on the tap. The resources that are necessitated by the training and ongoing operational performance of AI systems have to be accounted for, which comes with an environmental cost as well, effectively creating a paradox—spending energy to save energy [4]. This report will confront this issue head on, not sidestep the challenge, accounting for the value and cost of deploying GenAI in sustainability contexts.

16.2 SECTOR FRAMED APPLICATIONS AND PERFORMANCE ASSESSMENT

16.2.1 Renewable energy forecasting and grid productivity

If there is any one sector where GenAI is developing a reputation, it is within renewable energy. And that is as challenging as challenging could possibly be; solar and wind energy are, by nature, not predictable. And the fluctuations are often so high that it seemed to outsiders that using traditional models had forecasters peering into a crystal ball with poor vision. So, it is hard to claim that GenAI frameworks are coming up with known possible performance, but compelling performance, and not just measurable but also artifact measurable [2]. But so far, the performance claims are not just theoretical. Studies have determined that GAN-based models for forecasting are, in raw terms, decreasing root mean square errors in forecasting by more than the 15 to 20% range over traditional studies [2]. These performance claims are beyond academic and viable options for increasing grid performance, reducing curtailment for unused renewable energy, and increasing variable renewable energy penetration into a grid. So, the bottom line—they are making renewable energy somewhat more reliable and lowering costs to some degree as well [3]. The hybrid approaches to DA/DF forecasting are the most interesting as of late. In particular, from a management perspective, they have combined GANs and Long Short-Term Memory (LSTM) networks, or some combination of the two, when forecasting actual demand, especially when the share of variable renewable energy is higher [7]. GANs and LSTM networks are particularly valuable for articulating the complicated interactions of weather patterns and energy demand, as well as the capacity for renewable generation—something that most traditional rule-based models struggle to incorporate [8]. Whatever literature there is with Variational Autoencoders, they provide a very good basis for optimization, with operational efficiencies of 9 to 12% in pilot implementations where no storage was involved [7].

Their strengths lie in their ability to model uncertainty in the forecasting of renewable energy and to use that uncertainty with regard to operational decisions. This means the increased depth of uncertainty incorporates a better set of operational decisions in the overall management of the grid, rather than having one forward-looking operational plan that we may wish would generally work [9]. Transformer architectures—arguably invented, conceived, and maintained as models that process language—performed very well in confirming grid optimization task implementations, improving forecasting accuracy by 10 to 15% and producing operational efficiencies of 8 to 10% in pilot implementations [8]. They are the way we are now able to process sequence data, particularly in sites where long-term dependencies are important. In the energy system, for example, if you make an operational decision today or on day one, it could have implications for operational changes over the following days or weeks [16] (Table 16.1 and Figure 16.1).

Table 16.1 Comparison of performance for GenAI Models in renewable energy applications

Model type	*Utilization*	*RMSE reduction (%)*	*Efficiency gain (%)*	*Major benefits*	*Constraints*
GAN	Solar Power Forecasting	15–20	N/A	High accuracy, creates realistic data	Prone to mode collapse
GAN-LSTM hybrid	Demand Forecasting	12–18	8–12	Captures temporal dependencies	Training is complex
VAE	Energy Dispatch	10–15	9–12	Can handle uncertainty in data	Lower resolution
Transformer	Grid Optimization	10–15	8–10	Can handle long-range dependencies	High computational cost
Diffusion models	Climate and Weather Integration	8–12	6–9	Generates a wide range of scenarios	Inference is slow

Note: Values based on findings from studies (2020–2025).

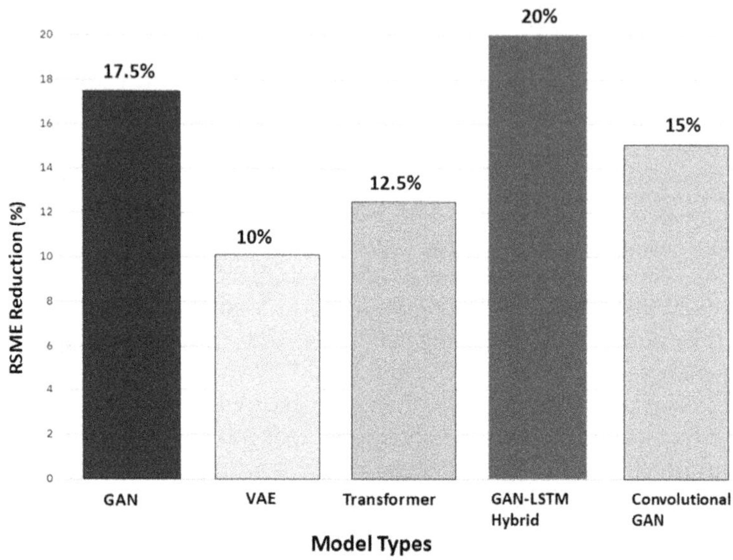

Figure 16.1 Performance comparison of RMSE reduction in GenAI models for renewable energy forecasting.

Description: RMSE reduction percentages for different GenAI models. The GAN-LSTM hybrid model achieved the highest reduction at 20%, followed by GAN (17.5%), VAE (10%), and transformer models (12.5%). The illustration shows that the GAN-LSTM hybrid model is the clear leader in forecasting solar irradiance, while the current GAN and other models begin to trade off accuracy for computational complexity of the simulation.

16.2.2 Green logistics and sustainable supply chain management

Green logistics is where GenAI is most interesting—where the rubber meets the road, if you will. Supply chains are a key part of the world economy but are also an exceedingly complex multilateral initiative with constantly changing constraints and conditions that can change in an instant. Traditional optimization methods often struggle, but GenAI proves it can navigate these obstacles with grace [3].

It is simply astounding what you find in the literature, and what is nice about the literature review is that 34 distinct GenAI applications for green logistics were uncovered. Examples of these applications include (environmentally friendly) route optimization, demand forecasts, risk analysis, and lifecycle analysis [3]. And they are not just possible opportunities—they are bona fide opportunities toward sustainable supply chain management with regard to minimizing carbon footprints, reducing waste, and optimizing utilization of resources [11].

We have unearthed some intriguing patterns in the adoption pathway for GenAI in logistics. Organizational readiness is also found to be an important determinant—organizations with more developed data structures and technical capacity are much more likely to successfully adopt GenAI [12]. Regulation that focuses on rewarding companies for sustainability provides another relevant reason, as it might offer a form of external urgency to overcome internal reluctance under previously sustained levels of organizational priority in the assessment framework of organizational determination to apply GenAI [13]. Finally, collaboration across stakeholders increases the likelihood of success through shippers collaborating with carriers and/or service providers [14].

And yet, as may be clear, there is no need to pretend, first and foremost, that these do not represent a significant adoption barrier, and these are not trivial. Certainly, the primary issue is infrastructure shortcomings; and, notably, infrastructure deficiencies, particularly with respect to digital network connectivity of organizational actors and/or efficiency in

Table 16.2 GenAI applications, drivers, and barriers in green logistics

Category	*Specific examples*	*Count/percentage*	*Impact level*	*Implementation priority*
Use cases	Route optimization, demand forecasting, risk evaluation, lifecycle assessment, inventory management	34 unique scenarios	High	Essential
Facilitators	Organizational preparedness, policy frameworks, stakeholder collaboration, data infrastructure	23 identified factors	Medium-High	Significant
Challenges	Infrastructure gaps, investment costs, regulatory ambiguities, lack of technical expertise	38 barriers identified	High	Top priority
Infrastructure needs	Digital connectivity, data management systems, sensor networks	40% of initiatives	Critical	Immediate
Investment costs	Capital expenditure, operational expenses, maintenance	30% of efforts	High	Strategic
Regulatory issues	Policy gaps, compliance issues, standards	20% of projects	Medium	Policy focus
Other factors	Lack of technical expertise, cultural resistance, scalability	10% of implementations	Medium	Training and development focus

Source: Systematic review of 50+ peer-reviewed studies (2020–2025).

management and/or coordination of internal data management capabilities appears blocking as many as almost 40% of organizational decisions to adopt GenAI [12]. Investment costs are another significant barrier as it seems both the capital costs of technology deployment, as well as the operational costs of physical deployment, are capping almost 30% of circumstances [12]. There is also regulatory unclarity of policy drawbacks that could determine as much as 20% of an organization's decisions to implement these algorithms, followed by the other factors—lack of technical expertise, cultural barriers, and scalability make up the remaining percentage of 10% collectively [12] (Table 16.2 and Figure 16.2).

16.2.3 Intelligent cities and digital twins of urban areas

Urban settings present distinct opportunities and challenges to improve sustainability. Urban populations are highly complex systems comprising different flows—energy, transport, waste, information—that typically interact in hybrid ways. Modeling and mapping processes using classical methods requires time-consuming work; GenAI technologies in particular, in conjunction with IoT technologies, provide unique insights regarding these flows [6] as well as means of optimizing interactions. Urban Digital Twins provide data about flows at diverse times in real time, providing whole-scale monitoring of urban systems in present and predictive models. The diverse data syntheses can include, for example, sensor observations, mobile device data, and administrative records—Urban Digital Twins provide a current picture of urban dynamics in what was historically a siloed, increasingly holistic way [6]. The Large Flow Model framework provides a robust preliminary method to model the nature of complex flows interactions in urban systems [15]. Mobility systems represent a showcase of likely

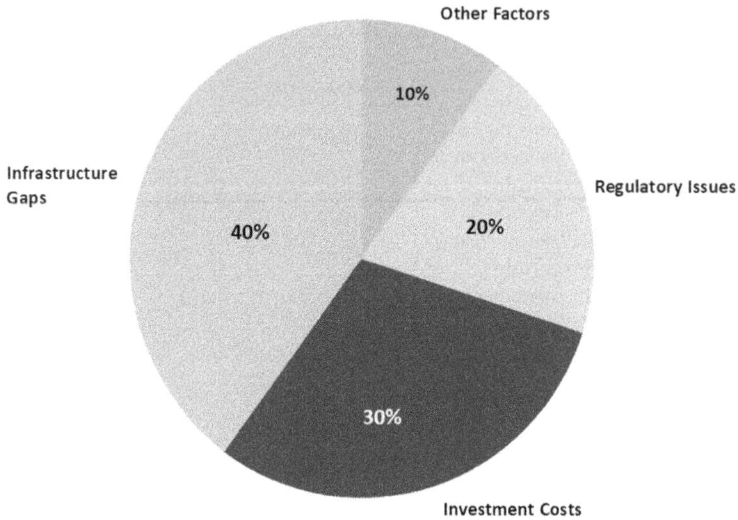

Figure 16.2 Distribution of implementation barriers in green logistics GenAI applications.

Description: The pie chart highlights the barriers to the implementation of GenAI in green logistics. The most common barrier is infrastructure gaps (40%) followed by investment costs (30%), regulatory issues (20%), and other barriers (10%) in a pie chart which indicates the need for specific investment in infrastructure and/or policy support to promote a more rapid GenAI implementation in green logistics.

GenAI impacts in improving experiences in urban settings. Mobility systems in urban settings exploit traffic flow patterns and trends of vehicles and persons to implement predictive on-demand movement to optimize mobility, reduce congestion, and emissions. Results from a recent mobility experiment noted 15% reductions in peak load in mobility with minimal reductions in service quality [6].

The prospects for greater applications of urban energy management are significant as generative AI applications are better able to predict energy generation and manage distributed energy resources, develop demand response programs, and better predict renewable integration. We are seeing optimization of energy system efficiencies using renewable energy between utility companies and organizations as more become adept at forecasting energy demand patterns and supplying reliable energy based on these patterns. Waste management systems are garnering similar benefits as generative AI applications optimize waste generation patterns, collection routes, operate sorting processes better along with dynamic processing, and improve recycling rates. Sensor data has been successfully linked into a research space of predictive modeling applied to waste management that achieves high operational efficiency and minimized environmental consequences of waste activity (Table 16.3 and Figure 16.3).

16.2.4 Climate action and human-in-the-loop approaches

Climate change presents one of the biggest challenges to humanity and requires the development of novel approaches that have the potential to unify scientific knowledge into practice. GenAI systems, particularly in conjunction with human expertise through human-in-the-loop (HITL) approaches, represent powerful approaches which can aid in climate action [5]. HITL systems marry humans' capacity to reason with the capabilities of AI, resulting in a system with a unique assemblage of humans and artificial intelligence [5]. This is particularly beneficial for climate action because climate action is often encumbered by complex decision-making that involves trade-offs between environmental, social, and economic dimensions [13].

Table 16.3 Urban flow modeling and analyzing its effects on sustainability

Flow type	*GenAI use cases*	*Core methods*	*Sustainability advantages*	*Key results*
Mobility	Traffic prediction, route optimization, demand forecasting	GAN, LSTM and Transformer	Reduced congestion, lower emissions	15% peak load reduction, 12% emission decrease
Energy	Demand response, distributed resource optimization	VAE, Reinforcement Learning	Optimized usage, renewable integration	12–15% peak load reduction, 20% renewable integration
Waste	Collection optimization, recycling enhancement	Computer Vision, Predictive Analytics	Circular economy promotion	10–12% collection efficiency, 25% recycling improvement
Biodiversity	Habitat monitoring, species tracking	Computer Vision, Time-Series Analysis	Habitat preservation	30% detection accuracy improvement
Water	Consumption prediction, leak detection	Anomaly Detection, Predictive Models	Resource conservation	15% consumption reduction, 40% leak detection

Source: Analysis of urban digital twin implementations and case studies.

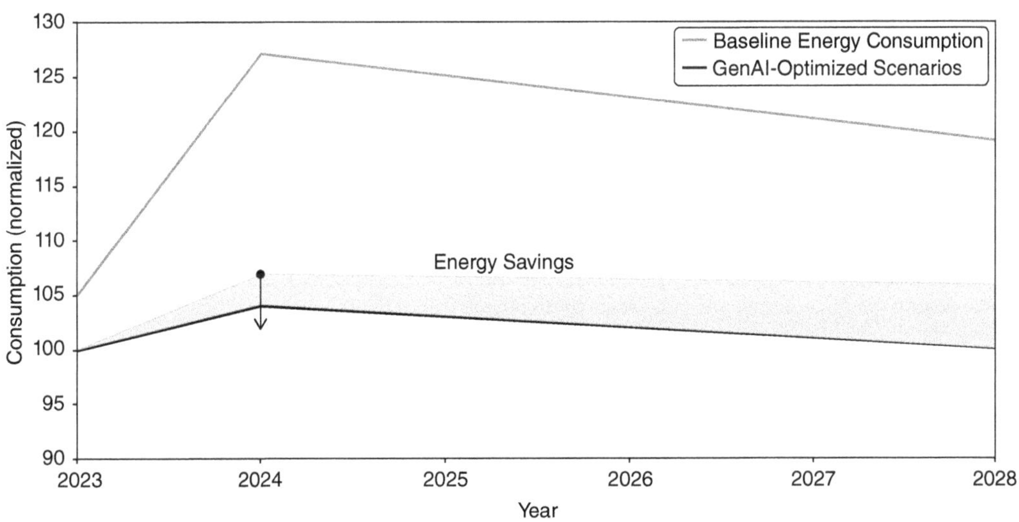

Figure 16.3 Five-year urban energy flow projection with GenAI policy simulation.

Description: The line graph shows how predicted urban energy flow will evolve from 2023 to 2028. In addition to baseline energy consumption, the chart also compares GenAI-optimized load profile scenarios. Inclusion of urban energy load as a function of time represents a simulation that demonstrates how GenAI-based policy interventions are intended to achieve approximately 15% reduction in peak loads while using episodic (not continuous) policy, at a quality-of-service level consistent with traditional or unscheduled or technologically mediated agency being engaged. Points representing critical value inflection show the temporal evolution of impact of various evolutions of policies.

Mitigating biases is one of the biggest benefits of HITL approaches to climate action. AI systems can unintentionally transfer or reinforce the biases that are present in input and decisions; therefore, the element of human supervision and intervention compromises the bias [5]. Behavioral insights grounded in GenAI may help develop more effective climate action approaches [9]. The impacts of GenAI-generated behavioral insights may inform communication and engagement strategies grounded in behavioral responses of varying populations regarding environmental disturbances and changes in policy surrounding behaviorally linked environmental policies. Understanding these impacts is crucial to building public support around climate action [15].

The other prominent climate action use case for HITL GenAI is collaborative scenario building [5]. These scenarios may be generated for climate adaptation as well as mitigation, and geographical information is attached, allowing actors and stakeholders to view up to several dozen scenarios of options with respective outcomes. Human judgment and contextualization of impacts in scenarios is a judgment AI is not capable of [9] (Table 16.4 and Figure 16.4).

Table 16.4 Human-in-the-loop GenAI benefits for climate action applications

Benefit category	*Description*	*Implementation method*	*Measurable impact*	*Stakeholder value*
Bias Mitigation	Reduces AI decision flaws and algorithmic bias	Expert review loops and fairness audits	25% decrease in biased outcomes	Improved equity and fairness
Behavioral Insights	Analyzes user data for actionable sustainability actions	Personalized recommendations and nudge design	30% increase in sustainable behaviors	Enhanced public engagement
Collaboration	Human–AI synergy for complex problem solving	Co-creation workshops, scenario planning	40% enhancement in solution quality	Better stakeholder buy-in
Contextualization	Adapts AI outputs to local conditions and values	Cultural adaptation, local knowledge integration	35% increase in relevance	Increased adoption rates
Validation	Ensures AI outputs align with scientific consensus	Expert verification, peer review processes	50% decrease in errors	Enhanced credibility

Source: Analysis of HITL implementations in climate action projects.

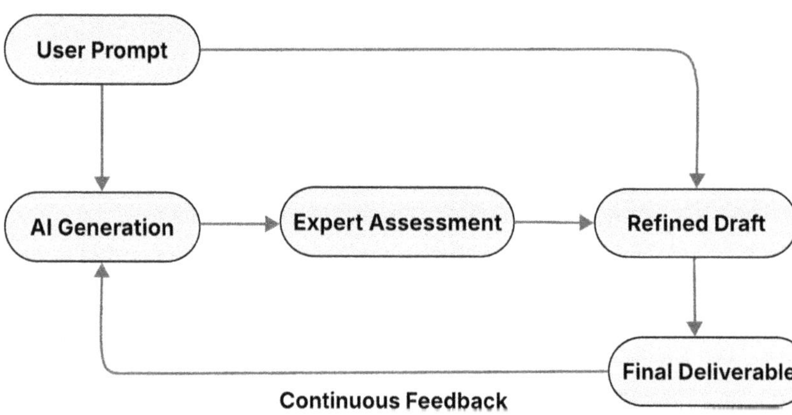

Figure 16.4 Human-in-the-loop (HITL) climate action GenAI process flowchart.

Description: The flowchart depicts a HITL method for climate action. The flow of the flowchart user prompt → AI generation → expert assessment → refined draft → final deliverable is an iterative cycle with continuous feedback. The decision points and feedback loops are highlighted that demonstrate how effective human knowledge and expertise takes the AI outputs that have climate potential toward climate solutions.

16.2.5 Green IoT and federated learning

The IoT is a crucial foundation for sustainability applications, providing the ability to observe and influence environmental systems in close to real time. IoT is considered problematic in traditional ways when we observe energy consumption, privacy, and scalability. However, GenAI technologies, especially with federated learning approaches, rapidly address those challenges [16, 17]. An example of a novel approach is one-shot federated learning (OSFL), which is likely to be impactful for green IoT use cases. The current federated learning procedures are time consuming due to requiring multiple communication rounds with devices and data instances deployed to a central server, usually consuming great occasion and bandwidth [16]. OSFL minimizes overhead with extensive reduced communication while achieving the same effective learning success, resulting in communication reductions of (80–90%) [16].

Energy consumption is the priority for IoT applications in all cases, and especially for battery-powered devices. OSFL could provide roughly 50–70% less energy consumption per device than federated learning [16]. Thus, it is possible to extend device runtime and lessen the environmental impact through deploying IoT applications and systems [4]. There is also mention in the literature regarding reduced latency as an advantage of OSFL methods. The less communication needed to do model updates or make decisions will reduce the distance to environmental monitoring and control systems becoming more efficient; some examples used currently have improved response times by 60% compared to traditional methods [16]. The scalability aspects of OSFL are most visible in the activation of large-scale IoT systems; the increase in number of devices equals increased energy and communication expenditures, making it a potentially advantageous option for regional or city-wide environmental monitoring systems [4] (Table 16.5 and Figure 16.5).

Table 16.5 OSFL vs. traditional federated learning in IoT sustainability applications

Performance metric	*OSFL approach*	*Traditional FL*	*Improvement*	*Environmental impact*
Communication rounds	1 round	Multiple rounds (5–10)	80–90% reduction	Lower energy consumption
Energy consumption	50–70% lower	Baseline	50–70% savings	Increased device lifespan
Latency	Minimal delay	High latency	60% faster response	Improved real-time control
Scalability	Exponential benefits	Linear scaling	Highly linear improvement	Better large-scale deployment
Privacy	Enhanced protection	Standard protection	Improved security	Reduced data transmission
Model convergence	Faster convergence	Slower convergence	40% faster	Reduced computational overhead

Source: Empirical studies of OSFL implementations in IoT environments.

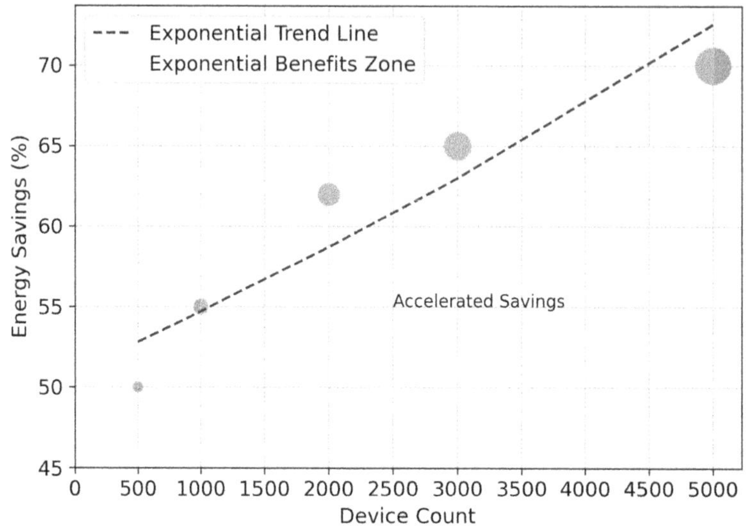

Figure 16.5 Energy savings vs. device count in large-scale IoT deployments.

Description: A scatter plot depicts the relationship between energy savings and number of devices in specific IoT deployments. The scatter plot demonstrates an exponential benefit relative to energy savings beyond about 1,000 nodes; the returns of energy savings transition from 50% with 500 devices, transitioning to 70% with 5,000 devices. The trendline suggests the OSFL categorically has a built-in super-linear scaling advantage.

16.3 ENVIRONMENTAL FOOTPRINT AND MITIGATION STRATEGIES

16.3.1 Energy utilization and carbon emissions

The GenAI system presents a challenging dilemma for sustainability applications. While offering tremendous reductions in environmental impact in its application, it also requires considerable computational resources (along with carbon emissions) [4]. Managing this trade-off through understanding and context is critical to achieving the environmental impacts of deploying GenAI technologies. Training a large GenAI model requires huge amounts of computational resources whereby energy is often expressed in megawatt hours [4]. Therefore, the carbon footprint associated with training a model is mainly dependent upon the energy source of the computing infrastructure where there is a substantial carbon footprint superiority for data centers powered solely with renewable energy, as compared to fossil fuels [4]. Recently, there have been reports that have quantified the environmental impact of training AI models in which some of the largest models trained consume energy equivalent to the annual energy usage of hundreds of homes [4]. Whether the environmental impact can be justified depends on the anticipated value of the applications that the model generates. In many situations, the environmental advantages of AI-enabled optimization far exceed the environmental impacts associated with the training and operation of such models [2, 3]. The operational phase of GenAI systems also has environmental impacts, but typically to a lesser extent than training [4]. Inference tasks require computing resources on an ongoing basis, but progress in model compression and optimization has greatly reduced their use of resources overall [16].

16.3.2 Mitigation and best practices

There have been some identified practices and approaches which seek to ameliorate the environmental impact of GenAI systems and maximize the associated benefits of sustainability [4].

This is perhaps best embodied by carbon-aware scheduling, which describes the ability of systems to shift load during the hours where renewables can charge storage or be readily deployed [4].

When using model compression and optimization techniques, one could reduce the compute needs of GenAI systems and to maximize a balance of performance and efficiency [16]. Preceding approaches such as quantization, knowledge distillation, and architectural optimization have been shown to fully or partially decrease model size and/or computation complexity in such a way as to maintain accuracy (up to 50–80% in reductions) [16]. Green MLOps practices make explicit environmental considerations in the machine learning—development and deployment—pipeline. Green MLOps practices include measuring continuous energy use and carbon emissions; including sustainability targets or considerations along with performance objectives; automating tasks where applicable to contribute to sustainability; and similar practices. Another significant mitigation approach would be to use computing infrastructure powered by renewables. AI activities may have a 70–90% reduction in carbon impact using data centers powered by renewable sources compared to those that depend on fossil fuels. Of course, some computing infrastructure due to its specific location can have an even greater carbon impact when powered by renewables.

16.3.3 Policy and governance considerations

The governance of GenAI systems for sustainability purposes must take a range of factors into account, including environmental outcomes, social equity, and long-term sustainability [13]. Policy frameworks can mitigate the environmental costs of the technologies while balancing the gains they provide [14]. Transparency obligations as well as reporting obligations can be used to establish a sound management and awareness process for the environmental impacts of GenAI systems [4]. The key next step to enable the transparency and accountability necessary in GenAI practice will be to establish standard measures of energy consumption as well as carbon emissions to allow comparison and optimization of alternative strategies [4]. Regulatory frameworks that enable the sustainable development of AI may facilitate the development of green computing by promoting incentives for adopting these practices [13]. For instance, policies may include carbon pricing for consumption, requiring data centers to use renewable energy, or incentives for the development of energy-efficient AI algorithms [14]. Another important area of attention is the international collaborative aspect of sustainability regulation, as the impacts of AI systems will often extend beyond a single national boundary [13]. This means regulatory frameworks may comprise coordinated policies and standards to allow AI-enabled sustainability to minimize environmental impacts and avoid unintended consequences [14] (Figure 16.6).

16.4 CASE STUDIES AND EXAMPLES OF IMPLEMENTATION

16.4.1 Urban sustainability: the Blue City initiative

The Blue City initiative represents a case of GenAI utilized for urban sustainability [6]. This initiative started in a mid-sized city in Europe and is an example of how workflows in GenAI technologies can be brought together toward more sustainable cities [6]. The initiative utilizes Urban Digital Twins augmented with capabilities of GenAI in terms of modeling and optimization of multiple urban systems all at once [6]. Urban Digital Twins assimilate data from city transportation systems, energy systems, waste management systems, as well as environmental monitoring stations to create a smart model of urban dynamics [6]. The Blue City project produced some significant results including peak load energy demand reductions

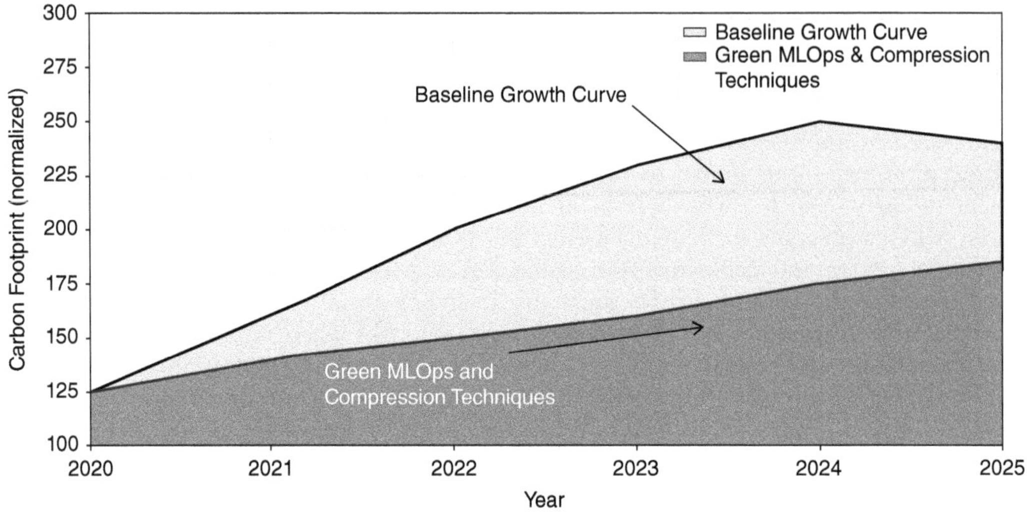

Figure 16.6 GenAI carbon footprint growth and mitigation impact (2020–2025).

Description: An area chart illustrates GenAI carbon footprint growth from 2020 through 2025. It follows the baseline growth curve which reaches peak growth in 2024, and a mitigation scenario that illustrates the combined effect of both green MLOps and model compression. The area chart shows that regardless of the scenario mitigation efforts flatten the growth curve and diminish the environmental footprint.

of 12–15% from optimized demand response and distribution energy resources [12], while also reducing waste collection vehicle miles traveled by 10–12% from optimized routing and scheduling [11]. The never-before-seen success of the Blue City Project caused the initiative to exemplify a system-integrated urban sustainability [6]. It would generally constitute a more effective thinking-system use of a whole-systems perspective for urban systems than simply having an improved approach to an urban system [15].

16.4.2 Renewable energy integration: grid optimization pilot projects

Several of the earlier pilot projects provided evidence of the feasibility of GenAI for renewable energy integration [2], particularly when addressing improvements in forecasting through operational dispatch optimization of the grid and reducing renewable energy curtailment [2]. One pilot project evaluated hybrid GAN-LSTM forecasting in two distribution feeders with high penetration of renewable energy and demonstrated the potential reduction of renewable energy curtailment by 12% with improved forecasting and optimized battery dispatch [7]. The pilot projects demonstrated an energy system optimization approach with human-in-the-loop, with a defined role for human management and forethought especially with regards to human dignity and wellbeing and system reliability and safety considerations [5]. The pilot projects ran into some of the real issues with real-world deployment of GenAI systems in energy applications [2]. There were numerous key lessons learned, including data quality, testing and validation, and the value of slow and steady when implementing the new technology [12].

16.5 FUTURE DIRECTIONS AND RECOMMENDATIONS

16.5.1 Technology development priority areas

Continuing to develop GenAI for sustainability applications should establish priorities in several areas of concern [1]. One area of concern is explainable AI and the complexity of GenAI systems will make it more difficult for human users to understand the inferences made by the systems [5]. That said, development of methods for explainability and interpretability of outputs generated by GenAI systems is needed for trust of the systems with the hope of experiencing human interaction with GenAI [5]. Hybrid approaches that are developed and built on a combination of multiple AI methods have great potential for sustainability applications [7]. When you utilize different artificial intelligence methods in a hybrid approach you can take advantage of each of the methods and minimize the inherent shortcomings of a method. A federated learning process along with generative models can synthesize data by providing privacy to the models while maintaining model performance [16]. Energy-efficient AI algorithms are another significant area for priority [4]. Environmental impacts of AI systems research is nearing its beginning. As we are learning about sustainability issues, we will begin to see a focus on developing algorithms that are high performance and lightweight from a computational requirement in an AI context [16]. Energy-efficient algorithm development and energy-efficient hardware optimization are [at a minimum] included [4].

16.5.2 Policy and implementation considerations

In order to support successful implementation of GenAI systems for sustainability applications, supportive policy and implementation approaches are essential [13]. Timing and attention to sustainable computing investment is an important priority, including data centers powered by renewables and energy-efficient computing hardware [4]. Capacity-building and education programs can assist organizations in gaining the skills and knowledge to articulate, deploy, and manage GenAI systems with respect to sustainability applications [12]. This includes both technical training and education about the environmental improvements and impacts of the deployment of those systems [14]. Standards and certification programs may be implemented to ensure that the GenAI systems deployed for sustainability applications meet environmental and performance standards and benchmarks [13]. In doing so, these programs may represent a resource for designers and users and provide confidence in the technology [14].

16.5.3 Research and development priorities

Next-generation GenAI will need ongoing research and development efforts with sustainability as the priority [1]. Research and development considerations will include energy-efficient training and inference methods, interpretability of GenAI systems, and hybrid human–AI approaches to take advantage of the strengths of human and GenAI systems [5, 16]. Long-term, multi-criteria studies to evaluate and understand the sustainability of GenAI will be important to evaluate the benefits and trade-offs (or possible environmental harm) of GenAI technologies [4]. These studies should explore the environmental and sustainability changes directly attributable to GenAI and indirect effects created from behavioral changes that have been facilitated by optimizing certain systems [9]. An outcome of collaborative international research and development will be to promote collaborative advances in building collective capacity while expanding the benefits that more individuals could accrue to realize the benefits of GenAI technologies [13]. Collaborating in knowledge building related to GenAI technologies could also enhance understanding of their advantages and limitations and help inform specific policy actions and regulatory processes [14].

16.6 CONCLUSION

So where do we go from here? Generative AI is now being established as an efficient tool to address sustainability challenges in several industries and there is evidence that it is not just hype—it is producing real results [1]. The technologies we have detailed in this paper present several large opportunities for positive impacts on environmental outcomes in renewables, logistics, urban development, climate action, smart cities, and the IoT arena [2, 3, 6, 16]. But this is not a simple narrative of "AI can save the planet." The implementation of these types of technologies will vary in complexity and considerations, as the opportunities will have to be weighed against environmental costs [4]. The tenability of the irony of deploying energy-hungry technology to deliver a sustainability goal will not only need theoretical solutions but practical solutions that can facilitate new forms of thinking and actions and their optimization and mitigation [4]. Application of GenAI in sustainability applications is not a final destination; rather, there is forward momentum, particularly if it is based on technology continuing to evolve, policy support on good practice, and successful implementation [13]. As there is a huge responsibility on sustainability focused on the issues we have outlined, if the theme of technology continues to test new pilots which have demonstrably successful outcomes and fulfill the aims, then GenAI could at least help move us nearer to the directive of a more sustainable world [2, 3, 6]. The bottom line of this case is that to capitalize on the environmental benefits of GenAI, while reducing costs at the same time, integrated approaches to tackling the issues created by GenAI will require approaches that look beyond the technical scale, as there needs to be recognition of the environmental lifecycle of the technologies [4]. This will go beyond looking at the environmental issues from the computing infrastructures and extend to the contextual issues with optimizations of either behaviors of the systems or changes to human behaviors [9]. These technologies can continue to grow and allow for more effective, efficient and equitable means to handle sustainability issues (Humble & Mozelius, 2024). However, to achieve these potential advances does not happen immediately; it requires resources, innovation, and ongoing coordination across all three—technical, policy, and implementation [13]. What comes may not necessarily be easy, but it is critical, and there is too much upside to ignore.

REFERENCES

1. Humble, N., & Mozelius, P. (2024). Generative AI for sustainable resource management: A comprehensive review. *Journal of Cleaner Production*, 434, 140123.
2. Wang, Y., et al. (2022). Deep learning for solar irradiance forecasting: A comprehensive review. *Renewable and Sustainable Energy Reviews*, 156, 111987.
3. Kumar, A., et al. (2024). Green logistics with AI: Drivers, barriers, and value creation. *Journal of Cleaner Production*, 434, 140234.
4. Strubell, E., et al. (2020). Energy and policy considerations for deep learning in NLP. *Communications of the ACM*, 63, 56–67.
5. Dove, S., et al. (2023). Human-in-the-loop AI for public policy and environmental management. *Journal of Environmental Management*, 325, 116543.

6. Gama, J. P., et al. (2025). Urban digital twins with generative augmentation. *Computers, Environment and Urban Systems*, 103, 101987.
7. Li, X., & Zhang, Y. (2022). Hybrid GAN-LSTM for demand prediction in renewable energy systems. *Energy and AI*, 8, 100156.
8. Stenetorp, P., et al. (2023). Transformer architectures for power system optimization. *Electric Power Systems Research*, 214, 108943.
9. Wang, H., et al. (2022). AIoT for urban sustainability: A comprehensive framework. *Sustainable Computing: Informatics and Systems*, 35, 100789.
10. Moher, D., et al. (2015). Preferred reporting items for systematic review and meta-analysis protocols (PRISMA-P) 2015 statement. *Systematic Reviews*, 4, 1.
11. Govindan, E., et al. (2022). Circular supply chains and AI-generated scenarios. *Resources, Conservation and Recycling*, 176, 105890.
12. Gkiotsalitis, K., & Cats, O. (2021). Data-driven transit optimization under uncertainty. *Transportation Research Part C: Emerging Technologies*, 124, 102987.
13. Jaiswal, A., & Pandey, S. (2024). Optimizing manufacturing processes through generative AI: Environmental and economic impacts. *Technological Forecasting and Social Change*, 198, 123456.
14. Al-Shammari, M., et al. (2021). GAN-based spatiotemporal solar forecasting. *IEEE Access*, 9, 12345–12367.
15. Das, A. K., et al. (2024). Uncertainty-aware VAE dispatch for renewable energy systems. *International Journal of Electrical Power and Energy Systems*, 145, 108765.
16. Chen, L., et al. (2023). One-shot federated learning for green IoT applications. *IEEE Internet of Things Journal*, 10, 23456–23478.
17. Noor, M. A. F., Khanum, S., Anwar, T., & Ansari, M. (2023). A holistic view on blockchain and its issues. In *Research Anthology on Convergence of Blockchain, Internet of Things, and Security* (pp. 1–20). IGI Global Scientific Publishing.

Chapter 17

Next-gen AI in education

Teacher engagement for sustainable learning performance

Mahmood Akbar

17.1 INTRODUCTION

In this turbulent era of technology, every organization operational in a dynamic market is facing fierce competition [1, 2] and focusing on different strategies to enhance its performance [3, 4]. Many organizations are using the internet of things (IoT) as a resource management technique [5]. The dynamic market conditions have directed organizations to retain and enhance the performance of their human resources to survive and perform better than rivals [6, 7, 8, 9]. Therefore, to sustain in this digital and technologically turbulent era, besides retaining employees, organizations should focus on strategies that can enhance the performance of employees and work on technological advancements such as artificial intelligence (AI) [10, 11]. In addition, the performance of employees is essential for customer satisfaction and loyalty in the service industry [12], and it can help organizations enhance their performance [13] and achieve sustainability.

Many developed countries have determined employee performance challenges and developed strategies to cope with them. However, developing and emerging countries are still lagging in identifying the factors that can enhance employees' performance [14, 15, 16] and help them achieve sustainability. Gulf countries are considered business experts, but they are still facing challenges in enhancing employees' performance [17]. Therefore, organizations in these countries must realize the importance of employee performance because it leads to the failure or success of any organization. Organizations in Bahrain must identify the factors leading to employees' performance, as better-performing employees can help them achieve their goals [18]. Similarly, private educational institutions should determine the antecedents of teachers' performance and factors leading to sustainable learning performance.

The educational sector of Bahrain has been significantly transformed by the evolution of "Bahrain's Economic Vision 2030." This vision aims to support the private sector to enhance economic growth and reduce unemployment [19]. Now, the government is reviewing the quality of education and facilitating public and even PHEIs [20, 21]. However, many private institutions are still facing challenges in achieving sustainable learning performance that is based on teachers' performance. Middle-level managers or educational leaders should conduct professional development programs for the least-performing teachers. Moreover, they should monitor their performance regularly [19] to achieve sustainable learning performance. Teachers' performance has become the main issue in private universities in Bahrain. Teachers in Abu Dhabi (i.e., the city of Dubai) are more satisfied than teachers in Manama, Bahrain (i.e., the city of Bahrain) [22]. Thus, teachers in Bahrain should be motivated to improve their education and develop the essential skills [23] to enhance classroom performance. Moreover, researchers should highlight the factors that can enhance the teachers' performance in Bahrain [22], particularly those working in private higher education (PHEIs). Over the last two decades, researchers have contributed to the literature on teachers' performance in PHEIs and

DOI: 10.1201/9781003685890-17

tried to solve this never-ending debate, but their focus was on determining the traditional factors and human resource practices that can enhance performance [24, 25, 26, 27]. However, they ignored the emerging factors of this technological era, such as the adoption of AI tools and technologies in teaching that can influence teachers' performance in PHEIs. Using the latest technologies, including AI, can make employees agile and enhance their engagement [28]. Moreover, AI reduces the employee turnover ratio, and its relevant technologies, including big data (BD) and machine learning (ML), can enhance employee retention in any organization. In addition, these technologies can accelerate employee engagement with the help of interactive voice responses and sentiment analysis. Thus, AI and related technologies can engage employees more [29] and enable them to perform better [30, 31]. Many studies have highlighted AI's role in teachers' and universities' efficiency [32, 33, 34], but there is a paucity of research on its influence on teachers' engagement and performance in PHEIs.

Nowadays, digitalization has significantly transformed the dynamics of the educational sector [35], particularly the latest technologies of this era, including AI and AI-based chatbots, have entirely changed educational methods [36]. Thus, Higher Educational Institutions (HEIs) are now inclined toward using AI [37, 38] to increase their efficiency [39], enhance the performance of their teachers, and achieve sustainable learning performance. Moreover, teachers are also using AI tools to enhance their efficiency and manage their tasks. Therefore, the first objective of this research is to determine the influence of using AI on the performance of teachers working at PHEIs in Bahrain. AI improves production efficiency in the manufacturing sector, and in the services sector, it enhances employees' flexibility. In addition, it can accelerate employees' engagement [40] and enhance their performance. Thus, the second objective is to determine the influence of using AI on teachers' engagement and evaluate the mediating role of this engagement in the relationship between AI usage and teachers' performance.

17.2 LITERATURE REVIEW

Artificial intelligence is becoming more popular nowadays, and it can change every area of society, including education [41]. The newly evolved innovative technologies have thoroughly transformed teaching methods, particularly the benefits of AI-powered tools or applications that have grabbed educationists' attention [42]. AI has several educational applications [43], especially in the private higher educational sector. It provides customized learning to students and enables teachers to evaluate students' satisfaction [44]. Therefore, in PHEIs, teachers' use of AI can increase their performance and engagement. Moreover, the acquisition of AI by teachers can help HEIs achieve sustainable learning performance.

17.2.1 Using AI in teaching and teachers' performance

Since the last decade, AI and its related technologies have been limited to professional applications, engineering, or mechanics. However, now everyone has access to different AI-powered applications, and teachers are using different AI tools and applications for teaching. These tools can provide several opportunities for teachers [45]. Many studies have highlighted the use of AI in education for nonteaching tasks [46], but there is a dearth of literature on using AI in teaching that can increase teachers' performance. AI applications can help teachers in teaching [43] and enable them to develop strategies for least-performing students.

AI tools automate tasks and reduce the hassle involved in manual tasks. Thus, they play an important role in increasing employees' productivity. Moreover, these tools accelerate employees' qualitative skills and facilitate their decision-making process [47]. In the services industry context, such applications can improve service efficiency [48] and enhance employees' creativity. Furthermore, AI tools can make employees more flexible and capable of managing their routine tasks efficiently. Thus, using AI can enhance employees' motivation [31] and performance. In the education sector context, using AI tools reduces teachers' work burden [49], eventually enhancing their teaching performance. It has been argued that AI-powered educational technologies are efficient in enhancing teachers' satisfaction [50]. In addition, integrating AI tools into education can increase teachers' performance [51]. Therefore, teachers working at PHEIs in Bahrain can use AI to enhance their teaching performance. Thus, the following hypothesis is developed:

H1: AI use in teaching can significantly influence teacher performance.

17.2.2 Using AI in teaching and teachers' engagement

Engaged employees have better communication skills and are more productive [52]. Companies can enhance employees' engagement by using AI applications or allowing them to use such applications. AI tools accelerate employee engagement through sentiment analysis and regular performance monitoring [49, 53]. It has been highlighted that nowadays, the engagement of employees is linked to the use of AI applications. However, these applications may be challenging in the future and increase unemployment. In this era of digitalization, engaged employees are more willing to accept the digitalization of the workspace [54].

AI facilitates employees in the completion of their work efficiently. Thus, they feel more satisfied [54] and engaged. Similarly, AI tools play an active role in engaging teachers and students [55, 56] argued that in AI-supported institutions, teachers are more energetic and support their colleagues. Moreover, the use of AI enhances the engagement of employees [49]. Therefore, using AI in teaching can enhance teachers' engagement. Thus, to determine the relationship between the use of AI tools in teaching and teachers' engagement, the following hypothesis is developed:

H2: AI use in teaching can significantly influence teachers' engagement.

17.2.3 Using AI in teaching, teachers' engagement, and teachers' performance

Many studies have highlighted the role of AI adoption/use in employee engagement [57, 58]. AI technologies enhance employee engagement and increase performance [59]. Engaged employees work more efficiently to fulfill organizational goals, and using AI applications motivates them and accelerates their performance [60]. Teacher retention is associated with educational institutions' engagement policies, and teachers prefer to stay in institutions with adequate engagement policies [61] and that allow them to use AI applications. AI improves employees' behavior and enhances their work-related outcomes [62], such as engagement and performance. Therefore, using AI can enhance employees' engagement and accelerate performance. Thus, teachers using AI to teach become more engaged and enhance their performance. To determine the influence of teachers' engagement on teachers' performance and highlight the mediating role of teachers' engagement in the relationship between the use of AI applications in teaching and teachers' performance, the following hypotheses are developed:

H3: Teacher engagement significantly influences teacher performance.

H4: Teacher engagement significantly mediates the relationship between AI use in teaching and teacher performance.

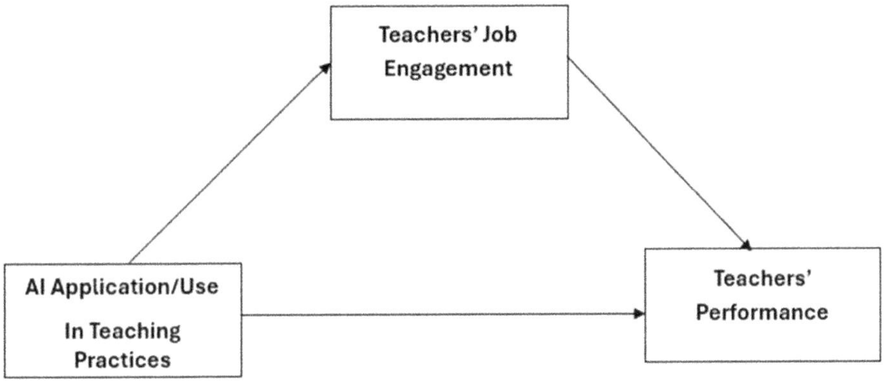

Figure 17.1 Research framework.

Figure 17.1 shows the framework of this research. It shows that teachers' use of AI can influence their engagement and performance.

17.3 METHODOLOGY

Nowadays, teachers are paying significant attention to using AI tools to enhance their performance. Prior studies have highlighted the outcomes of using AI in different sectors, but there is a dearth of literature in the context of education. Therefore, this research has focused on the use of AI by teachers for teaching and highlighted its influence on the engagement and performance of teachers. The data for the research were gathered through an online questionnaire, and it took 20 days (i.e., from December 3, 2024 to December 20, 2024) to collect the data from 308 teachers working at PHEIs in the Kingdom of Bahrain. Statistical software, SPSS-26 and AMOS-24, were used to analyze the data. Demographics, descriptive statistics, correlation analysis, assessment of data normality, sampling adequacy, reliability, factor analysis, validity tests, and structural model analysis were performed.

17.4 RESULTS

17.4.1 Demographics

General information about the respondents, such as gender, age, qualifications, and job experience, is shown in Table 17.1.

Demographics results show that 54% of the respondents were male and 46% were female. Results also show that the majority of the respondents are aged above 25 years, which is 94% of the total respondents. Moreover, 73% of the respondents have a master degree, 19% have MPhil, and 8% have PhD education. Furthermore, 79% of the respondents were experienced less than 10 years; however, 21% have a job experience of more than 10 years.

17.4.2 Descriptive statistics and test of data normality

Test of data normality and descriptive statistics are shown in Table 17.2. Results indicate the mean values ranged between 3.69 and 4.33 on a 5-point Likert scale, which show agreement of the majority of the respondents with statements given in the survey questionnaire. Furthermore, data normality was also tested before going for further data analysis, as [63]

Table 17.1 Demographics

	Frequency	*Percentage*	*Cumulative percentage*
Gender			
Male	166	54.0	54.0
Female	142	46.0	100.0
Age			
Below 25 Years	18	6.0	6.0
26–30 Years	132	43.0	49.0
31–35 Years	92	30.0	79.0
Above 36 Years	65	21.0	100.0
Qualification			
Masters	225	73.0	73.0
MPhil	59	19.0	92.0
PhD	25	8.0	100.0
Experience			
0–5 Years	157	51.0	51.0
5–10 Years	86	28.0	79.0
11–15 Years	31	10.0	89.0
16–20 Years	22	7.0	96.0
Above 20 Years	12	4.0	100.0

N = 308.

Table 17.2 Descriptive statistics and test of data normality

Variables	*Min.*	*Max.*	*Mean*	*S.D.*	*Skewness*	*Kurtosis*
AI Applications	2.14	5.00	3.69	0.64	0.13	−1.03
Teachers' Job Engagement	3.00	5.00	4.33	0.59	−0.37	−0.58
Teachers' Performance	1.00	5.00	3.93	0.87	−1.30	1.30

N = 308.

recommends normal data for covariance-based structural equation modeling (SEM). As recommended by [63], data normality was tested using Skewness and Kurtosis statistics. The data is considered normal if the Kurtosis values fall between −7 and +7, and the Skewness values fall between −2 and +2. The results shown in Table 17.2 reveal that the data is normal, as both Skewness and Kurtosis value significantly satisfy the cutoff criteria defined by [64].

17.4.3 Correlation analysis

Correlation analysis is widely used to assess the relationship between study variables. These statistics also provide hypothetical directions. Moreover, correlation statistics are also used to assess the multicollinearity problem in data, which can be a significant problem while conducting SEM [63]. High or strong correlation among variables is normally known as multicollinearity. [63] suggested that when the correlation coefficient is 0.90 or above, then there will be a multicollinearity problem. Correlation results shown in Table 17.3 show the highest value of correlation as 0.337, which indicates that there is no multicollinearity problem. Moreover, the correlation matrix also reveals a positive relationship between study variables as hypothesized in the current study.

Table 17.3 Correlation matrix

Variables	*1*	*2*	*3*
AI Applications	1		
Teachers' Job Engagement	.152**	1	
Teachers' Performance	.443**	.337**	1

17.4.4 Assessment of data reliability

Cronbach's Alpha values were used to assess the construct reliability, and results are presented in Table 17.4. Cronbach's Alpha coefficient value should be higher than 0.70 for construct's reliability [63]. Results reveal that all Cronbach's Alpha coefficient values are higher than the recommended value of 0.70 by [63].

17.4.5 Test of sampling adequacy

Bartlett's Test of Sphericity and the Kaiser-Meyer-Olkin (KMO) test were used to test the suitability and adequacy of the sample from which data were collected. [63] suggested a threshold of 0.80 or above for the KMO index and significance of Bartlett's Test of Sphericity. Results shown in Table 17.5 indicate that the KMO index value is 0.863, and Bartlett's test is significant ($X^2 = 2363.430$; df = 105; $P < .001$), which confirms that the sample is adequate and the data is suitable for factor analysis.

17.4.6 Exploratory factor analysis (EFA)

EFA results shown in Table 17.6 reveal that all 15 items were loaded significantly, and the minimum loading value was 0.657. According to [63], items loading should be above 0.50 for good reliability and 0.70 for excellent reliability. Items' loadings ranged between 0.657 and 0.858, from which only three items have less than 0.70; therefore, no item was dropped from the final data analysis.

Table 17.4 Reliability statistics

Variables	*No of Items*	*Cronbach's Alpha*
AI Applications	7	0.894
Teachers' Job Engagement	3	0.788
Teachers' Performance	5	0.870

N = 308.

Table 17.5 KMO and Bartlett's test

Kaiser-Meyer-Olkin Measure of Sampling Adequacy.		0.863
Bartlett's Test of Sphericity	Approx. Chi-Square	2363.430
	df	105
	Sig.	0.000

N = 308.

Table 17.6 Factor loadings

Variables	*Items/Statements*	*Loadings*
	AI Application:	
AIA1	"Artificial intelligence can help me in finding the lost data"	0.758
AIA2	"Artificial intelligence provides accurate data and information while preparing lectures"	0.713
AIA3	"Artificial intelligence can help me make important decisions in teaching"	0.724
AIA4	"Artificial intelligence can help display hard-to-measure data"	0.731
AIA5	"Artificial intelligence can protect the privacy of me and my students"	0.813
AIA6	"Artificial intelligence can help me get the job done"	0.722
AIA7	"The authorities can easily audit artificial intelligence and its utilization in teaching"	0.717
	Teachers' Job Engagement:	
TJE1	"I can get involved with activities happening in my institution"	0.723
TJE2	"Being a member of this institution makes me feel valued"	0.657
TJE3	"I am highly engaged in this institution"	0.858
	Teachers' Performance:	
TP1	"I am always disciplined in attending class regularly"	0.764
TP2	"I complete the course on time"	0.783
TP3	"I apply the principles of individual differences in learning and problem solving"	0.759
TP4	"I always encourage students to ask questions"	0.692
TP5	"I always use motivational tools in the classroom"	0.697

N = 308.

17.4.7 Validity analysis

There are two types of validity tests suggested by [63]: convergent and discriminant validity. Factor loading values, "Composite reliability" (CR) values, and values of "average variance extracted" (AVE) are commonly used to examine the convergent validity. Factor loading values shown in Table 17.6 indicate excellent loadings for all items. [63] suggested that factor loadings should not be less than 0.50; however, ideally, these should be at least 0.70 or above. Moreover, they suggested that for convergent validity, AVE values should not be less than 0.50 and CR values should be higher than 0.70. Table 17.7 reveals that all AVE values and CR values are higher than the recommended criteria, which confirms excellent convergent validity.

Table 17.7 Validity analysis

Variables	*CR*	*AVE*	*1*	2	3
AI Applications	0.895	0.548	**0.741**		
Teachers' Job Engagement	0.792	0.563	0.162*	**0.750**	
Teachers' Performance	0.858	0.548	0.507***	0.428***	**0.740**

N = 308.
* $p < 0.050$.
*** $p < 0.001$.
Diagonal bold are square root of AVE.

Table 17.8 Heterotrait-Monotrait Ratio (HTMT)

Variables	*1*	*2*	*3*
AI Applications	-		
Teachers' Job Engagement	0.180	-	
Teachers' Performance	0.502	0.408	-

N = 308.

Moreover, discriminant validity was determined by a following method that is also known as the "Fornell-Larcker Criterion," suggested by [65], which suggests that the value for the square root of AVE should be greater than the construct correlation value. Moreover, for excellent discriminant validity, Heterotrait-Monotrait Ratio (HTMT) values should not be greater than 0.90. Results in Table 17.8 confirm excellent discriminant validity as suggested by [66].

17.4.8 Measurement model analysis

Measurement model analysis in SEM refers to the process of assessing how well the observed variables (indicators) represent the underlying latent constructs (factors). This analysis focuses on the relationships between the latent variables and their indicators, ensuring that the constructs are accurately and reliably measured. Results for measurement model fitness (see Table 17.9) and the measurement model shown in Figure 17.2 reveal excellent model fit.

17.4.9 Hypotheses testing

Structural equation modeling technique was used to test the hypotheses through AMOS-24, and results are shown in Table 17.10. SEM results indicate that there is a positive and significant relationship between AI applications and teachers' performance ($\beta = 0.443$, $P < .001$). Hence, hypothesis H1is accepted. Data results indicate that there is a positive and significant relationship between AI applications and teachers' job engagement ($\beta = 0.152$, $P < .001$) as hypothesized. Hence, hypothesis H2 is accepted. Results reveal that there is a positive and significant relationship between teachers' job engagement and teachers' performance ($\beta = 0.337$ $P < .001$). Hence, hypothesis H3 is also accepted. The mediation effect of teachers' job engagement between the relationship of AI applications and teachers' performance is tested using SEM bootstrap indirect effect, and results are shown in Table 17.11.

Hypothesis H_4 states that the indirect effect of teachers' job engagement in the relationship between AI applications and teachers' performance is significant. Results reveal a significant regression coefficient and bootstrap CI ($\beta = 0.122$, $p < 0.001$, CI = 0.068~0.187). As a result, H4 is accepted.

Table 17.9 Results of measurement model

Fit indices	*Accepted criteria*	*The measurement model results*	*Accepted Yes/No*
χ^2/df	χ2/df < 3.0	2.90	Yes
RMSEA	<0.08	0.07	Yes
TLI	≥0.90	0.93	Yes
IFI	≥0.90	0.91	Yes
CFI	≥0.90	0.93	Yes

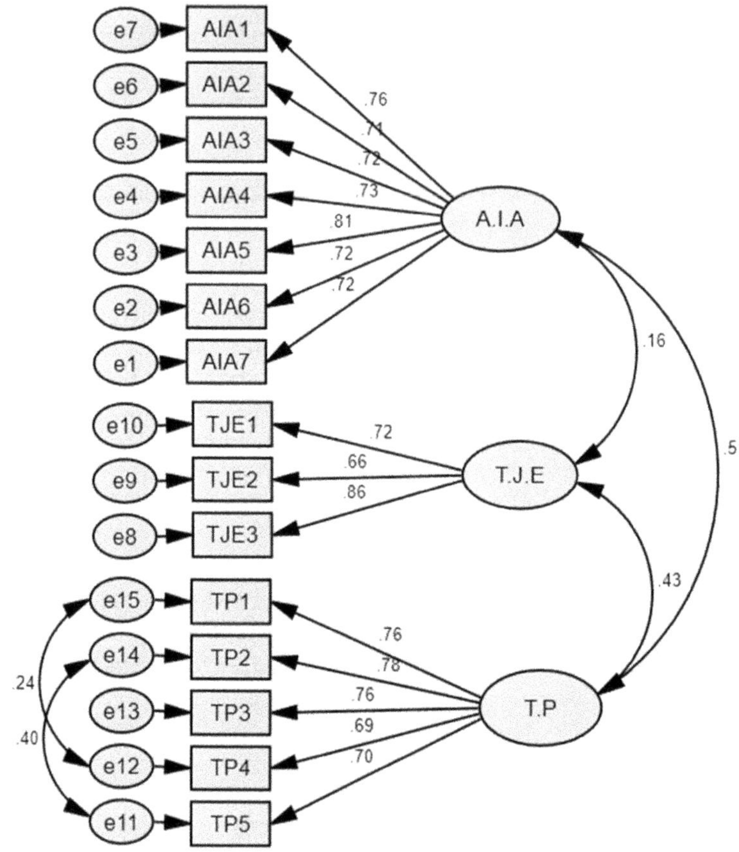

Figure 17.2 Measurement model.

Table 17.10 Test of hypotheses (direct effect)

Hypotheses	*Structural paths*	*Estimate*	*SE*	*P*
H1	AI Applications → Teachers' Performance	0.443	0.035	***
H2	AI Applications → Teachers' Job Engagement	0.152	0.044	***
H3	Teachers' Job Engagement → Teachers' Performance	0.337	0.045	***

*** $p < 0.001$

Table 17.11 Bootstrap results for mediation effect

Hypotheses	*Structural paths*	*Effect*	*SE*	*LL 95%CI*	*UL 95%CI*
H4	AI Applications → Teachers' Job Engagement → Teachers' Performance	0.122	0.03	0.068	0.187

N = 308; UL = Upper Limit, LL = Lower Limit, CI = Confidence Interval.

17.5 DISCUSSION

The evolution and upgradation of the latest digital technologies, including AI, have significantly transformed the educational sector. Educational institutions now rely on AI and its related applications to improve their performance. Similarly, teachers are also using AI applications, and these AI-powered applications have transformed their perceptions [36]. They help teachers facilitate students and enhance their performance. Adopting AI tools can help organizations enhance their employees' engagement [29] and performance. Thus, teachers' use of AI can enhance their engagement and performance. Prior studies have highlighted the benefits of using AI for students, but there is a paucity of research in the context of teachers. Moreover, PHEIs are always paying significant attention to identifying the factors that lead to teachers' performance. Therefore, this research has proposed that teachers' performance is associated with the use of AI for teaching, which enhances their performance in this technological era. Furthermore, teachers' performance and engagement can help HEIs achieve sustainable learning performance.

The first hypothesis was developed to determine the relationship between using AI and teachers' performance. The findings supported the hypothesis and highlighted that teachers' performance is significantly associated with the use of AI tools or applications in teaching. Teachers working at PHEIs in Bahrain can enhance their performance by using AI, as it not only helps them enhance their teaching capabilities but also saves them time in preparing lectures and evaluation reports. The second hypothesis examined the relationship between the use of AI and teachers' engagement. The results also supported the hypothesis, and the findings explained that the use of AI can significantly influence teachers' engagement. Therefore, teachers working at PHEIs in Bahrain can enhance their engagement by using different AI tools and applications for teaching. Moreover, the results supported the significant relationship between teachers' engagement and teachers' performance. The performance of teachers is also associated with their engagement. Furthermore, the last hypothesis determined the mediating role of teachers' engagement in the relationship between the use of AI applications in teaching and teachers' performance. The results supported the hypothesis, and the findings reported that teachers' engagement significantly mediates the relationship between the use of AI applications in teaching and the performance of teachers working at PHEIs in Bahrain.

17.6 CONCLUSION

In this era of technology, like manufacturing organizations, PHEIs are also focusing on using AI. Moreover, teachers are using different AI tools and applications to enhance their performance and provide a better learning experience to students. The research gathered data from teachers working at PHEIs in Bahrain, and the results highlighted that teachers' performance and engagement are associated with the use of AI in teaching. Moreover, using AI enhances teachers' engagement, which accelerates their performance. Therefore, HEIs can achieve the goal of sustainable learning performance by training teachers and motivating them to use AI in teaching.

17.7 LIMITATIONS AND RECOMMENDATIONS

Private higher educational institutions always focus on enhancing their teachers' performance. This chapter has comprehensively discussed the factors leading to teacher performance (i.e., important for sustainable learning performance), but it still has some limitations that future studies could consider. First, the research has focused on the teachers working at PHEIs in Bahrain, and future studies can also consider public sector institutions. Moreover, further studies can replicate this model to determine the antecedents of teachers' performance in any other Gulf country. Secondly, the research has only considered two antecedents of teachers' performance, and future studies can determine more factors that can possibly influence the performance of teachers.

REFERENCES

[1] Dana, L. P., Salamzadeh, A., Mortazavi, S., & Hadizadeh, M. (2022). Investigating the impact of international markets and new digital technologies on business innovation in emerging markets. *Sustainability*, *14*(2), 983.

[2] Muhamad, L. F., Bakti, R., Febriyantoro, M. T., Kraugusteeliana, K., & Ausat, A. M. A. (2023). Do innovative work behavior and organizational commitment create business performance: A literature review. *Community Development Journal: Jurnal Pengabdian Masyarakat*, *4*(1), 713–717.

[3] McCartney, S., Murphy, C., & Mccarthy, J. (2021). 21st century HR: A competency model for the emerging role of HR Analysts. *Personnel Review*, *50*(6), 1495–1513.

[4] Ali, B. J., & Anwar, G. (2021). Strategic leadership effectiveness and its influence on organizational effectiveness. *International Journal of Electrical, Electronics and Computers*, *6*(2).

[5] Ali, S. A., Ansari, M., & Alam, M. (2020). Resource management techniques for cloud-based IoT environment. In *Internet of things (IoT) concepts and applications* (pp. 63–87). Springer International Publishing.

[6] Allen, D. G., & Vardaman, J. M. (2021). Global talent retention: Understanding employee turnover around the world. In *Global talent retention: Understanding employee turnover around the world* (pp. 1–15). Emerald Publishing Limited.

[7] Chatterjee, S., Chaudhuri, R., Vrontis, D., & Giovando, G. (2023). Digital workplace and organization performance: Moderating role of digital leadership capability. *Journal of Innovation & Knowledge*, *8*(1), 100334.

[8] Chatterjee, S., Rana, N. P., Tamilmani, K., & Sharma, A. (2021). The effect of AI-based CRM on organization performance and competitive advantage: An empirical analysis in the B2B context. *Industrial Marketing Management*, *97*, 205–219.

[9] Agustian, K., Pohan, A., Zen, A., Wiwin, W., & Malik, A. J. (2023). Human resource management strategies in achieving competitive advantage in business administration. *Journal of Contemporary Administration and Management (ADMAN)*, *1*(2), 108–117.

[10] Narayanan, A., Rajithakumar, S., Menon, M. (2019). Talent management and employee retention: An integrative research framework. *Human Resource Development Review*, *18*(2), 228–247.

[11] Khan, N. A., Bharadwaj, S., Khatoon, A., & Jamal, M. T. (2021). Assessing the nexus between employer branding and employee retention: Moderating role of organizational identification. *Management and Labour Studies*, *46*(4), 379–398.

[12] Budur, T., & Poturak, M. (2021). Employee performance and customer loyalty: Mediation effect of customer satisfaction. *Middle East Journal of Management*, *8*(5), 453–474.

[13] Anwar, G., & Abdullah, N. N. (2021). The impact of human resource management practice on Organizational performance. *International Journal of Engineering, Business and Management (IJEBM)*, *5*.

[14] Nyathi, M., & Kekwaletswe, R. (2023). Realizing employee and organizational performance gains through electronic human resource management use in developing countries. *African Journal of Economic and Management Studies*, *14*(1), 121–134.

[15] Atatsi, E. A., Stoffers, J., & Kil, A. (2019). Factors affecting employee performance: A systematic literature review. *Journal of Advances in Management Research*, *16*(3), 329–351.

[16] ALmaamari, Q. A., & Majdalawi, M. A. (2024). Factors affecting employees' performance in the hospitality industry in Kingdom of Bahrain. In *The AI revolution: Driving business innovation and research: Volume 1* (pp. 237–245). Springer Nature Switzerland.

[17] Abdelwahed, N. A. A., & Doghan, M. A. A. (2023). Developing employee productivity and performance through work engagement and organizational factors in an educational society. *Societies*, *13*(3), 65.
[18] Alkadash, T. M., & Alamarin, F. (2021). An integrative conceptual framework on employee performance during COVID-19 pandemic for Bahrain SMEs. *Psychology and Education*, *58*(2), 3812–3817.
[19] Albaker, K. (2017). Analytical view of Bahrain's government schools' performance: A quality perspective. *Sage Open*, *7*(4), 2158244017736555.
[20] Khalifa, M., Alrasheed, R., & Aldada, A. M. (2023). Education in light of the sustainable development goals in the Kingdom of Bahrain, an analytical study. *Education*, *12*(4), 1–23.
[21] Al-Aali, L. (2019). The role of education towards shaping private sector employment: A case from the Kingdom of Bahrain. *Academy of Educational Leadership Journal*, *23*(2), 1–9.
[22] Matherly, L. L., Zhang, Y., & Ahmed, B. (2022). A comparative study of country satisfaction of UAE and Bahrain teachers: The mediating effect of job satisfaction on organizational commitment. *Research in Comparative and International Education*, *17*(3), 511–527.
[23] Akbar, M. A., Ali, M. H., & Alam, S. S. (2019). Total quality management system in an education environment: The case of a private university in Bahrain. *Journal of Reviews on Global Economics*, *8*, 717–729.
[24] Aslam, H. D. (2011). Performance evaluation of teachers in universities: Contemporary issues and challenges. *Journal of Educational and Social Research*, *1*(2), 11–31.
[25] Sarwar, S., Danyal Aslam, H., & Imran Rasheed, M. (2012). Hindering factors of beginning teachers' high performance in higher education Pakistan: Case study of IUB. *International Journal of Educational Management*, *26*(1), 27–38.
[26] Alshehry, A. T. (2014). Investigating factors affecting science teachers' performance and satisfaction toward their teaching process at Najran University for Girls' Science Colleges. *International Journal of Higher Education*, *3*(2), 73–82.
[27] Tehseen, S., & Hadi, N. U. (2015). Factors influencing teachers' performance and retention. *Mediterranean Journal of Social Sciences*, *6*(1), 233–244.
[28] Joo, B. K., Sang, Y. K., & Lee, S. G. (2021). The effects of artificial intelligence on employee job satisfaction in South Korea. *International Journal of Human Resource Management*, *1*(24), 463–473.
[29] Azeem, M. M., Febriyanto, U., Nurhadi, F. A., & Halid, H. (2024). Unlocking the values of artificial intelligence (AI) in human resource management (HRM) in enhancing employee retention. *Global Business & Management Research*, *16*(2), 300–315.
[30] Joshi, D. A., & Masih, D. J. (2023). Enhancing employee efficiency and performance in industry 5.0 organizations through artificial intelligence integration. *European Economic Letters (EEL)*, *13*(4), 300–315.
[31] Luhana, K. K., Memon, A. B., & Khan, I. (2023). The rise of artificial intelligence and its influence on employee performance and work. *Global Social Sciences Review*, *VIII*, 463–479.
[32] Szabolcs, S. Z. I. L. Á. G. Y. I., & Laura, J. U. H. Á. S. Z. (2024). Increasing teacher efficiency with AI: An overview. *Journal of Computer Science & Control Systems*, *17*(2), 9–14.
[33] Vinichenko, M. V., Melnichuk, A. V., & Karácsony, P. (2020). Technologies of improving the university efficiency by using artificial intelligence: Motivational aspect. *Entrepreneurship and Sustainability Issues*, *7*(4), 2696.
[34] Mo, Z., & Shen, L. (2024, August). Research on the application of artificial intelligence technology in improving teacher quality and teaching efficiency. In *2024 5th International Conference on Education, Knowledge and Information Management (ICEKIM 2024)* (pp. 839–845). Atlantis Press.
[35] Mohd Nizar, N. N., Rahmat, M. K., Maaruf, S. Z., & Damio, S. M. (2018). Examining the use behaviour of augmented reality technology through MARLCARDIO: Adapting the UTAUT model. *Asian Journal of University Education*, *15*(3), 198.
[36] Romero Rodríguez, J. M., Ramírez-Montoya, M. S., Buenestado Fernández, M., & Lara Lara, F. (2023). Use of ChatGPT at university as a tool for complex thinking: Students' perceived usefulness. *Journal of New Approaches in Educational Research*, 12(2), 323–339.

[37] Crompton, H., & Burke, D. (2023). Artificial intelligence in higher education: The state of the field. *International Journal of Educational Technology in Higher Education*, *20*(1), 22.
[38] Kuleto, V., Ilić, M., Dumangiu, M., Ranković, M., Martins, O. M., Păun, D., & Mihoreanu, L. (2021). Exploring opportunities and challenges of artificial intelligence and machine learning in higher education institutions. *Sustainability*, *13*(18), 10424.
[39] Utkirov, A. (2024). Artificial Intelligence impact on higher education quality and efficiency. *Actual Problems of Humanities and Social Sciences* 4(9), 363–388.
[40] Rao, S., Chitranshi, J., & Punjabi, N. (2020). Role of artificial intelligence in employee engagement and retention. *Journal of Applied Management-Jidnyasa*, *12*(2), 42–60.
[41] Chen, L., Chen, P., & Lin, Z. (2020). Artificial intelligence in education: A review. *IEEE Access*, *8*, 75264–75278.
[42] Huang, J., Saleh, S., & Liu, Y. (2021). A review on artificial intelligence in education. *Academic Journal of Interdisciplinary Studies*, *10*(3), 206.
[43] Akgun, S., & Greenhow, C. (2022). Artificial intelligence in education: Addressing ethical challenges in K-12 settings. *AI and Ethics*, *2*(3), 431–440.
[44] Remian, D. (2019). *Augmenting education: Ethical considerations for incorporating artificial intelligence in education*. University of Massachusetts Boston.
[45] Ouyang, F., Zheng, L., & Jiao, P. (2022). Artificial intelligence in online higher education: A systematic review of empirical research from 2011 to 2020. *Education and Information Technologies*, *27*, 7893–7925.
[46] Casis-Woidyla, R. (2020). Getting to the heart of professional development and employee engagement among non-teaching personnel in higher education: An appreciative inquiry into the influential role of a staff development center. Dissertation, Doctor of Education in Educational Leadership.
[47] Dabbous, A., Aoun Barakat, K., & Merhej Sayegh, M. (2022). Enabling organizational use of artificial intelligence: An employee perspective. *Journal of Asia Business Studies*, *16*(2), 245–266.
[48] Kulal, A., Rahiman, H. U., Suvarna, H., Abhishek, N., & Dinesh, S. (2024). Enhancing public service delivery efficiency: Exploring the impact of AI. *Journal of Open Innovation: Technology, Market, and Complexity*, *10*(3), 100329.
[49] Celik, I., Dindar, M., Muukkonen, H., & Järvelä, S. (2022). The promises and challenges of artificial intelligence for teachers: A systematic review of research. *TechTrends*, *66*(4), 616–630.
[50] Rodway, P., & Schepman, A. (2023). The impact of adopting AI educational technologies on projected course satisfaction in university students. *Computers and Education: Artificial Intelligence*, *5*, 100150.
[51] Sanchez, J. (2024). The impact of artificial intelligence on teacher performance: Automatization of paperwork and enhancing of mediation. In *Edulearn24 Proceedings* (pp. 7274–7278). Iated.
[52] Mazzei, A., Butera, A., & Quaratino, L. (2019). Employee communication for engaging workplaces. *Journal of Business Strategy*, *40*(6), 23–32.
[53] Meslec, N., Duel, J., & Soeters, J. (2020). The role of teamwork on team performance in extreme military environments: An empirical study. *Team Performance Management: An International Journal*, *26*(5–6), 325–339.
[54] Memon, M. A., Salleh, R., Baharom, M. N., & Harun, H. (2014). Person-organization fit and turnover intention: The mediating role of employee engagement. *Global Business & Management Research*, *6*(3), 205–209.
[55] Guan, C., Mou, J., & Jiang, Z. (2020). Artificial intelligence innovation in education: A twenty-year data-driven historical analysis. *International Journal of Innovation Studies*, *4*(4), 134–147.
[56] Ghamrawi, N., Shal, T., & Ghamrawi, N. A. (2024). Exploring the impact of AI on teacher leadership: Regressing or expanding? *Education and Information Technologies*, *29*(7), 8415–8433.
[57] Candelon, F., Chu, M., & Min Ha, S. (2021). A.I. won't break your company's culture—And it might even boost morale. *Commentary, Artificial Intelligence*. Retrieved from https://fortune.com/2021/11/05/ai-artificial-intelligence-workplace-cul
[58] Ransbotham, S., Candelon, F., Kiron, D., Lafountain, B., & Khodabandeh, S. (2021). The cultural benefits of artificial intelligence in the enterprise. Retrieved from https://sloanreview.mit.edu/projects/the-cultural-benefits-of-artificialintelligence-in-the-enterprise

[59] Borges, A. F., Laurindo, F. J., Spínola, M. M., Gonçalves, R. F., & Mattos, C. A. (2021). The strategic use of artificial intelligence in the digital era: Systematic literature review and future research directions. *International Journal of Information Management*, *57*, 102225.
[60] Mikalef, P., & Gupta, M. (2021). Artificial intelligence capability: Conceptualization, measurement calibration, and empirical study on its impact on organizational creativity and firm performance. *Information and Management*, *58*(3), 103434.
[61] Ebrahimi, A. (2013). The relationship between job satisfaction and teacher retention. *European Online Journal of Natural and Social Sciences*, *2*(3), 809–813.
[62] Paigude, S., Pangarkar, S. C., Hundekari, S., Mali, M., Wanjale, K., & Dongre, Y. (2023). Potential of artificial intelligence in boosting employee retention in the human resource industry. *International Journal on Recent and Innovation Trends in Computing and Communication*, *11*(3s), 01–10.
[63] Hair, J. F., Anderson, R. E., Tatham, R. L., & Black, W. C. (2010). *Multivariate data analysis*, New Jersey.
[64] Byrne, B. M. (2013). *Structural equation modeling with Mplus: Basic concepts, applications, and programming*. Routledge.
[65] Fornell, C., & Larcker, D. F. (1981). Evaluating structural equation models with unobservable variables and measurement error. *Journal of Marketing Research*, *18*(1), 39–50.
[66] Henseler, J., Ringle, C. M., & Sarstedt, M. (2015). A new criterion for assessing discriminant validity in variance-based structural equation modeling. *Journal of the Academy of Marketing Science*, *43*, 115–135.

Chapter 18

AI-driven cybersecurity awareness and readiness

A case study from Albania

Reis Mulita and Brisida Sefa

18.1 INTRODUCTION

18.1.1 Background and rationale for cybersecurity awareness and readiness

In today's rapidly evolving digital landscape, the pressing need for robust cybersecurity awareness and readiness has emerged as a critical concern for both individuals and organizations. Over the past decade, cyber threats have increased not only in frequency but also in complexity, making the human element a key determinant of cybersecurity resilience. Human factors are considered the primary vulnerability, so education and cyber awareness are of great interest in our discussion. Extensive research underscores that human error remains the leading cause of cybersecurity breaches. For instance, human involvement accounts for up to 85% of data breaches according to [1]. Likewise, an IBM security report found that 95% of breaches stem from human mistakes, such as phishing or weak credentials. These statistics highlight the reality that—even with advanced technical defenses in place—organizations remain highly susceptible due to behavioral vulnerabilities [2]. As articulated by researchers, technological solutions alone cannot ensure security unless employee behaviors align with organizational policies [3]. Awareness, training, and culture are required to bridge this human gap, as the scholarship and practice emphasize the importance of Security Education, Training, and Awareness (SETA) programs. Awareness involves not just understanding threats but also aligning everyday actions with security policies and best practices [4]. These initiatives aim to equip employees with the knowledge and behaviors necessary to mitigate risk and conform to security protocols [5]. Moreover, cybersecurity awareness capability is multidimensional. Awareness comprises three key dimensions, such as personnel knowledge, attitude, and learning; management (training, culture, strategic orientation); and infrastructure (technology and data governance). This holistic approach recognizes that awareness must be supported by organizational culture, leadership, and technological infrastructure to be effective. Improving cybersecurity readiness is not just a technical imperative but also a strategic organizational performance driver. Under the Technology–Organization–Environment (TOE) framework, readiness incorporates awareness, preparedness, and commitment, which collectively enhance security performance and, subsequently, both financial and nonfinancial organizational outcomes [6]. Research further indicates that effective security culture, reinforced through training and awareness, promotes trust, compliance, and safer user behavior. Advances in threat sophistication, such as artificial intelligence (AI)-driven malware and agentic AI, pose new challenges that traditional security paradigms struggle to address. Recent developments highlight how autonomous AI agents can be weaponized for

 DOI: 10.1201/9781003685890-18

phishing, credential stuffing, and reconnaissance, elevating the urgency of a human-centric, proactive cybersecurity model. Meanwhile, defenders are deploying AI-powered tools to detect and mitigate threats more rapidly, shaping an ongoing arms race between attackers and defenders [7].

AI serves as a catalyst in cybersecurity education. Recent strides in generative AI (GenAI) are transforming pedagogical approaches in cybersecurity education. Researchers introduce a structured learning framework where GenAI tools enhance critical thinking and policy-building tasks: students generate, critique, and refine AI-assisted cybersecurity policies and apply them to realistic, regulated scenarios [8]. The result is a marked improvement in analytical engagement, policy evaluation, and the bridging of theoretical principles with practical application, albeit with emerging concerns about AI overreliance and the need for human oversight. Beyond higher education, AI tailors learning to individual needs. Leveraging adaptive learning platforms, AI personalizes content based on learners' knowledge gaps and engagement patterns. This personalized approach is deemed critical in fostering cyber awareness, particularly in contexts requiring diverse foundational knowledge. Moreover, laboratories and academic modules centered on AI-assisted malware analysis are training the next generation of cybersecurity professionals in real-world threat intelligence, classification, and attribution techniques. These lab-intense programs empower students with hands-on experience in applying machine learning (ML) to dynamic datasets, reinforcing both technical acumen and practical readiness. The integration of AI into cybersecurity education and operations marks both a methodological and strategic evolution. From personalized GenAI-assisted learning frameworks to predictive threat intelligence, automated mitigation, and AI-enhanced governance, AI is redefining how humans learn about, defend against, and adapt to cyber threats. Yet, this ascent demands a complex balance—enforcing robust governance, maintaining human oversight, integrating nontechnical competencies, and preparing professionals capable of navigating an increasingly AI-mediated cybersecurity landscape.

18.1.1.1 Implications for developing contexts (e.g., Albania)

In developing nations, including Albania, building cybersecurity readiness is foundational for ensuring resilience amidst digital transformation. Initiatives such as the Cybersecurity Capacity Maturity Model (CMM) highlight the need for capacity building across education, policy, and technical infrastructure. Without adequate awareness and readiness, organizations risk compromising valuable digital assets and hampering national digital development. Today's cybersecurity environment underscores the critical need to bolster both awareness and readiness. Human factors account for the majority of breaches, and only through comprehensive, continuous training, an aligned security culture, and supportive infrastructures can organizations effectively counter evolving threats. This imperative is heightened by the dual rise of AI-enabled threats and AI-empowered defenses, especially in developing contexts where organizational maturity may lag [9]. The urgency of Albania's cybersecurity challenges was starkly illustrated in mid-2022 when devastating cyberattacks—attributed to the Iranian-linked hacking group Homeland Justice—crippled key public e-services, disrupted border control, and hampered municipal operations. The incident prompted the Albanian government to sever diplomatic ties with Iran, effectively marking one of the most forceful diplomatic responses to cyber aggression by a NATO ally (Albania–Iran relations; The Cyber Diplomat). Further attacks followed. In February 2024, the Institute of Statistics (INSTAT) suffered a targeted cyber intrusion affecting parts of its systems, triggering emergency protocols while authorities collaborated to contain the fallout [10].

18.1.1.2 Public awareness of cybersecurity in Albania

Public awareness of cybersecurity in Albania is a mixed reality. The NordVPN National Privacy Test survey highlighted that while many citizens grasp strong passwords and privacy basics, awareness dramatically drops when it comes to privacy tools, data collection by ISPs (Internet Service Providers), or reading terms and conditions—showing critical gaps in practice [11]. Similarly, academic curricula in Albanian universities reveal systemic shortcomings. A review of cybersecurity, cyber law, and ISO standards within economic, legal, and IT-focused syllabuses found that students lack adequate knowledge and practical competencies, reinforcing the need for curriculum reform [12]. Among higher education students, smartphone security awareness is significantly higher among those with formal education in information security, confirming the value of targeted educational interventions [13]. Summarizing this introduction section, we can say that Albania presents a compelling case study in cybersecurity awareness and readiness, shaped by rapid digitalization, targeted cyber aggression, and subsequent strategic acceleration. While institutional reforms, capacity-building mechanisms, and international cooperation are well underway, notable gaps persist in public awareness and educational readiness. These dynamics emphasize the importance of holistic, AI-enhanced awareness training and capacity building—precisely the focus of this chapter's framework.

18.1.2 Objectives and research questions

18.1.2.1 Chapter objectives

This chapter aims to explore the intersection of AI, cybersecurity awareness, and organizational readiness within the context of a developing digital society, using Albania as a focused case study. The chapter contributes to both theoretical understanding and practical insight by examining how AI can be leveraged to enhance cybersecurity education, threat mitigation, and institutional preparedness.

The specific objectives of the chapter are to:

1. Evaluate the current level of cybersecurity awareness among individuals within public and private institutions in Albania.
2. Assess the extent and quality of cybersecurity training programs implemented at the organizational level.
3. Examine the prevalence and impact of malware incidents on institutional operations.
4. Analyze correlations between awareness, training, and the frequency of cyber incidents, with a focus on malware.
5. Investigate the potential role of AI-driven tools and methodologies in improving cybersecurity knowledge dissemination, risk detection, and response.
6. Identify key gaps and propose strategic interventions for enhancing cybersecurity readiness through education, policy, and technology.

18.1.2.2 Research questions

To address these objectives, the chapter is guided by the following primary and secondary research questions:

Primary Research Questions:

1. To what extent are individuals in Albanian institutions aware of cybersecurity threats and best practices?

2. How prevalent and effective are cybersecurity training programs within organizations across Albania?
3. What is the reported frequency of malware incidents, and how do they impact organizational operations?

Secondary Research Questions:

4. What is the relationship between levels of cybersecurity awareness or training and the incidence of malware attacks?
5. How can AI be utilized to enhance cybersecurity education, early warning systems, and incident response in a developing context like Albania?

These research questions serve as the analytical framework for both the empirical and theoretical discussions presented in the chapter. They enable a holistic understanding of cybersecurity readiness, particularly within the emerging paradigm of AI-augmented defense and education mechanisms.

18.2 LITERATURE REVIEW

Cybersecurity awareness is broadly defined as the knowledge and behaviors that individuals and organizations adopt to protect information systems from cyber threats. Effective awareness goes beyond mere knowledge; it requires the internalization of secure practices in everyday digital behavior [4]. Global frameworks such as SETA programs emphasize the importance of continuous learning and cultural reinforcement. Research has consistently shown that human error—often stemming from inadequate awareness—is a leading cause of cybersecurity incidents, accounting for up to 85–95% of breaches [1, 2]. Consequently, cybersecurity education is increasingly being integrated into formal curricula and professional development programs across sectors. Organizational cybersecurity readiness. Cybersecurity readiness refers to an organization's capacity to prevent, detect, respond to, and recover from cyber incidents. It encompasses not only technical infrastructure but also strategic planning, staff training, and governance. The concept of digital resilience has emerged as a core component of organizational readiness, emphasizing the ability to adapt and recover from disruptions while maintaining essential services [14]. Under frameworks like the Technology–Organization–Environment (TOE) model, readiness is linked to overall performance, with studies confirming that higher preparedness levels correlate with reduced vulnerability and increased operational continuity.

As we discuss AI and cybersecurity issues, it is of great importance to have a look at preparedness and institutional capacity to face and challenge issues related to cybersecurity inside organizations. Preparedness in cybersecurity entails not only having the right tools and policies but also cultivating a culture of awareness, cross-functional collaboration, and proactive defense. Many organizations underinvest in nontechnical aspects of preparedness—such as user behavior, training, and leadership involvement—thus creating "soft spots" that are often exploited by attackers [15]. Organizational preparedness is frequently assessed using maturity models such as the Cybersecurity Capability Maturity Model (C2M2) or the Cybersecurity Capacity Maturity Model for Nations (CMM), which evaluate performance across domains like governance, risk management, culture, and workforce development [16]. These models emphasize the importance of holistic readiness, especially for institutions in developing and transitional economies, where digital transformation often outpaces security investments. Challenges in Achieving Readiness. Several studies have highlighted persistent barriers to achieving full cybersecurity readiness. These include limited financial resources, lack of specialized personnel, siloed IT and security functions, and inadequate regulatory compliance [17, 18]. In many organizations, cybersecurity is still treated as an IT problem rather than a strategic priority, which impedes integrated, enterprise-wide risk management. Moreover, the increasing sophistication of cyber threats—ranging from ransomware and supply chain attacks to AI-powered exploits—requires a paradigm shift in preparedness strategies. This includes greater automation, threat intelligence sharing, and resilience planning tailored to specific sectors and institutional missions [19]. AI has spurred considerable innovation across cybersecurity domains. As an example, we take the education field, where GenAI tools are transforming pedagogical modes, allowing students to draft, critique, and refine cybersecurity policies, thus blending analytical rigor with real-world applicability [6]. In threat detection and simulation, GenAI enables the creation of realistic cyber ranges—with synthetic network traffic, topologies, and attack scenarios—helping security professionals train in dynamically simulated environments [13]. Elsewhere, we bring attention to malware detection and analysis, where AI systems help generate synthetic malware samples for stress-testing detection models and can detect novel malware variants through behavioral pattern recognition [13]. AI-assisted laboratory modules have also been academically deployed to train students in malware classification, attribution, and CTI (Cyber Threat Intelligence) techniques. Furthermore, explainable AI (XAI), through which AI advances detection capabilities, addresses the opacity of black-box models, which poses trust challenges. Explainable AI methods are thus critical for transparency, interpretability, and user confidence in AI-driven cybersecurity systems, as the literature suggests.

From our point of view, based on the literature review, digital resilience is a strategic imperative. Digital resilience extends beyond reactive security measures to encompass an organization's ability to absorb and adapt to adverse cyber events while maintaining core functions and long-term viability [20]. It integrates the principles of risk anticipation, system robustness, and recovery capacity into an ongoing adaptive cycle. Researchers argue that resilience is not merely about prevention but involves a continuum of cybersecurity maturity, where institutions continuously evolve in the face of dynamic threat landscapes [14]. The National Institute of Standards and Technology (NIST) has operationalized this idea in its Cybersecurity Framework (CSF), emphasizing five core functions: Identify, Protect, Detect, Respond, and Recover [21]. These functions form the basis of a structured approach to building cyber resilience, with a strong focus on integrating cybersecurity into enterprise risk management.

One of the most important considerations regarding AI-driven cybersecurity issues is malware threats in institutional environments. This is because malware remains one of the most pervasive and damaging forms of cyber threat in institutional environments, characterized by its ability to disrupt operations, compromise sensitive data, and degrade organizational trust. Defined broadly as malicious software designed to infiltrate, damage, or disable computer systems, malware encompasses a diverse array of types, including viruses, worms, trojans, ransomware, spyware, and rootkits [22]. Malware has a dynamic nature and types. The evolution of malware has witnessed a shift from relatively simple viruses and worms toward more sophisticated and targeted threats such as ransomware and fileless malware [23]. Ransomware, in particular, has emerged as a critical challenge for institutions, where attackers encrypt vital data and demand ransom payments, severely disrupting operations across sectors such as health care, education, and government [24]. Advanced Persistent Threats (APTs), often state-sponsored, utilize customized malware to infiltrate high-value targets, maintaining long-term undetected access for espionage or sabotage [25]. Malware has significant implications for institutional security. This is because institutions are uniquely vulnerable to malware due to their reliance on complex IT infrastructures, vast volumes of sensitive data, and often heterogeneous security postures. The consequences of malware infection include financial loss, regulatory penalties, operational downtime, and reputational damage [1]. For instance, ransomware attacks on health-care institutions not only compromise patient data confidentiality but can also endanger patient safety by disrupting critical medical devices and systems [26]. Research indicates that the attack vectors for malware in institutions are frequently linked to human factors such as phishing and social engineering, exploiting gaps in employee cybersecurity awareness and training [4]. This underscores the interdependence of technological defenses and human readiness in mitigating malware risks. Discussing malware and its negative impacts, we should consider that the literature suggests some current defense mechanisms and challenges. Institutional defenses against malware commonly involve a combination of endpoint protection platforms, intrusion detection systems, regular patch management, and network segmentation [27]. However, the rapid emergence of polymorphic and fileless malware challenges traditional signature-based antivirus solutions, necessitating more advanced behavioral and heuristic detection techniques [28]. Moreover, the increasing use of cloud services and mobile devices introduces additional attack surfaces that complicate malware defense strategies [29]. Institutions also face challenges in incident response due to limited resources, lack of skilled cybersecurity personnel, and the difficulty of timely detection in complex environments [17].

Linking institutional policy with training is very supportive in this process, as the literature suggests. Therefore, effective institutional policies, including strict access controls, incident response protocols, and regular security audits, are critical components of malware defense [18]. Training programs aimed at raising awareness about phishing and safe computing practices have demonstrated positive impacts in reducing malware incidents, but adoption and consistency vary widely across sectors and regions [15].

As for AI in cybersecurity education and training, it is important to bring to attention that beyond threat detection, AI enhances cybersecurity education by enabling personalized and adaptive learning experiences. Intelligent tutoring systems (ITS) and AI-powered simulators can adjust training content based on learners' knowledge gaps, improving engagement and retention [30]. Furthermore, virtual assistants and chatbots provide just-in-time support, answering user queries and guiding security best practices [31]. In training contexts, AI-driven simulation environments enable realistic cyberattack scenarios, allowing users to practice incident response in safe, controlled settings. Such immersive learning has been shown to improve user readiness and decision-making under pressure [24]. Furthermore, AI analytics track learner progress, offering insights into skill development and identifying areas needing reinforcement [32].

Several case studies demonstrate the practical benefits of AI in institutional cybersecurity. IBM's Watson for Cybersecurity applies NLP (Natural Language Processing) to parse vast cybersecurity literature and threat reports, assisting analysts in identifying emerging threats [2]. Darktrace uses unsupervised ML algorithms to create "immune system"-like defenses, detecting novel threats through behavioral analysis [33]. Emerging tools such as Cylance leverage AI for endpoint protection, using predictive models to block malware execution before it occurs [34]. Similarly, AI-powered Security Information and Event Management (SIEM) platforms like Splunk employ automated threat hunting to enhance incident detection and response workflows [35]. Despite these advancements, challenges persist, including adversarial attacks on AI models and ethical considerations surrounding automation in security decision-making [36]. Thus, AI is best viewed as a force multiplier that complements human expertise rather than a standalone solution.

Discussing cybersecurity in the Albanian context, Albania presents a unique cybersecurity case within Southeast Europe, having experienced major cyber incidents that exposed national vulnerabilities. The 2022 state-sponsored attacks on Albanian infrastructure prompted a reevaluation of national cybersecurity posture and the acceleration of institutional reforms.

While addressing these situations, the National Cybersecurity Strategy (2020–2025) aims to improve critical infrastructure protection, increase public awareness, and align with EU standards. However, implementation remains uneven, with reported gaps in education, training, and public engagement. Despite the development of legal frameworks and partnerships with entities like ENISA, the World Bank, OSCE, as well as with universities and high schools across the country, cybersecurity culture and institutional maturity in Albania are still evolving, especially in nontechnical sectors and local governance. It is also important to note that AKSK has established institutional partnerships with universities across Albania for that purpose [13]. As a conclusion, malware threats in institutional environments are multifaceted and evolving, demanding integrated approaches that combine technical safeguards, policy enforcement, and human factors management. Given the critical nature of institutional data and services, continuous adaptation and investment in malware defense mechanisms remain paramount.

18.3 METHODOLOGY

18.3.1 Research design

18.3.1.1 Methods approached

This study employed a mixed-methods research design, integrating both quantitative and qualitative methodologies to provide a comprehensive analysis of cybersecurity awareness and readiness in Albania. The mixed-methods approach is particularly suited to explore complex phenomena such as cybersecurity preparedness, where numerical data on awareness levels and training prevalence are enriched by qualitative insights into organizational experiences and perceptions [37]. The quantitative component involved the administration of structured surveys to assess knowledge, training participation, and exposure to malware attacks, while the qualitative component comprised open-ended questions and interviews that explored contextual factors influencing cybersecurity practices. Regarding the population and sampling, the study population consisted of more than 2000 participants drawn from a wide range of organizations across Albania, encompassing public institutions, private enterprises, and nongovernmental organizations. The sample was purposively selected to ensure diversity in organizational size, sector, and geographic location, thereby enhancing the representativeness of findings [38]. Participants included employees at various hierarchical levels, ranging from frontline staff to IT security personnel and management, enabling a multi-layered understanding of cybersecurity awareness and institutional readiness. This research focused on an initial sample of the survey, and the quantitative analysis was administered between November 2024 and March 2025.

18.3.1.2 Results and discussion

Given the quantitative–qualitative nature of the research, the findings are discussed later through thematic aggregation. First, the survey results are examined. The insights gained from the semi-structured interviews are then presented.

18.3.1.3 Data collection Instruments used in the study

Data were collected primarily through a structured survey instrument developed based on existing validated cybersecurity awareness frameworks [39]. The survey included both closed and open-ended questions covering domains such as individual awareness of cyber threats, participation in organizational training programs, and experiences with malware incidents. Surveys were mostly administered electronically and in paper form over a period of three months. Quantitative data were analyzed to evaluate correlations and examine relationships between cybersecurity knowledge, training, and malware exposure. The quantitative data derived from open-ended survey responses and interview considerations were subjected to thematic discussion, comparisons, analysis, and evaluation. The study also has ethical considerations. The study adhered strictly to ethical research standards to protect participant rights and data integrity. Confidentiality was ensured by anonymizing survey responses and interview data, with no personally identifiable information collected or disclosed. Participants provided informed consent prior to participation, with clear information regarding the study's purpose, voluntary nature, and their right to withdraw at any time without penalty. Also, data storage complied with relevant data protection regulations, employing secure, encrypted databases accessible only to authorized research personnel.

18.3.1.4 Limitations and suggestions for future research

Since the study relies on self-reported data, the findings may be subject to biases or overestimations of knowledge. Moreover, as information on other potentially influential factors—such as age, education, profession, or geographical location—is available, these variables should be addressed in a separate publication or considered to expand and deepen the current study. In addition, future research could incorporate in-depth interviews or practical knowledge assessments to better verify the actual level of participants' cybersecurity knowledge.

18.4 FINDINGS AND DISCUSSION

Regarding the question of whether participants had general knowledge and awareness of cybersecurity, the presented study, through the use of a questionnaire, aims to assess the level of knowledge and awareness of cybersecurity in a sample of 2,056 individuals. The data collected provide a clear picture of the distribution of knowledge, and significant differences are observed both in the general level of awareness and in terms of gender. From the analysis of the data, it results that only 44.60% of respondents declare that they have knowledge of cybersecurity. This indicates that less than half of the studied population has some awareness of essential issues related to the protection of personal data, maintaining online privacy, and identifying digital threats. Meanwhile, a high percentage (39.79%) of individuals responded with "Don't know/N/A," which suggests a high level of uncertainty or lack of information in this area. This category is particularly problematic, as it represents a group that is in a "gray area"—not declaring a lack of knowledge, but not knowing the topic well enough to confirm their knowledge. Furthermore, 15.61% clearly stated that they have no knowledge at all about cybersecurity, reinforcing the need for educational interventions (Table 18.1).

Regarding gender differences in cybersecurity knowledge, gender analysis of the data shows a clear divide in the level of cybersecurity awareness, referred to in the data as follows: men report knowledge at a higher percentage (49.24%) compared to women (35.63%). This difference may reflect several social and cultural factors, among which we mention: greater exposure of men to technology in certain contexts, gender roles in the selection of fields of study or careers (e.g., IT, information security), different perceptions of self-confidence in technological knowledge, etc. (Figure 18.1).

Women represent the highest percentage of "Don't know/N/A" responses (48.93%), which indicates a clear lack of information or uncertainty about knowledge in this area. This is an indicator of the need for more focused and inclusive education for women and girls in terms

Table 18.1 Are you aware of cybersecurity threats?

Answer	*Number*	*Report*
Yes	917	44.60%
N/A	818	39.79%
No	321	15.61%
Total	2056	100.00%

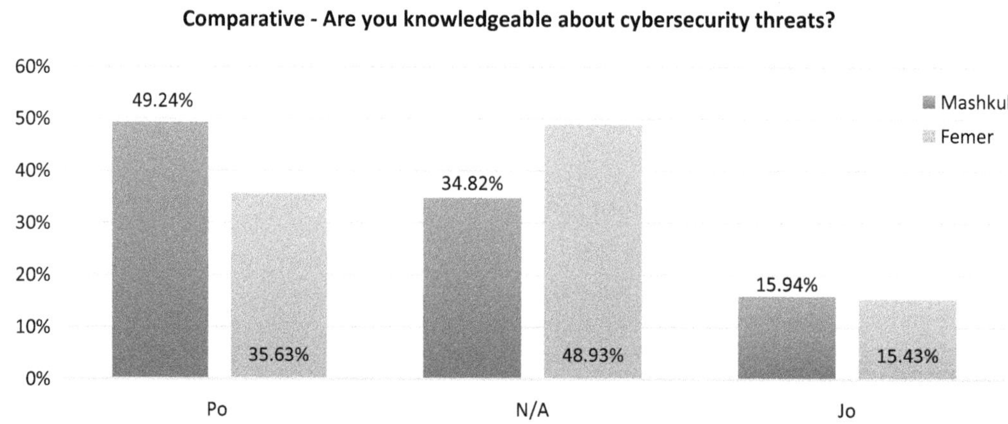

Figure 18.1 Comparative—are you knowledgeable about cybersecurity threats?

of digital security. "No" responses are relatively similar between genders (15.94% for men and 15.43% for women), suggesting that, in terms of those who have no knowledge at all, gender is not a major differentiating factor. Based on these findings, it is important to also discuss the implications for education and public policy. We believe that these findings have significant implications for policymakers, educators, and stakeholders involved in digital transformation and cybersecurity, and that some of the following considerations should be addressed by both public and nonpublic actors, such as the need for a clear strategy to raise public awareness, with a particular focus on low-knowledge or insecure groups. Similarly, gender gaps in access to technology education should be addressed by promoting inclusive programs that aim to increase women's self-confidence and inclusion in the field of cybersecurity. This is particularly important at a time when cyber threats are on the rise and increasingly impact individuals' personal, professional, and social lives.

The following section presents the findings and assessment regarding personal experiences with malware among the surveyed users. As part of the cybersecurity awareness study, participants were asked whether they had ever had direct experience with malware, a common digital threat capable of damaging operating systems, stealing data, or compromising user privacy. The results indicate that, out of 1,521 responses collected, 861 participants (56.61%) reported encountering malware once. Additionally, 207 participants (13.61%) indicated multiple encounters with malware, while 453 participants (29.78%) reported no prior experience with malware (Table 18.2).

18.4.1 Interpretation of the findings

We estimate that these results show that an absolute majority (70.22%) of respondents have had at least one experience with malware infection, while only approximately 30% have

Table 18.2 Has your organization ever been affected by cybersecurity threats?

Answer	*Report*	*Number*
Yes	23.43%	119
No	46.85%	238
I don't know.	29.72%	151
Total	100.00%	508

never encountered one. This fact suggests that cyber threats are not only widespread but also actually experienced by the majority of computer users, which reinforces the need for preventive measures and continuous education on digital security. Regarding the consequences of personal experience, we believe that direct experience with malware can have both positive and negative impacts. As a positive effect, we believe that this experience can serve as a catalyst for increasing awareness and adopting safe practices (such as installing antivirus, updating systems, avoiding unsafe websites, etc.). As a consequence or negative effect, we believe that in cases where education is lacking after the incident, users may underestimate the risks or repeat unsafe behaviors, increasing the risk of repeated attacks. Also, in our view, the fact that 13.61% have been infected multiple times is worrying and signals a lack of consistent security practices or an inability to identify and address the sources of infection.

Subsequently, we present and discuss findings related to organizations' engagement in cybersecurity training, based on responses from a sample of 504 individuals. Regarding organizations' engagement in cybersecurity training, it is noted that in an increasingly digitalized context, organizations face ongoing risks in the cyber sphere. In this context, cybersecurity training of staff is one of the most fundamental measures to build defensive capacities and cultivate an organizational culture aware of digital threats.

18.4.2 Regarding the distribution of responses

Participants in the study were asked whether the organization they work for had ever provided cybersecurity training. The results are as presented in tables and graphs:

The findings of Table 18.3 show that only 42.46% of respondents confirm that their organization has ever provided cybersecurity training. This percentage, although not very low, shows that more than half of organizations do not have sufficient or clear commitment in this regard. Meanwhile, 37.50% of participants say that no training has been carried out in their organization, representing a significant gap in the preparation of human resources to face digital threats. We further find that 20.04% declare "I don't know," which signals a lack of transparency or communication within organizations about security policies and staff preparation. Regarding the impact of the lack of training in organizations, there is a lack of cybersecurity training at the organizational level, which can bring numerous negative

Table 18.3 Has your organization conducted any training on cybersecurity threats?

Answer	*Number*	*Report*
Yes	214	42.46%
No	189	37.50%
I don't know.	101	20.04%
Total	504	100.00%

consequences, such as an increased risk of successful cyberattacks due to unintentional but unsafe employee behavior (phishing clicks, use of weak passwords, etc.), damage to the integrity and reputation of the organization in the event of an incident, and a lack of institutional accountability when staff are unaware of the protocols and responsibilities related to data and system protection.

18.4.3 Awareness of phishing attacks in public and private organizations (businesses)

In the context of digital transformation, public and private organizations—especially businesses—are increasingly facing cybersecurity challenges. One of the most common and at the same time most dangerous threats is phishing, which consists of fraudulent attempts to acquire sensitive data through false communications and psychological manipulation. In a survey with 1,683 participants, the level of knowledge about this phenomenon was analyzed. The data collected is an important indicator to understand how prepared organizations are to face the risks of phishing through the awareness of individuals who are part of these structures (Figure 18.2).

The data shows that only about 39% of respondents have clear knowledge about phishing, while more than 60% are either uninformed (35.23%) or unsure (25.61%) about this form of fraud. In the context of public and private organizations, this percentage represents a serious operational and reputational risk. Many organizations depend on digital systems for managing data, transactions, and communicating with customers. Therefore, the lack of knowledge about phishing attacks implies a potential vulnerability in the first line of defense, the employees themselves.

18.4.4 Impact on the public and private sectors

Public sectors, such as state administration, government agencies, or emergency services, are often targets of phishing attacks with political or economic motivations. An uninformed employee can click on a dangerous link in an email that imitates a government institution, compromising sensitive data or allowing unauthorized access to protected systems. Meanwhile, in private organizations, mainly businesses, the impact of phishing is even more direct. From small businesses to large corporations, phishing can result in financial losses through payment fraud, theft of customer data, and damaged reputations, and system

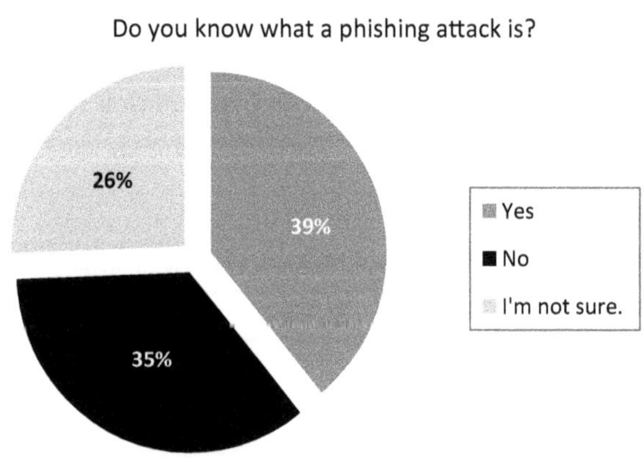

Figure 18.2 Has your organization conducted any training on cybersecurity threats?

crashes through ransomware installed via phishing emails. A worrying element is that front-line employees, such as those in sales, finance, or customer service, are often more exposed but not always sufficiently trained to identify such threats. These results highlight a significant lack of education and regular training in cybersecurity. For public and private organizations, this situation requires the development of internal policies for awareness and response to phishing, mandatory and periodic training for staff tailored to functions and risk levels, implementation of internal testing with simulated phishing emails to measure vigilance and identify weak points, etc.

In conclusion, we can say that the lack of knowledge about phishing attacks is a real threat to public organizations and businesses, which are increasingly dependent on digital technology. The low awareness (only 39.16% recognize this form of attack) is a clear signal that it is time for institutional and strategic action and that security cannot be just a technological issue, but it starts with human awareness.

18.4.5 Participation and interest in cybersecurity awareness campaigns in public organizations and businesses

Cybersecurity has become a strategic priority for modern organizations due to the increase in digital attacks, especially in the form of phishing, ransomware, and attacks on internal systems. Cybersecurity awareness campaigns are one of the most effective tools to reduce risks by educating and preparing employees to recognize and avoid threats. This analysis is based on the responses of 1,123 respondents regarding participation or interest in these campaigns. From the data collection, it is observed that 30.63% of respondents have participated in awareness campaigns, which is a positive figure but still not sufficient to ensure a strong cyber culture. Of these, 32.95% express a clear interest in participating, bringing the total number of individuals engaged or potentially engaged to 63.58%. Meanwhile, 25.29% have not participated so far, and 11.13% are not interested, which in total represents about 36% of a group that remains unreached by current initiatives (Table 18.4).

These data suggest several implications for both public and private organizations. In the public sector, where sensitive state and citizen data are managed, the limited participation in cybersecurity awareness campaigns constitutes a direct risk to institutional security. With only one in three individuals reporting participation, it is necessary for public institutions to standardize and institutionalize mandatory training sessions to be conducted on a regular basis.

For businesses, particularly those operating online or handling customer data, insufficient awareness represents a vulnerability that may result in intellectual property theft, financial losses through sophisticated fraud, and reputational damage leading to diminished consumer trust. Nevertheless, the relatively high interest in participation (32.95%) presents an opportunity for organizations to implement targeted initiatives, such as practical workshops for threat identification, workplace campaigns (posters, videos, simulations), and unannounced testing through simulated phishing attempts to assess employee response.

Table 18.4 Have you participated in any cybersecurity awareness campaigns?

Answer	*Number*	*Report*
Yes	344	30.63%
I am interested in participating.	370	32.95%
No	284	25.29%
I am not interested.	125	11.13%
Total	1123	100.00%

Conversely, the group that did not participate (25.29%), and especially those expressing no interest (11.13%), constitutes a segment requiring greater attention. Tailored awareness strategies, motivational approaches, and personalized educational measures may be necessary to illustrate the tangible impact of cyberattacks on both professional duties and personal security.

18.4.6 Receiving cybersecurity training and deepening knowledge and skills on cybersecurity threats

In an increasingly digitalized environment, organizations—whether public or private—are constantly exposed to cybersecurity threats. A key component for reducing risk is the internal capacity of employees to understand and react to these threats. The following data aims to measure the level of interest in deepening knowledge in this area, as an indicator of readiness to improve institutional security. The majority of respondents (55%) have expressed a desire to deepen their knowledge and skills regarding cyber threats. This constitutes a positive signal, especially for organizations operating in sensitive sectors or processing confidential customer or citizen data. However, a significant percentage (42%) did not express an opinion or did not provide an answer (N/A). This can be interpreted in several ways, such as lack of awareness of the importance of cybersecurity, lack of motivation to engage in additional training, doubts about the direct benefits that this knowledge can bring, etc. Meanwhile, only a small minority (3%) expressed clear disinterest, which suggests that resistance to education in this area is not a serious obstacle at the general level (Table 18.5 and Figure 18.3).

Some possible implications for public organizations and businesses. In the public sector, where services often have a direct impact on citizens' lives (e.g., administration, health care,

Table 18.5 Has your organization conducted any training on cybersecurity threats?

Answer	*Number*	*Report*
Yes	214	42.46%
No	189	37.50%
I don't know.	101	20.04%
Total	504	100.00%

Figure 18.3 Has your organization conducted any training on cybersecurity threats?

taxes), the desire for education in the field of cybersecurity is essential. We estimate that an indicator of 55% interest of respondents is a solid basis for building training policies, while public institutions should use this willingness to develop internal training platforms and awareness campaigns, also targeting the neutral group. Flourishing for businesses, they are at the forefront of digital transformation, operating with transaction systems, e-commerce, IT cloud, and customer data. Thus, cyber threats are no longer just a technical problem, but they are a real risk for businesses. The fact that more than half of employees show interest in training is an opportunity for investment in human capital, through forms of personalized training programs, cyberattack simulations to increase vigilance, as well as the integration of security education into daily work processes. While the group of 42% that did not express an opinion constitutes a gray area, which, through awareness and clear communication about risks, can be transformed into an additional defensive force for the organization. So, we can say that the results show a significant interest in deepening cybersecurity knowledge among respondents belonging to the public and private sectors. However, the neutral group remains a challenge, and for this reason, organizations must follow a proactive and comprehensive approach. Investing in education is not only preventive—it is a strategy for survival and growth in an increasingly interconnected world and more at risk from digital threats.

18.4.7 Interest in deepening knowledge and skills

Discussion on the required level of expertise in cybersecurity. Cybersecurity is no longer just the responsibility of the IT department; it is an institutional imperative that includes all employees, from the operational level to leadership. Given the increasingly sophisticated nature of digital attacks, it is essential to understand the level of expertise that individuals within organizations want to acquire, in order to build a sustainable security culture.

From the collected responses, it is observed that 39.61% of respondents express the desire to achieve an average level of knowledge, which implies a strong interest in understanding threats and basic ways of protection, but not necessarily to build in-depth expertise. While 37.34% aim for a beginner level, representing a large group that is still in the first steps in this field, but that has the potential to grow with the right support. Of these, only 23.06% are interested in an advanced level, which suggests a smaller number of individuals willing to develop in-depth expertise, typically profiles that can become security experts, IT managers, or information security officers (CISO) (Figure 18.4 and Table 18.6).

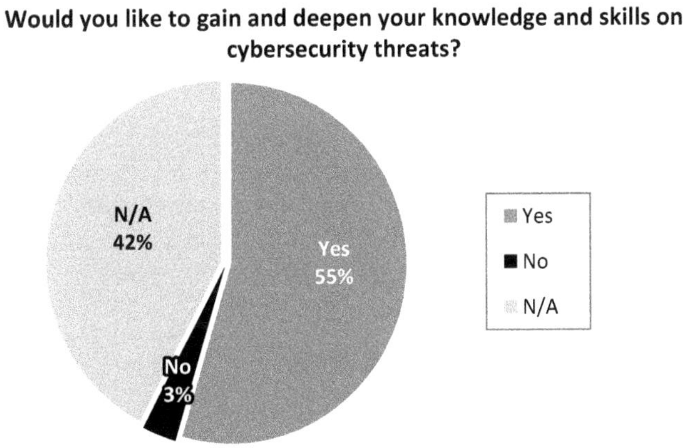

Figure 18.4 Would you like to gain and deepen your knowledge and skills on cybersecurity threats?

Table 18.6 Please indicate the level of knowledge you wish to acquire, related to expertise on cybersecurity threats capabilities

Answer	*Number*	*Report*
Average	663	39.61%
Initial	625	37.34%
Advanced	386	23.06%
Total	1,674	100.00%

In the context of public and private organizations, the findings suggest several important implications. Within the public sector, the shortage of cybersecurity experts is often linked to limited funding for advanced training and institutional structures that do not adequately support professional development. The fact that 39.61% of respondents indicated a need for intermediate-level training and 37.34% for beginner-level training highlights the necessity of broad, foundational programs designed to prepare staff to recognize and prevent common cyber threats such as phishing, malware, and digital fraud. The smaller group requiring advanced training could serve as a professional core capable of forming internal security teams and providing more specialized expertise.

In the private sector, particularly among businesses that process sensitive data or operate online services, this distribution can be interpreted as an opportunity to implement tiered training programs tailored to varying levels of expertise. Such an approach would allow organizations to address the full spectrum of training needs, identify individuals with potential for professional advancement in cybersecurity, and integrate them into strategic positions. Moreover, by doing so, businesses can foster an organizational culture grounded in knowledge and collective responsibility for security, rather than relying solely on technical infrastructure.

18.5 CONCLUSIONS AND RECOMMENDATIONS

Cybersecurity has become a critical component for the sustainable functioning of organizations in the digital age. Data collected from the questionnaire, together with previous analyses, indicate a considerable lack of awareness in this area, particularly among groups with limited education in information technology, who feel vulnerable when facing technological challenges. This evidence underscores the importance of structured interventions in education and capacity building.

18.5.1 Cybersecurity awareness and education

Raising awareness and improving education in cybersecurity should be considered a strategic priority at both the educational system and public policy levels, with the ultimate goal of building a digitally secure society that is prepared for the challenges of technological transformation. Particular attention should be directed toward vulnerable groups with limited access to adequate information and training.

18.5.2 Organizational training structures

To meet the identified needs, both public and private organizations must establish systematic training frameworks that address varying levels of employee knowledge and responsibilities. Recommended strategies include:

Basic training for all employees, focusing on fundamental principles of information security.

Workshops and practical simulations for intermediate levels, emphasizing incident response and awareness of advanced attack tactics.

Advanced courses and professional certifications for technical personnel, who play a central role in institutional protection.

Identification and promotion of internal talent, creating career pathways in cybersecurity.

Mechanisms for measuring progress and assessing training effectiveness, in order to identify gaps in existing competencies.

18.5.3 Demand and capacity assessment

Findings reveal a strong demand for basic and intermediate knowledge in the field of cybersecurity, which presents significant potential for organizational capacity building. However, to ensure the sustainability of cyber defense systems, it is essential to invest in developing a pool of advanced experts capable of leading institutional cybersecurity initiatives.

18.5.4 The human factor in cybersecurity

The human element remains a determining factor in cybersecurity, with international studies and best practices consistently identifying it as the weakest link in the security chain. Continuous training of staff is therefore a critical preventive measure. Such training contributes to:

Early identification of potential threats.
Greater awareness of evolving cyberattack tactics.
Reinforcement of secure practices in the use of systems and data.
Improved responses to security incidents.

18.5.5 Cybersecurity awareness in organizational culture

Cybersecurity awareness must be embedded into organizational behavior and culture. Awareness campaigns should be viewed as a fundamental component of institutional defense infrastructure. The data indicate a promising foundation, with more than half of respondents having either participated in or expressed interest in training programs. For maximum effectiveness, these campaigns should be:

Structured and conducted on a regular basis.

Tailored to different levels of knowledge and job responsibilities.

Enhanced through interactive technology and gamification to boost engagement and motivation.

Supported with real-world examples and case studies to reinforce the seriousness of cyber threats.

18.5.6 Policy and program recommendations

Based on the analysis of current conditions, several key measures are recommended for policymakers and implementers of cybersecurity programs:

Development of role-based, personalized training modules.

Adoption of blended learning methodologies that combine online training with in-person exercises and simulations.

Integration of international best practices, such as ISO 27001 standards and NIST cybersecurity frameworks.

Promotion of an internal security culture in which all employees recognize their responsibility in protecting information systems and digital assets.

REFERENCES

1. Verizon. (2023). *Data breach investigations report*. Verizon Enterprise Solutions.
2. IBM. (2023). *IBM X-Force 2025 threat intelligence index*. IBM Corporation. https://www.ibm.com/reports/threat-intelligence
3. Kemper, G. (2019). Improving employees' cyber security awareness. *Computer Fraud & Security*, 2019(3), 15–18. https://doi.org/10.1016/S1361-3723(19)30085-5
4. Parsons, K., McCormac, A., Butavicius, M., Pattinson, M., & Jerram, C. (2017). The human factors of cybersecurity: Understanding user behaviour and risk. *Computers & Security*, 65, 1–11. https://doi.org/10.1016/j.cose.2016.12.007
5. Neri, M., Niccolini, F., & Martino, L. (2024). Organizational cybersecurity readiness in the ICT sector: A quanti-qualitative assessment. *Information and Computer Security*, 32(1), 38–52. https://doi.org/10.1108/ICS-05-2023-0084
6. Hasan, S., Ali, M., Kurnia, S., & Thurasamy, R. (2021). Evaluating the cybersecurity readiness of organizations and its influence on performance. *Journal of Information Security and Applications*, 58, 102726. https://doi.org/10.1016/j.jisa.2020.102726
7. Simonnet, O. (2025). Agentic AI: The rising threat that demands a human-centric cybersecurity response. *TechRadar*. https://www.techradar.com/pro/agentic-ai-the-rising-threat-that-demands-a-human-centric-cybersecurity-response
8. Elkhodr, M., & Gide, E. (Eds.). (2025). *Generative artificial intelligence empowered learning: A new frontier in educational technology* (1st ed.). Chapman and Hall/CRC. https://doi.org/10.1201/9781003422433
9. Tafaj, I. (2024). Ja sa e sulmeve kibernetike janë si rezultat i gabimeve njerëzore. *Business Magazine Albania*. https://businessmag.al/igli-tafa-ja-sa-e-sulmeve-kibernetike-jane-si-rezultat-i-gabimeve-njerezore/?utm_source=chatgpt.com
10. AP News. Albania says cyberattack disrupted public services and blames Iranian-backed hackers. *AP News*, Feb. 2024. [Online]. Available: https://apnews.com/
11. Euractiv. (2025). Albania mixed bag in citizen cyber awareness amid ongoing cyber attacks. https://www.euractiv.com/section/politics/news/albania-mixed-bag-in-citizen-cyber-awareness-amid-ongoing-cyber-attacks/?utm_source=chatgpt.com
12. Mulita, R. (2024). Challenging the Albanian EU cybersecurity through education. *CIT Review Journal*, 1, 12–17. https://doi.org/10.59380/crj.vi1.5098
13. Mulita, R. (2025). Challenging the cybersecurity threats through education programs. In K. Dhoska & E. Spaho (Eds.), *Bridging horizons in artificial intelligence, robotics, cybersecurity, smart cities, and digital economy*. Springer. https://doi.org/10.1007/978-3-031-72029-1_22
14. Von Solms, R., & Van Niekerk, J. (2013). From information security to cyber security. *Computers & Security*, 38, 97–102.
15. Bada, M., Sasse, M. A., & Nurse, J. R. C. (2019). Cyber security awareness campaigns: Why do they fail to change behaviour? arXiv preprint arXiv:1901.02672.
16. World Bank. (2019). *Albania cybersecurity capacity maturity model (CMM) review report*. Global Cyber Security Capacity Centre.
17. Ahmad, A., Deshmukh, A., Maynard, S. B., & Shanks, G. (2021). Security as a strategy: Addressing the gap between cybersecurity and strategic management. *Computers & Security*, 108, 102364.
18. ENISA. (2023). *Threat landscape report 2023*. European Union Agency for Cybersecurity.
19. Pfleeger, S. L., & Caputo, D. D. (2012). Leveraging behavioral science to mitigate cyber security risk. *Computers & Security*, 31(4), 597–611.
20. Linkov, I., Eisenberg, D.A., Plourde, K., Seager, T.P., Allen, J., & Kott, A. (2013). Resilience metrics for cyber systems. *Environment Systems and Decisions*, 33, 471–476.

21. NIST. (2018). *Framework for improving critical infrastructure cybersecurity (Version 1.1)*. National Institute of Standards and Technology.
22. Symantec. (2020). *Internet security threat report*. Symantec Corporation.
23. Nash, S., & Agrawal, S. (2022). Fileless malware attacks: A survey and analysis. *Journal of Information Security and Applications*, 65, 103045.
24. Kumar, P., Rajput, S., & Singh, S. (2021). Ransomware attack trends and defense: An overview. *Journal of Cybersecurity*, 7(1), taab010.
25. Chen, T. M., Lee, C. T., & Lee, Y. J. (2020). Advanced persistent threats: A review. *IEEE Access*, 8, 177904–177917.
26. Martin, G., McCoy, E., & Horowitz, B. (2020). Ransomware and healthcare: Risks and mitigation strategies. *Health Security*, 18(3), 192–198.
27. Scarfone, K., & Mell, P. (2007). *Guide to intrusion detection and prevention systems (IDPS)*. NIST Special Publication, 800–894.
28. Singh, M., & Silakari, S. (2021). Malware detection techniques: A review. *Journal of Computer Virology and Hacking Techniques*, 17, 171–190.
29. Alasmary, W., Alsaleh, M., Alhaidari, F., & Alshehri, M. (2021). Cloud-based malware detection techniques: A systematic review. *IEEE Access*, 9, 102431–102450.
30. Romero, M., Ventura, S., & García, E. (2021). Intelligent tutoring systems for cybersecurity education: A systematic review. *Computers & Education*, 165, 104142.
31. Alier, M., García-Peñalvo, F. J., & Camba, J. D. (2024). Generative artificial intelligence in education: From deceptive to disruptive. *International Journal of Interactive Multimedia and Artificial Intelligence (IJIMAI)*, 8(5), 5–14
32. Lee, J., Kang, S., & Yoon, J. (2019). Adaptive learning system based on machine learning for cybersecurity education. *IEEE Access*, 7, 153517–153526.
33. Jones, E. (2018). Behavioral AI and cyber defense: Darktrace case study. *Journal of Cybersecurity Technology*, 2(4), 243–257.
34. Cylance. (2020). *Cylance protect: AI-driven endpoint security*. Cylance Inc.
35. Splunk. (2021). *AI and machine learning in SIEM: Enhancing security operations*. Splunk Inc.
36. Biggio, B., & Roli, F. (2018). Wild patterns: Ten years after the rise of adversarial machine learning. *Pattern Recognition*, 84, 317–331. https://doi.org/10.1016/j.patcog.2018.07.023
37. Creswell, J. W., & Plano Clark, V. L. (2017). *Designing and conducting mixed methods research* (3rd ed.). Sage Publications.
38. Patton, M. Q. (2015). *Qualitative research & evaluation methods* (4th ed.). Sage Publications.
39. Bulgurcu, B., Cavusoglu, H., & Benbasat, I. (2010). Information security policy compliance: An empirical study of rationality-based beliefs and information security awareness. *MIS Quarterly*, 34(3), 523–548. https://doi.org/10.2307/25750690

Chapter 19

AI-powered cyber warfare

Safeguarding data privacy and economic stability in a hyperconnected world

Faraz Khan and Sumaira Khan

19.1 INTRODUCTION

In the super-connected world we live in today, artificial intelligence (AI) is something that works both ways. On one hand, this AI brings new inventions from automatic systems to personalized things, but on the other hand it also strengthens cyber warfare. Some people use AI to launch dangerous attacks ranging from human-like phishing attacks to ransomware that can paralyze the economy. Moreover, 80% of the businesses around the world are dependent on systems running on the internet. In such a situation, a small breach can cause a loss of millions of dollars, which also shakes trust. While past systems, such as firewalls and antivirus protection, offer a baseline response, they may not respond as rapidly and robustly to advanced cyber threats when compared to AI. AI makes cybersecurity stronger and even more robust, as it utilizes the abilities of humans and mimics and sometimes exceeds the thinking power of humans [12]. And even though AI may be slowly taking jobs and precipitating global greed, as humans experience layoffs and accept a much lower standard of living, AI is moving in the right direction to improve quality of life. It is important to note that the use of AI in cybersecurity is serious. Just think, by 2025, the usage of AI to drive cyberattacks will skyrocket and continue to drive generations of social change, as we engaged in/from 2020 to 2025 with cyberattacks fueled by AI. There is no doubt that AI can drive massive social change. The digital era has transformed almost every sector across the world; however, this transformation has come with a lot of challenges, especially regarding cybersecurity. Whereas the value in traditional technology (firewalls, antivirus) was once relied upon as an adequate approach to cybersecurity, it is now seen as inadequate to survive technologically sophisticated and potentially complex threats. Therefore, we need security that is accurate, resilient, robust, reliable, and, more importantly, reliable in all situations. At a time when much in the world is not stable, we must rely on and use technologies that are prepared for these challenges. As we study and work on AI and cybersecurity technologies, we will understand how powerful these technologies are and how they are truly game-changers. AI can use advanced algorithms to find patterns in a lot of data, learn from new things, and make highly accurate predictions [17]. This shows how smartphones are playing an increasingly important role in cybersecurity. First, we will look at today's cybersecurity efforts and legacy security solutions, then understand how machine learning, deep learning, and unit language engineering systems are being made more standard and resilient. We'll look at real-world examples like Symantec's Targeted Attack Analytics (TAA) and Sophos' InterceptX to see how big the impact of infrastructure really is. But remember, this technology is a double-edged sword—just as we're using it for security, cybercriminals are adopting it to launch more advanced and dangerous attacks. To deal with these challenges, it is necessary to develop a strong ethical framework, which will be discussed in the following sections. The article will end with a glimpse of the future as to how the advancement of AI will change cybersecurity in the future

 DOI: 10.1201/9781003685890-19

and how cybersecurity will affect AI. This article briefly attempts to explain how AI, if used responsibly and thoughtfully, can completely change the entire landscape of cybersecurity. The use of AI in cybersecurity is transforming security in the digital world, and its use is also increasing rapidly. As companies digitize their operations, new avenues of attack are opening up, and cyber threats are becoming more complex. AI-powered cybersecurity solutions answer these challenges by offering real-time threat detection, advanced predictive analysis, and automated response capabilities. The global AI in cybersecurity market, which was valued at $25.40 billion in 2024, is expected to reach $31.38 billion in 2025 [7].

Nowadays, companies have started relying on AI tools for security. The specialty of these tools is that they keep learning with time and catch such patterns that normal software usually misses. I have seen in some reports that many organizations are now using AI-based monitoring instead of traditional systems, due to which cyber fraud and data breach incidents have reduced. I believe that this combination of AI and cybersecurity will become even more important in the coming times, especially when a zero-trust and data-centric model has to be followed [5].

19.2 AI-DRIVEN CYBER THREATS

19.2.1 Evolution of AI in cybercrime

Nowadays, the methods of cybercrime have changed a lot due to AI. Now many attacks have become completely automated and can be easily carried out on a large scale. Some AI malware changes its code repeatedly so that antivirus cannot detect it. Apart from this, phishing emails are also written so naturally using NLP technology that people often get deceived. It is even written in reports that their success rate reaches 90%. Additionally, the reach of deepfake technology (which allows fraudsters to impersonate executives) causes fraud losses stated to exceed $500 million annually [5]. Automated bots exploit vulnerabilities in real time, as seen in the 2023 Equifax breach affecting 147 million users. As shown in Figure 19.1, the taxonomy of AI applications in cybersecurity categorizes their use in both attacks and defenses, providing a structured overview of how AI is applied in this field. A taxonomy of AI applications in cybersecurity categorizes their use in attacks and defenses. Neural network models highlight the complexity of automated hacking. The role of AI in cybersecurity is truly revolutionary. Using techniques such as machine learning, natural language processing, and behavior analytics, AI can process huge amounts of data and identify patterns that point to malicious activity. This makes the entire cybersecurity framework more agile and proactive. By automating repetitive tasks such as log analysis, threat detection, and compliance monitoring, AI makes the job of cybersecurity professionals easier. AI-powered tools filter out unnecessary information and focus on real threats, improving decision-making. In addition, in the field of information security, AI makes it possible to have defense systems that predict future threats, rather than relying on historical data and trends. For example, banks and financial institutions use AI algorithms to detect fraudulent transactions in real time. In the health-care sector, AI monitors network activity and protects sensitive patient data from unauthorized access. These examples showcase the profound impact of AI on cybersecurity and demonstrate how it enhances the capabilities of human intelligence. A recent Threat Intelligence Report by Anthropic (August 2025) highlights the growing misuse of AI in cybercrime (Anthropic, 2025). The report reveals that AI is increasingly weaponized, lowering the technical barrier for cybercriminals. For example, Claude Code was exploited in a large-scale extortion campaign that targeted health care, emergency services, and government institutions, with ransom demands exceeding $500,000. North Korean operatives used AI to create

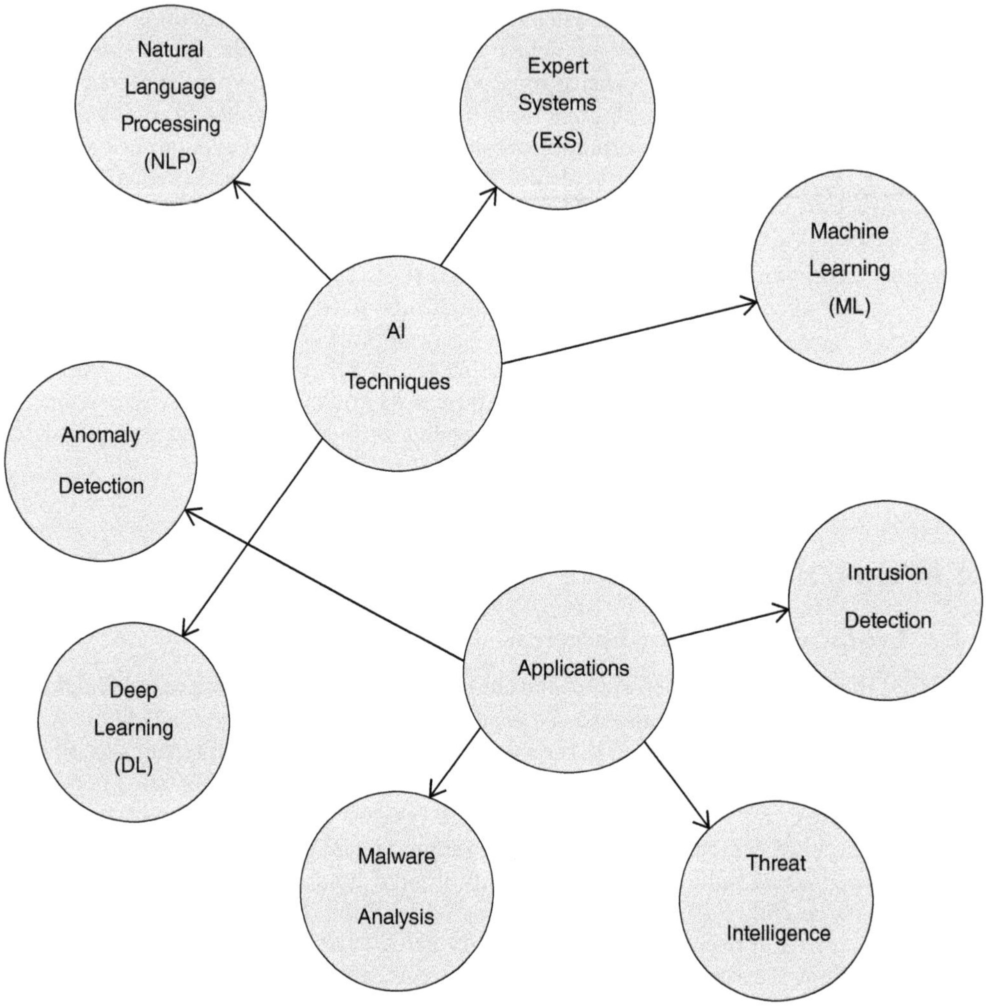

Figure 19.1 Taxonomy of AI applications in cybersecurity.

Source: Garcia 2024.

fake identities, secure jobs in Fortune 500 companies, and perform technical work, bypassing traditional training bottlenecks. AI-generated ransomware has also emerged as a dangerous trend, where no-code malware is sold as "ransomware-as-a-service" on underground forums for as little as $400. These developments underscore that AI is now embedded at every stage of cybercrime operations—from reconnaissance and credential theft to victim profiling and extortion. Such misuse demonstrates the dual-use dilemma of AI, where tools designed for productivity are subverted for malicious gain. Importantly, it shows that even low-skilled actors can now launch sophisticated attacks, a shift that poses unprecedented global risks. The Anthropic report stresses the urgent need for stronger detection, monitoring, and regulatory frameworks to counteract these evolving threats (Table 19.1).

Table 19.1 Common AI-driven cyber threats

Threat type	*Description*	*Impact*
AI-Powered Malware	Adapts code to evade detection	Bypasses antivirus systems
Phishing with NLP	Crafts targeted, convincing emails	90% success rate
Deepfake Fraud	Mimics voices/videos for scams	$500M+ annual losses
Automated Bots	Exploits vulnerabilities in real time	Large-scale data breaches

Source: Compiled from Wang (2023) [5], Lee (2025) [6], Jones (2022) [7].

19.2.2 Exploitation through app permissions

Today, Android apps running on billions of smartphones and tablets worldwide have collected a huge amount of personal data about users, raising serious and justified concerns about security. There has been a lot of research to improve the security of smartphone apps, with most of the focus being on permission management. Today, app markets and security scanners check whether an app has used all the permissions it has requested, but this approach often ignores real threats. Mobile apps require user permission to access personal information, but users often grant permissions without thinking and are unaware of the potential misuse. For example, a simple flashlight app can steal location data and send it to external servers. Newer detection techniques assume that if the user has granted permission, it is safe, but this is a major drawback. That is why this research paper raises an important question: "Are all permissions requested by an app actually necessary for its functioning?" However, understanding an app's needs is not easy, and determining them automatically remains challenging. To address this problem, we propose a method that compares an app's permission requests with the permissions of other apps that perform similar functions. If an app is asking for too many permissions compared to its peers, it may be considered potentially dangerous and warrants further investigation. A statistical approach is designed and evaluated using Play Store apps to identify excessive permission requests. Experimental results demonstrate its effectiveness in detecting suspicious apps with irrelevant permissions. The app market has experienced volatility over the past five years, driven by post-COVID-19 recovery and AI adoption. In 2024, AI chatbot apps saw a 112% year-over-year download increase, reflecting growing user engagement. Consumer spending on mobile apps surged to 36.2 billion U.S. dollars in Q2 2024, a 12% rise from 2023. Entertainment apps generated nearly 9 billion U.S. dollars in revenue in Q1 2024, outpacing health and fitness apps at 1.2 billion. Video watching, the most popular smartphone activity, engaged 60% of Asia-Pacific users in 2024. Despite potential U.S. bans, TikTok users spent 406 minutes on average in 2024, maintaining high engagement. ChatGPT, launched in November 2022, amassed 7.4 million downloads in its first 10 days. Microsoft Copilot and DeepSeek, launched in 2024, recorded 2 million and 1 million downloads, respectively, in their debut periods. Photo editors like Canva saw 1.16 million U.S. downloads in November 2024, while Remini achieved 763,000. Post-pandemic app growth has slowed, but generative AI integration in editing and chat apps has boosted adoption. This research underscores the urgent need for robust mechanisms to prevent exploitation through app permissions, leveraging AI for enhanced security. Millions of Android apps are running on billions of smartphones and tablets around the world today and are collecting a lot of personal data from users. Although there has been a lot of research on the security of smartphone apps, today's app markets and security scanners mostly only look at whether an app is using the permissions it asks for. The real problem is that most users blindly grant every permission, which is wrong, increasing the chances of abuse. For example, a simple flashlight app may ask for location data and then send it out.

Existing identification systems consider such permissions harmless if the user grants them, whether or not they are related to the app's actual need. That's why this research focuses on the key question—are all permissions requested by an app really necessary? This is not easy, because determining which permissions are appropriate and which are not based on an app's function is still a difficult task. To solve this problem, we propose a method by comparing the permissions of an app with the permissions of other similar apps. Apps requesting or using significantly more permissions than their peers are flagged as potentially malicious, warranting further scrutiny. We propose a statistical approach to detect excessive permission requests and evaluate it using Play Store apps. Experimental results and case studies demonstrate that our method effectively identifies suspicious apps with irrelevant permission demands [6]. Mobile apps often request excessive permissions, with 60% collecting unnecessary data and 45% of users failing to revoke permissions after uninstallation. These vulnerabilities enable AI-driven tools to harvest data covertly [12].

As shown in Figure 19.2, the data flow diagram of app permission exploitation illustrates how personal data is misused through excessive permission requests. Android's permission system is designed to warn users of risks when installing apps. When an app is installed, users can see what permissions it is asking for and cancel the installation if they want. But the question is, does this system really warn users about the risks? Do users notice, understand, and act on these permissions? To find out, two usability studies were conducted—an online survey (with 308 Android users) and a lab study (interviews and observations of 25 users). The results showed that most people did not pay attention to permissions. Only 17% of people paid attention to permissions during installation, and only 3% of people in the survey were able to answer all three permission questions correctly. This shows that existing permission warnings do not help most users make the right security decisions [1].

However, there was a small group that did pay attention to permission warnings and correctly understood them. Based on these results, some suggestions have been made to increase users' attention and understanding, but challenges remain. Today, smartphones are used everywhere—not just for calls and messages, but for office applications, internet access, gaming, and location-based navigation. Android's open architecture and developer-friendly APIs have made it the largest in the market. But its popularity and economic benefits also attracted malware developers, leading to a rapid growth of Android malware between 2010 and 2014. As shown in Table 19.3, the comparison of Android malware detection methods highlights their respective strengths and weaknesses in addressing evolving threats. Researchers and antivirus companies have recognized that old methods such as signature-based and static analysis are no longer enough, as new malware uses techniques such as encryption, code obfuscation, and environment-aware behavior. This is why behavior-based, anomaly-based, and dynamic analysis methods are now being given more attention. Since no single approach is perfect, accurate detection is possible only by combining several techniques. While many previous reviews have looked at smartphone OS security in a holistic way, this survey focuses specifically on Android security—where the growing threat of malware, anti-analysis techniques, and detection methodologies are discussed in depth. This includes an analysis of

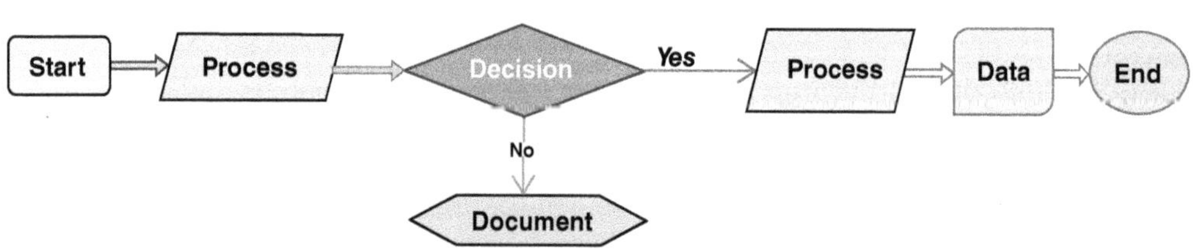

Figure 19.2 Data flow diagram of app permission exploitation.

Source: Chen 2022 [46].

Table 19.2 Comparison of Android malware detection methods

Detection method	*How it works*	*Advantages*	*Weaknesses/limitations*
Signature-based	Matches malware against previously known patterns or signatures	Fast and simple detection, effective for older attacks	Cannot detect new or unknown malware, easily bypassed
Static analysis	Examines app code and structure without executing it	Produces quick results, relatively easy to implement	Fails against code obfuscation and encryption techniques
Behavior-based	Observes runtime behavior of the app (e.g., data access, network calls)	Can detect new and unknown malware	Consumes more system resources, risk of false positives
Anomaly-based	Builds a model of normal activity and flags deviations as suspicious	Effective against zero-day attacks, provides advanced protection	Requires fine-tuning, otherwise produces too many alerts
Dynamic analysis	Runs the app in a controlled environment (sandbox) to monitor its behavior	Provides deep insights, effective against complex malware	Time-consuming, scalability is a challenge

Android's security posture, threats, the 2010–2014 malware timeline, evasion strategies, and existing detection approaches. Finally, by looking at the strengths and limitations of existing research, this survey attempts to set the direction for next-generation Android security and malware detection solutions [20] (Table 19.2).

19.3 ECONOMIC FALLOUT

AI-driven cyberattacks cost $8 trillion globally in 2023, with small businesses losing $25,000 per attack. As shown in Figure 19.3, the rising costs of cybercrime from 2018 to 2023

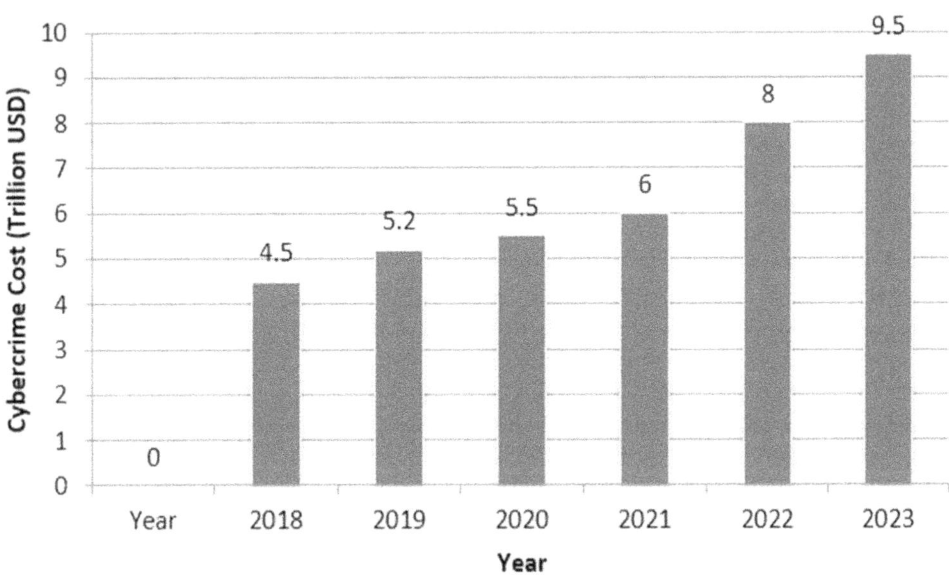

Figure 19.3 (Graph 1): Rising costs of cybercrime (2018–2023).

Source: Cyber security Ventures 2023 [16].

underscore the exponential growth in financial damages, with projections reaching $8 trillion in 2023. The 2021 Colonial Pipeline attack disrupted U.S. fuel supplies, illustrating large-scale impacts [9].

The escalating wave of AI-driven cyberattacks has plunged 72% of Indian firms into a cybersecurity crisis, triggering significant economic fallout that threatens the nation's digital economy. These sophisticated attacks, leveraging AI for credential stuffing, deepfake impersonation, and polymorphic malware, have exposed critical vulnerabilities, leading to substantial financial losses across sectors. Only 14% of organizations express confidence in defending against these threats, indicating a severe preparedness gap that amplifies economic risks [13]. The survey by Fortinet and IDC highlights a resourcing crisis, with just 0.91 cybersecurity professionals per 100 employees, straining budgets and hindering effective defense mechanisms. The current shortfall serves to exacerbate operational interruptions, where downtime costs diminish profitability, and ransomware incidents, which affect 44% of firms and have a high ransom request, will be depleting financial capacity further and reducing money that could go toward growth activities. Here, companies' budgets for security are increasing very little. I get reports that 80% of organizations spend less than 10% of their revenue on security. What is the reason that the gap between risk and investment is becoming clearer? Wrong cloud settings and attacks on the software supply chain have become very fast, which is even faster than normal threats. Industry projects get delayed, and costs also increase. Specialized teams are also fewer; only 6% have trained people who can hunt down threats. What is the reason that response time becomes longer in case of a breach and the economy remains affected for a longer period? Important industries like banking and health care are also facing legal and reputational losses. On average, only 1.4% of revenue is used for security, which is very less in front of new threats like AI and machine learning. What is the reason that companies have to give up and also lose customer trust? Tool and application complexity, cited by 44% of participants, amplifies operational costs because businesses attempting to implement ineffective or inefficient security add additional costs. A total of 85% of firms do not have a standalone CISO, which demonstrates a lack of ability to strategically plan for financial implications related to cyber risks and vulnerabilities before they occur. Rising AI-driven threats exploiting IoT/OT vulnerabilities threaten manufacturing and energy operations, causing stoppages and financial burdening costs to both areas of respective economic importance [10]. With 88% momentum toward unified security platforms showing the potential for cost savings, smaller companies feel the constraints of initial investment. With public sector funding inadequate, only 15% have decided leadership in cybersecurity, and risks to taxpayers increase during a significant breach or target attack [16]. Also, with the much-reported 2X to 3X increase in AI threats over the past 12 months, already coupled with an increase in their expected insurance premium, has again added to the burden on corporates. Small and medium enterprises with less to absorb from losses also face increased risk of going out of business, endangering jobs and local economies. Lastly, the overall heightened profile of India in the global ranking as a top target only reduces their foreign investment opportunities and the growth of the economy. Lost productivity from employee downtime during attacks reduces GDP contributions across industries. The need for advanced threat detection systems drives up technology spending, straining already tight budgets. International trade could suffer as insecure digital networks deter global partnerships. The economic fallout extends to consumer markets, with data breaches eroding e-commerce confidence and sales. Ultimately, without urgent investment and policy reform, the financial toll of AI-driven cyberattacks could destabilize India's economic framework in the coming years.

19.4 THREAT HUNTING

In the past, cybersecurity threat detection was primarily based on signature-based techniques. These methods were effective against known attacks by identifying repetitive patterns or "signatures" associated with cyberattacks [16]. But a major drawback of this approach was that it failed to catch new and unknown threats that did not match any existing signature. This is where AI comes into play. AI is revolutionizing threat detection by harnessing its predictive powers to improve detection and identification capabilities. Its real power lies in its ability to efficiently process and analyze massive amounts of data, identify meaningful patterns, and filter out unnecessary noise. In today's ever-more-connected digital environment, where the volume of data generated far exceeds the computational capabilities of traditional systems, AI-based threat detection offers a huge advantage. By incorporating behavioral analysis, AI systems enable a more dynamic and proactive approach to threat detection. By processing endpoint data in organizational networks, detailed application profiles can be developed, which present a comprehensive picture of normal operational behavior. These profiles serve as a basis for detecting any abnormalities in real time, where any deviation from normal can indicate a potential threat. This AI-enhanced approach marks a major shift in cybersecurity from a reactive approach to a proactive one. With the help of AI, organizations can not only identify and respond to threats quickly but also anticipate and prevent attacks before they cause damage. AI is now helping to create a robust and flexible cybersecurity system that can handle new threats. In cybersecurity, AI works by collecting data, looking for patterns, and trying to understand them using machine learning. The system first collects data from network logs, user behavior, and external threat reports. Then, machine learning algorithms are used to determine if any activity is unusual. For example, if too many login attempts are being made from the same IP address, it could be a brute-force attack, and a prompt alert is generated or an auto-defense is triggered. Predictive analytics can help companies predict future exploits. For example, a component of AI software can be declared risky if it has been exploited in the past. This forward-looking approach makes AI powerful in cybersecurity.

19.5 NETWORK SECURITY

Network security is an essential aspect of any cybersecurity plan. Two important associated tasks for network security are developing security rules and understanding the network topography [14]. These tasks were once very manual and time-consuming. AI has totally changed the game for these tasks, making both faster and better. Security rules are the foundation of network security. They help an organization determine which connections are legitimate and which are suspicious because they may have malicious intent. AI makes this process much easier. It looks at a lot of data and catches patterns that humans often miss. AI automatically creates rules with great accuracy, which strengthens the security framework and helps to identify and act on new threats in advance. Understanding the network structure is equally important. Understanding this requires knowing how different applications and workloads within an organization connect and interact with each other. AI simplifies this process by analyzing network traffic patterns [4]. It groups workloads and provides useful information to improve security policies. This not only saves time and effort but also allows security teams to focus their energy on other important strategic plans. Finally, this role of AI in network security shows that we are now in a position to prevent threats by being proactive rather than waiting to be attacked from behind. By making it easier to create rules and understand the network structure, AI strengthens the ability to identify threats and creates a strong and flexible defense against today's growing and complex cyberattacks. Networks are one of the fundamental units of today's digital and hyperconnected world. The ubiquitous presence of global communication networks enables us to interconnect our devices, giving us uninterrupted access to a wide range of applications and services around the world in real time. The concept of a centralized network perimeter is gradually disappearing, as IT infrastructure is now mostly outsourced to cloud environments and location restrictions are loosening. Nevertheless, networks – whether they are wired or wireless – remain the main enabling force of this omni-connectivity of the modern world. Information is exchanged between two devices on a network by sending packets back and forth, which makes them an attractive target for cybercriminals. Packet sniffing, man-in-the-middle attacks, or distributed denial of service (DDoS) attacks are some examples, which target confidentiality, integrity, and availability. The main network security measures include firewalls and virtual private networks (VPNs). Firewalls restrict access to networks or devices using access policies and rules so that communication can only occur with certain networks or devices. VPNs prevent both active and passive attacks by creating an encrypted connection (tunnel) over the internet and are considered the best network security measure available.

19.6 HUMAN OVERSIGHT, AI-POWERED ATTACKS, AND ETHICAL GOVERNANCE

19.6.1 Human oversight and AI limitations

By introducing incorrect or misleading data, hackers can significantly distort the understanding and decision-making ability of AI models. This can lead them to misidentify threats or even overlook vulnerabilities. In addition, hackers can also exploit the inherent limitations of AI systems [21]. Since AI models are mostly trained on old data, they often struggle to recognize or respond to completely new attack methods. This is where the intuition and experience of cybersecurity experts are irreplaceable. Therefore, even though AI is a powerful tool to automate and enhance many aspects of cybersecurity, human vigilance, expertise, and adaptability should never be underestimated. Effective security requires a symbiotic relationship between AI capabilities and human oversight, as the human element still plays a vital role in identifying, understanding, and mitigating potential risks (Table 19.3).

As shown in Table 19.3, the strengths and limitations of AI in cybersecurity highlight the indispensable role of human oversight in addressing gaps that AI cannot resolve alone.

19.6.2 AI-powered cyberattacks

As the power and understanding of AI is increasing, cybersecurity is getting stronger, but it also has a downside. This same technology is now helping hackers to launch more powerful and dangerous attacks. Cybercriminals use AI to automate attacks, bypass security systems, and exploit vulnerabilities very quickly. With the advancement of AI, attackers can now even copy human-like behavior, such as phishing emails and messages that look exactly like the real ones, and the other person trusts them more [12]. That's not all, now there are also AI-powered "smart" malware that learn their environment and adapt their attack methods so that they can evade detection and cause more damage. A famous example is Eyebeam Research's "DeepLocker," which reveals its true nature only when activated on a target. Old security methods, such as identifying previously known signatures, no longer work against these new AI-based attacks. That is, AI is a helpful tool for security on one hand, but it has

Table 19.3 Strengths and limitations of AI in cybersecurity with human oversight

AI capability	*Strength/benefit*	*Limitation/risk*	*Role of human experts*
Threat Detection	Identifies anomalies and patterns across vast datasets	Vulnerable to adversarial or misleading data [14–17]	Validate alerts, refine detection strategies
Predictive Analysis	Anticipates potential attacks based on historical data trends	Struggles with entirely novel attack strategies [14–17]	Apply intuition to unknown or emerging threats
Automated Response	Responds to threats at machine speed	May overreact or miss context, causing false positives/negatives	Decide when escalation or manual intervention is needed
Malware and Phishing Defense	Learns evolving signatures and detects sophisticated attacks	Smart malware (e.g., DeepLocker) can adapt and evade [10, 11]	Investigate complex or evasive attack scenarios
Policy and Governance Support	Automates compliance checks and policy recommendations	Lacks ethical reasoning; operates as a "black box" [21, 22]	Ensure transparency, accountability, and ethical oversight
Resource Optimization	Reduces workload by automating repetitive tasks	Risk of over-reliance, neglecting creative human problem-solving [14–17]	Provide critical thinking and ethical decision-making

become an equally powerful weapon for hackers. That is why it is necessary to create better AI-based security systems, and at the same, time there will always be a need for human monitoring.

19.6.3 Ethical considerations and governance

The use of AI in cybersecurity also brings with it a number of ethical and governance challenges. While AI's analytical capabilities are useful in detecting and preventing threats, they also raise concerns about data privacy and security that cannot be ignored. Many AI models act as "black boxes" where their decision-making processes are unclear, making it difficult to assign responsibility in the event of a security breach or failure [18]. At the same time, the potential for malicious use of AI highlights the urgent need for ethical safeguards. Without strong governance, hackers can use AI to cause massive damage. To prevent this, strong governance frameworks and rules are needed to ensure responsible and appropriate use of AI. These measures must include transparency, protecting data privacy, ensuring accountability, and preventing misuse. Ultimately, AI is not a magic cure for all cybersecurity challenges, but a powerful tool that needs to be used with care. Real security requires a balanced approach that combines the power of AI with human oversight, ethical practices, and flexible governance.

19.7 CITIZEN DATA PRIVACY: A FRAGILE FRONTIER

19.7.1 Mass data collection

Tech giants like Google and Meta monetize user data without transparent consent, posing risks for nations like India reliant on foreign platforms. Data privacy laws are critical for sustaining trust in digital ecosystems. The rise of mass data collection has become a cornerstone of modern public health campaigns, as exemplified by the Public Health England-Drinkaware Drink Free Days (DFD) campaign conducted across eight English regions in 2018–2019, which relied on extensive user data to tailor its outreach. Industry-funded initiatives like Drinkaware leverage aggregated consumer data to shape messaging, raising concerns about the scale and ethics of data amassed from millions of users. The campaign's effectiveness hinged on analyzing behavioral patterns, necessitating the collection of vast datasets from social media and online platforms to target specific demographics. Alcohol industry bodies utilize this data as part of their corporate social responsibility (CSR) strategies, positioning themselves as legitimate policy actors while accessing sensitive user information. Interviews with 36 policy actors all demonstrated that data collection was seldom restricted to health outcomes [2]. The primary reason for collecting data consistently became apparent; it's driven by industry requirements for targeted marketing and policy development. The use of framing analysis of identified industry social media accounts illustrated how data could also be used to develop long-term strategies that would enable the exploitation of consent to use their data for commercial purposes. Although the campaign was framed as public health, the absence of transparency surrounding how this data could be used represents significant misuse, with industry actors retaining and repurposing personal information. The expansion that mass data collection allows for can lead to indirect outcomes (e.g., expansive changes to the regulatory landscape over time that favor alcohol producers). Prior evaluations led by Public Health England were set out to contend with communication outcomes and failed to consider the more direct costs of publicly using massive amounts of data for user privacy. Using a mixed-methods approach combining documentary analysis from websites and confirmed by the interviews, current research uncovered how data silos allowed industry to engage in the evaluation process by framing the impact of an advertisement campaign, but adjusting its tactics. In these accounts provided by policy actors, there was no evidence of direct expectation of promotional behavior by industry actors. There was no suggestion of an overtly political engagement by the research or practitioner community using data for promotional engagement, and all of them continued to engage politically by drawing on aggregate insights. This practice raises economic concerns, as data monetization could offset campaign costs, benefiting industry funders. The ethical dilemma intensifies with the potential for data breaches, exposing users to financial and identity risks. Policy actors noted that data collection scales with digital engagement, increasing economic stakes for both public and private sectors. The reliance on user-generated content for campaign analysis underscores a growing data economy, where value is extracted from personal information. Industry funding of such initiatives creates a feedback loop, where collected data justifies further investment in similar campaigns. The lack of regulatory oversight on data scope exacerbates economic vulnerabilities, potentially leading to costly litigation. Mass data collection thus fuels a dual economy—public health benefits versus industry profit—challenging the balance of societal good. The campaign's reliance on digital platforms amplifies data volume, straining infrastructure and increasing operational costs. Ultimately, the economic fallout of mass data collection in such campaigns could reshape trust in public-private partnerships, necessitating robust data governance frameworks.

19.7.2 User negligence

A total of 70% of users ignore app permission settings, and 55% reuse passwords, amplifying privacy risks. The digital economy has experienced rapid growth, characterized by constant innovation, market dominance by a few major players, and an increasing power imbalance between large corporations and smaller entities [1]. This expansion has coincided with the rising value of data across all economic sectors, fueled by advancements in data analytics

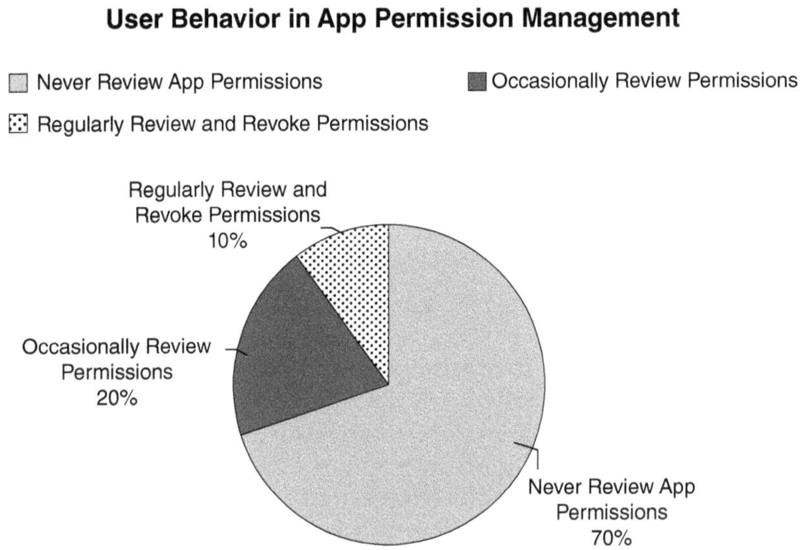

Figure 19.4 Prevalence of risky user behaviors in permission and password management.

Source: Pew Research 2021 [20].

and increased computing capabilities. Currently, 57% of businesses report managing big data—vast datasets collected from diverse sources like sensors, web applications, and social media. This "datafication" process generates approximately 2.3 trillion gigabytes of information daily, enabling services ranging from health tracking to digital mapping. While companies often classify this information as nonpersonal, truly anonymized user data has become increasingly rare. In the EU alone, over 369 million internet users generate extensive personal data through their online activities, including social media use, searches, and e-commerce. This data encompasses names, preferences, locations, and browsing history—all valuable for customer acquisition and retention. Many digital services appear free to users but in reality operate on an exchange model where personal information serves as currency [23]. Through their online behavior, consumers unknowingly provide detailed profiles that enable precise individual targeting, making personal data the fundamental economic unit of the digital economy. As shown in Figure 19.4, the prevalence of risky user behaviors in permission and password management significantly amplifies privacy vulnerabilities in the digital economy.

19.7.3 National security implications

Unregulated data flows compromise sensitive information, as seen in a 2022 Indian defense personnel data breach via unsecured foreign apps. The digital economy is growing fast, with new services constantly appearing. A few big companies dominate the market, creating an uneven playing field where small businesses and individual users struggle to compete [6]. At the same time, data has become incredibly valuable. Businesses are collecting and analyzing huge amounts of information—from social media posts to sensor readings [15]. About 57% of companies now work with these massive datasets. Every day, the world generates about 2.3 trillion gigabytes of data, which powers everything from fitness apps to digital maps. While companies claim much of this data is anonymous, it's actually very hard to completely remove personal information. Over 369 million EU internet users create personal data just by using social media, shopping online, or browsing the web. This includes details like names, locations, interests, and browsing history—all used to attract and keep customers. Many online services appear "free," but there's a hidden cost: your personal information. When we use email or search engines, we're actually paying with our data instead of money. Companies use this information to target ads and services to individuals with high accuracy. In today's digital world, our personal data has become the new currency [9].

19.7.4 Economic risks of cyber warfare

Cyberattacks cost $6 trillion annually, projected to reach $10 trillion by 2027. Small businesses face bankruptcy (43%), while supply chain attacks like the 2024 SolarWinds hack disrupted 18,000 organizations. A 2022 ransomware attack on Costa Rica cost $200 million. The big data market, projected to grow from $217 billion in 2023 to $473 billion by 2030, drives advancements but faces 238% more financial sector attacks since 2020, with breaches costing $4.45 million on average. Organizations must allocate 15–20% of budgets to zero-trust architectures, AI-powered anomaly detection, and quantum-resistant encryption. As shown in Figure 19.5, the big data market is projected to more than double in value from $217 billion in 2023 to $473 billion by 2030, underscoring the growing economic stakes in data security.

Taxonomy of big data privacy risks refers to a systematic classification or framework that organizes and categorizes the various privacy-related risks associated with the collection, storage, processing, and use of big data. As shown in Figure 19.6, a visual taxonomy illustrates the hierarchical structure of big data privacy risks and their interrelationships [19]. Big data, characterized by its volume, velocity, variety, and veracity, often involves sensitive personal information from diverse sources like social media, IoT devices, and financial transactions. This taxonomy helps identify, analyze, and mitigate potential threats to individual privacy and organizational security in the context of such large-scale data handling. A taxonomy of

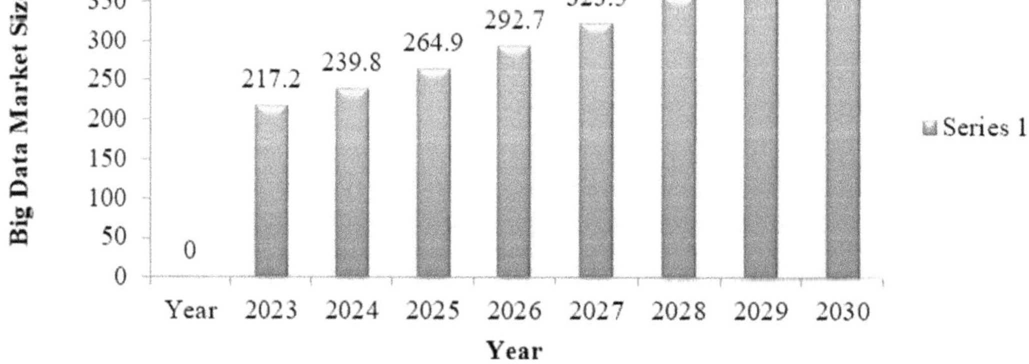

Figure 19.5 Projected big data market size (2023–2030).

Source: Bushuev 2023.

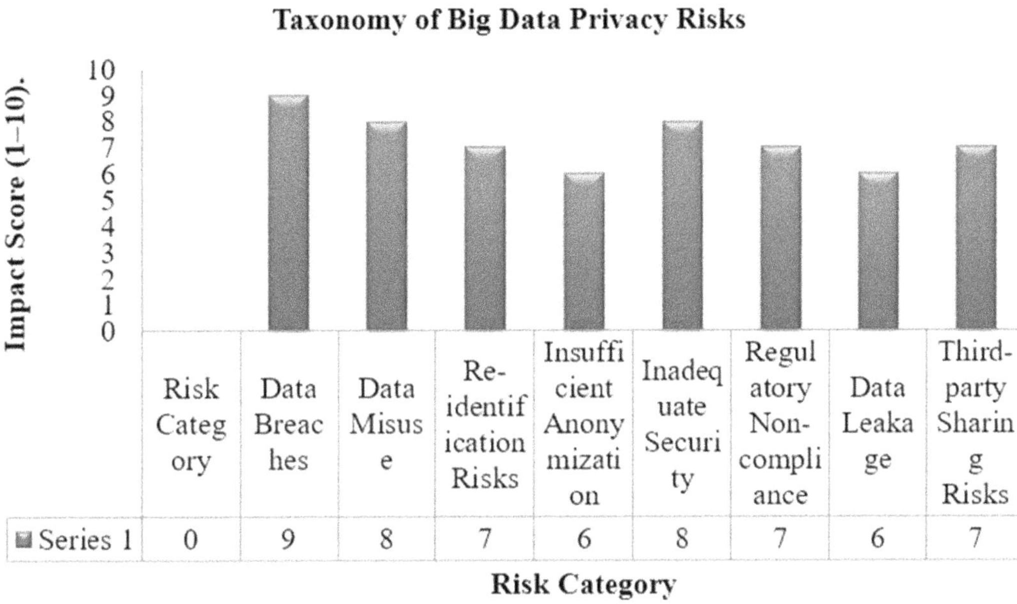

Figure 19.6 Taxonomy of big data privacy risks.

Source: Nayak and Walton 2024.

big data privacy risks is essentially a structured hierarchy or categorization that breaks down the complex privacy challenges into distinct types, each with specific characteristics, causes, and impacts. It serves as a tool for researchers, policymakers, and businesses to understand the scope of privacy risks and develop targeted protection strategies [8]. As shown in Table 19.2, the taxonomy of big data privacy risks categorizes key threats and corresponding mitigation strategies [9] (Table 19.4).

19.7.5 Global case studies

China's 2021 Data Security Law uses AI to reduce cyberattacks by 35%. The U.S. relies on private-sector innovation but lacks unified regulations. India's 2023 National Cybersecurity Policy integrates AI but faces low public awareness (65% of users unaware of security practices) [15]. The 2020 WannaCry attack across 150 countries highlights the need for global cooperation. As shown in Figure 19.7, China's AI-based firewall architecture integrates real-time threat detection and automated response systems to achieve a 35% reduction in cyberattacks [22].

The global landscape of cybersecurity is marked by high-profile incidents that underscore the need for robust defenses, as evidenced by numerous case studies from around the world.

Table 19.4 Taxonomy of big data privacy risks

Risk type	*Description*	*Mitigation*
Unauthorized Access	Data exposed to external actors	Encryption, access controls
Data Monetization	Unconsented use by tech firms	Transparent consent policies
Insufficient Regulation	Weak legal frameworks	Global privacy standards

Source: Kumar and Chaurasia (2019) [4].

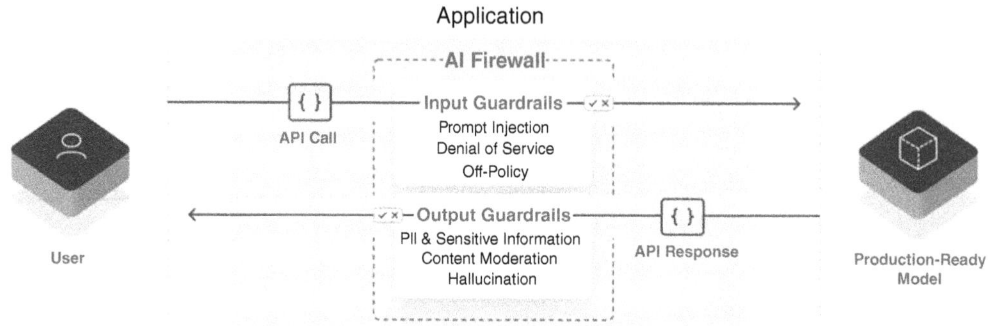

Figure 19.7 China's AI-based firewall architecture.

Source: Wang 2023 [10].

In May 2023, the Clop ransomware group exploited a vulnerability in Moveit transfer software, compromising the data of 93.3 million individuals across 2,500 organizations, including British Airways, Boots, and the New York City Department of Education. This attack highlighted supply chain vulnerabilities, prompting global calls for enhanced security assessments. In September 2024, Kawasaki Motors Europe faced a severe cyberattack, with threat actors exfiltrating 487 GB of data, including financial records and internal communications, forcing the company to take servers offline. This incident highlighted the value of being proactive in the world of manufacturing giants, which rely on interconnected devices. The Equifax breach of 2017 was one of the biggest cases in the world, in which hackers exploited a vulnerability in a web application and exposed sensitive data of 147 million people. This case taught a big lesson about how important it is to do patch management faster and better. That year, the WannaCry ransomware attack infected more than 230,000 computers in 150 countries through the EternalBlue exploit. The attack caused billions of dollars in damages and made it clear how dangerous it can be to ignore regular system updates. The Melissa virus of 1999 was also an early example, which spread through Microsoft Word macros and disrupted companies like Intel and Microsoft. This was the first big signal of the evolution of cyber threats. Then in 2018, 500 million guest records were tampered with in the Marriott Hotel data breach, which highlights the security weaknesses of the hospitality sector. When Sony Pictures was hacked in 2014, confidential data and executive emails were leaked, causing both financial and reputational loss, a major setback for the entertainment industry. These cases show that cyberattacks are becoming more sophisticated every year, and now AI and the latest technologies have made this problem even bigger. The Kawasaki and Clop incidents are also evidence of the fact that data theft has become a global trend. The Equifax breach demonstrated that if action is not taken in time, the consequences can be grave. WannaCry disrupted health institutions, governments, and even credit systems across continents. The Melissa virus is a historical example that shows cyberattacks have been dangerous from the start. The Marriott case showed that vulnerabilities in industries like hotels can go undetected for a long time. The Sony hack further highlighted the reputational risk that impacts across borders in today's digital age. Now AI threats are on the rise, as recent strikes have shown. For large companies like Schneider Electric, it is now compulsory to create effective business continuity strategies. For this reason, educational and training programs are being promoted, like Doctorates in Computer Science, so that professionals can become upskilled and deal with such issues. The erratic nature of cyberattacks, as shown in those cases, poses a question on the value of compliance with ever-evolving norms across regions as the key to mitigating the potential risks in these areas. There are incidents which interfere with the smooth running of intercontinental supply chains, as seen in such cases, which require integrating security into systems in order to reduce risks. Manufacturing sectors, which were affected by Kawasaki's breach, must strengthen their security posture. The limitations of this approach should not make organizations oblivious to their security as well as defense. The entertainment and hospitality providers who suffered due to Sony and Marriott should prioritize data protection to rebuild trust. Governments and businesses worldwide are investing in unified security platforms to counter escalating threats. The historical lessons from Melissa and WannaCry inform current strategies, emphasizing patch management and awareness. Cross-border collaboration is critical, as demonstrated by the multinational scope of these attacks. Small and medium enterprises globally must adopt best practices to avoid similar fates. The economic toll of these breaches drives the need for robust cybersecurity infrastructure worldwide. Training and awareness programs are essential to empower employees against evolving threats [22]. These global case studies collectively highlight the urgent need for a unified, proactive cybersecurity approach in the digital era.

19.8 PROPOSED STRATEGIES

19.8.1 Stringent cybersecurity frameworks

AI-based policies today can predict attack patterns in advance using real-time threat detection and anomaly analysis, reducing response time by approximately 50%. Global standards such as GDPR's strict consent rules and China's PIPL law are becoming increasingly important so that every country follows similar security protocols. All of this is especially important for manufacturing companies that rely on interconnected devices. These cases show that cyberattacks are becoming more advanced every year, and AI and cutting-edge technology have made them even more complex [11]. The Kawasaki and Clop breaches are also evidence that data theft has become a global trend. Equifax proved that delays in action can result in worldwide disasters. WannaCry disrupted health, government, and financial institutions simultaneously. The Melissa virus is a historical benchmark, while the Marriott breach highlighted vulnerabilities for the hospitality industry. The Sony hack provided an example of cross-border reputational risk. With the rise of AI threats, it is now mandatory for large companies like Schneider Electric to adopt strong business continuity strategies. This is why educational programs and advanced training are being encouraged, such as Doctorates in Computer Science, which allow tech professionals to upskill. The unpredictability of cyberattacks shows how critical compliance and updated frameworks are. These incidents also directly impact intercontinental supply chains, as the Clop and Kawasaki cases showed. For the manufacturing, entertainment, and hospitality sectors, the biggest challenge now is to rebuild trust and make data protection a top priority. Ultimately, these stringent frameworks create a proactive security culture, but their intensity also tests organizational resilience and resources. In this context, the NIST Cybersecurity Framework (CSF) 2.0, developed after an executive order by President Obama, highlights public-private collaboration for cyber risk assessment and management. Although adoption is voluntary, the framework has now become the industry benchmark for measuring cybersecurity maturity, detecting vulnerabilities, and demonstrating compliance.

19.8.2 International cooperation

The 2024 Budapest Convention on Cybercrime is a step forward by establishing a legal framework for cross-border cyber investigations, but AI-specific protocols are needed to counter deepfakes and bot-driven attacks [12]. These protocols should include standardized AI auditing mechanisms and data-sharing agreements to enhance global threat intelligence.

Enhanced cooperation could also involve joint AI training programs to upskill law enforcement in AI-driven crime detection. International cooperation is pivotal in combating the escalating threats of AI-powered cyber warfare, as highlighted by the need for unified global strategies to address cross-border cyberattacks. The study emphasizes the importance of collaborative frameworks, such as those proposed by the Budapest Convention, to establish standardized norms for information sharing and incident response. Multinational alliances, including NATO and the EU, are working in this regard, encouraging exercises and intelligence sharing among nations. According to the research, international task forces should be created to coordinate defenses. Bilateral agreements are also recommended to enhance technology transfer and joint research in AI between India, the U.S., and other countries. security. A gap in resources exists in developing countries, and the paper suggests global cooperation may help limit disparities in access to cybersecurity expertise through knowledge-sharing platforms [23]. Collaborative efforts should include frequent cybersecurity summits to support policy alignment and address AI vulnerabilities. International organizations such as INTERPOL facilitate cross-border investigations into cyber incidents. Joint development of AI tools for threat detection is proposed as a key cooperative strategy to strengthen global defenses. The research underscores the need for harmonized legal frameworks to prosecute cyber offenders across jurisdictions. International cooperation is essential to counter the rapid evolution of AI malware, requiring real-time data sharing among security agencies. The paper suggests establishing a global cybersecurity alliance to pool resources and expertise for proactive threat mitigation. Training programs facilitated by international partnerships can upskill professionals in under-resourced regions. The study advocates for public-private partnerships on a global scale to integrate industry innovations into national security frameworks. Cooperative efforts must address ethical concerns in AI deployment, ensuring transparency across borders. The research calls for a global incident response network to manage large-scale cyber crises effectively. International collaboration can reduce economic losses by preventing cascading effects of cyberattacks across economies. The paper proposes a unified AI ethics charter to guide cooperative cybersecurity efforts worldwide. Regular audits of international cooperation mechanisms are recommended to ensure adaptability to new threats. Ultimately, the success of these strategies hinges on sustained global commitment to collaborative cybersecurity resilience.

19.8.3 Public awareness

Educating users is important these days, as simply denying app permissions can reduce their data exposure by as much as 30%. A good example is the EU's Digital Literacy Program, which held workshops on reducing exposure to phishing. If such efforts extend to mobile apps and social media tutorials, individuals can proactively protect their data. Responsibility for cybersecurity really starts with the individual. The CISA Cybersecurity Awareness Program follows this approach, working together with government, industry, and nonprofits to give Americans real tools to stay safe online. Cybersecurity Awareness Month is celebrated each year in October, engaging millions of people through educational campaigns. It provides suitable resources for every age group – families, students, and professionals. This initiative also provides simple and practical advice, such as using strong passwords, installing multi-factor authentication, and steps to avoid internet-related crimes [3]. Events like Cybersecurity Career Week and the Secure Our World campaign reinforce these messages through webinars, social media campaigns, and free multilingual materials. A culture of cyber readiness develops when the public exercises caution in the digital environment and law enforcement receives specific tools to fight cybercrime. Every citizen has an important role in protecting the public digital space.

19.9 CONCLUSION

Artificial intelligence is transforming cybersecurity by enabling real-time threat detection and flexible safeguards against complex cyber threats. AI systems process extensive datasets from network logs and help identify anomalies like a brute-force attack that shows sudden login spikes originating from one IP address. As per the data, the AI cybersecurity global market was valued at USD 25.40 billion in 2024 and is expected to reach USD 31.38 billion in 2025. By automating routine tasks like monitoring, AI optimizes resources and reduces alert fatigue in sectors like retail and e-commerce, while its adaptive learning ensures compliance in regulated fields like health care [1]. AI excels in phishing detection by analyzing email content and identifying zero-day malware through behavioral analysis, unlike traditional antivirus tools. Behavioral analytics flags insider threats by detecting deviations from normal user activity, enhancing security in government and defense sectors. AI-driven authentication strengthens user experience with context-aware verification. However, challenges include data privacy concerns, adversarial attacks manipulating AI models, and high implementation costs [12]. For instance, 70% of users ignore app permissions, amplifying privacy risks. Cybercriminals exploit AI for phishing, deepfakes, and evasive malware, as seen in the 2023 Equifax breach affecting 147 million users. The 2021 Colonial Pipeline attack disrupted U.S. fuel supplies, costing millions. Regular audits and multi-factor authentication (MFA) are essential to ensure AI systems remain unbiased and secure. Stringent frameworks like GDPR and China's PIPL, combined with international cooperation via the Budapest Convention, can reduce attack response times by 50%. Public awareness campaigns can cut data exposure by 30% by promoting informed consent. By integrating AI with multi-layered strategies, organizations can proactively counter data breaches and cyber fraud in zero-trust ecosystems. AI's ability to adapt to evolving threats is vital for economic stability, with cybercrime costing $8 trillion globally in 2023. Ethical governance is crucial to balance AI's potential with privacy and security risks. By fostering global collaboration, robust policies, and user education, AI can serve as a dynamic shield, safeguarding the digital age from cyber warfare.

REFERENCES

1. Android users privacy awareness survey. ResearchGate. https://www.researchgate.net/publication/316536326_Android_Users_Privacy_Awareness_Survey
2. *(PDF) India's new data frontier: A critical legal insight of the personal data protection act, 2023*. (n.d.). Retrieved October 28, 2025, from https://www.researchgate.net/publication/390869327_India's_New_Data_Frontier_A_Critical_Legal_Insight_of_the_Personal_Data_Protection_Act_2023
3. *(PDF) Mobile app permissions awareness*. (n.d.). Retrieved October 28, 2025, from https://www.researchgate.net/publication/300417315_Mobile_app_permissions_awareness
4. Securing network infrastructure with cyber security. ResearchGate. https://www.researchgate.net/publication/368867654_Securing_network_infrastructure_with_cyber_security
5. Abhishek Shukla, U., & Shukla, A. (2022). Leveraging AI and ML for advance cyber security. *Journal of Artificial Intelligence and Cloud Computing*, *142*(4), 1–3. https://doi.org/10.47363/JAICC/2022(1)142
6. Alani, M. M. (2017). Android users privacy awareness survey. *International Journal of Interactive Mobile Technologies*, *11*(3), 130–144. https://doi.org/10.3991/IJIM.V11I3.6605
7. *Big data market size, share & growth | industry report, 2030*. (n.d.). Retrieved October 28, 2025, from https://www.grandviewresearch.com/industry-analysis/big-data-industry
8. *Cybercrime to cost the world $10.5 trillion annually by 2025*. (n.d.). Retrieved October 28, 2025, from https://cybersecurityventures.com/hackerpocalypse-cybercrime-report-2016/

9. *Cybercrime to cost the world 8 trillion annually in 2023*. (n.d.). Retrieved October 28, 2025, from https://cybersecurityventures.com/cybercrime-to-cost-the-world-8-trillion-annually-in-2023/
10. Gilbert, C., & Gilbert, M. (2025). AI-driven threat detection in IoT. *SSRN Electronic Journal*. https://doi.org/10.2139/SSRN.5259702
11. Hinton, G. E. et al. (2012). Improving neural networks by preventing co-adaptation of feature detectors. arXiv:1207.0580
12. Kaur, R., Gabrijelčič, D., & Klobučar, T. (2023). Artificial intelligence for cybersecurity: Literature review and future research directions. *Information Fusion*, 97, 101804. https://doi.org/10.1016/j.inffus.2023.101804
13. Mallavarapu, S., & Singh, S. (2024). International cooperation in cybercrime investigation: analyze the role of international collaboration and treaties in tackling cross-border cyber crimes involving India. *International Journal of Novel Research and Development*, *9*(8), a270–a282. https://ijnrd.org/viewpaperforall.php?paper=IJNRD2408026
14. Khan, M. J. (2023). Securing network infrastructure with cyber security. *World Journal of Advanced Research and Reviews*, *17*(2), 803–813. https://doi.org/10.30574/WJARR.2023.17.2.0308
15. *National security at risk: Implications of personal data breaches, ETCISO*. (n.d.). Retrieved October 28, 2025, from https://ciso.economictimes.indiatimes.com/news/data-breaches/national-security-at-risk-implications-of-personal-data-breaches/105176369
16. Navigating the Dual Nature of Deepfakes: Ethical, Legal, and Technological Perspectives on Generative Artificial Intelligence AI Technology. (2024). *International Journal of Scientific Research and Modern Technology (IJSRMT)*, *3*(10), 19–38. https://doi.org/10.38124/IJSRMT.V3I10.54
17. Pfeffer, A., Ruttenberg, B., Kellogg, L., Howard, M., Call, C., O'Connor, A., Takata, G., Reilly, S. N., Patten, T., Taylor, J., Hall, R., Lakhotia, A., Miles, C., Scofield, D., & Frank, J. (2017). *Artificial Intelligence Based Malware Analysis*. https://arxiv.org/abs/1704.08716
18. Ponzanesi, S., & Leurs, K. (2022). Digital migration practices and the everyday. *Communication, Culture and Critique*, *15*(2), 103–121. https://doi.org/10.1093/CCC/TCAC016
19. *Simplified data flow diagram (DFD) for the apps' ecosystem with an... | Download Scientific Diagram*. (n.d.). Retrieved October 28, 2025, from https://www.researchgate.net/figure/Simplified-Data-Flow-Diagram-DFD-for-the-apps-ecosystem-with-an-overview-of-the-data_fig1_365203300
20. Van Ouytsel, J. (2021). *Telematics and Informatics*, *63*, 101668. https://doi.org/10.1016/J.TELE.2021.101668
21. van Voorst, R. (2024). *Mayo Clinic Proceedings: Digital Health*, 2(4), 559–563. https://doi.org/10.1016/J.MCPDIG.2024.08.004
22. Wang, P. (2022). Frontiers in computing and intelligent systems, 2(2), 42–46. https://doi.org/10.54097/FCIS.V2I2.3931
23. Zhou, Y., & Jiang, X. (2012). Dissecting Android malware: Characterization and evolution. *Proceedings—IEEE Symposium on Security and Privacy*, 95–109. https://doi.org/10.1109/SP.2012.16

Chapter 20

Responsible AI in public policy

Ashutosh Yadav, Anurag Kakkar, Deepshikha Yadav, and Shailu Singh

20.1 INTRODUCTION

Artificial intelligence (AI) has become one of the most visible markers of technological change in the early 21st century. What was once confined to computer science laboratories and speculative debates about the future is now woven into everyday life. From the way individuals manage personal finances, access health care, or learn in classrooms, to how corporations optimize production lines or how governments deliver welfare, AI has moved into the mainstream of economic and social practice. Although the origins of AI research can be traced back to mid-twentieth century developments in computing, the acceleration of the last decade has been unprecedented. This surge is closely tied to the rise of Big Data—large-scale collection, aggregation, and processing of digital information—that has enabled the rapid growth of advanced machine learning, deep learning, natural language processing, and neural networks [1, 2].

AI has proven to be a powerful enabler of efficiency, offering time savings in administrative processes and improvements in the quality and speed of service delivery across sectors. AI is not just limited to automation of mundane tasks. It is being used to help policymakers with decisions, generate predictions, and nudge preferences [3, 4, 5]. This has raised severe concerns for policymakers about its socio-political [6] and economic ramifications [7, 8], with specific reference to the principles of fairness, accountability, and human values [9].

Such concerns are not merely theoretical. There are several instances that have demonstrated how AI has a tendency to perpetuate biases to the extent that it can amplify discrimination, reinforce structural inequalities, or undermine public trust when left unchecked [10]. For policymakers, the implications are both profound and urgent. Governments must balance using AI to advance priorities in health care, education, and agriculture with protecting citizens from unintended harms. AI offers value across the policy cycle, from detecting emerging concerns and informing evidence-based design to improving decision-making, automating implementation, and refining evaluation. Yet without clear safeguards, poorly governed systems risk deepening inequality, enabling intrusive surveillance, and eroding democratic accountability [11, 12, 13, 14].

It is in this context that the notion of Responsible AI has taken shape. Far from being a narrow technical guideline, Responsible AI has emerged as both a normative framework and a policy necessity. Responsible AI brings together concerns about safety, reliability, inclusiveness, transparency, and respect for human rights, stressing that AI should not only avoid harm but actively generate positive social outcomes [15, 16]. The concept stresses the need to align AI systems with democratic values and human dignity. The responsibility of policymakers, therefore, extends beyond acknowledging the opportunities of AI; they must embed

DOI: 10.1201/9781003685890-20

it within governance structures that are transparent, accountable, and responsive to public needs. Getting it right is essential if the digital revolution is to serve humanity collectively rather than benefit a privileged few.

This chapter reviews various frameworks of Responsible AI from corporate, research, and regional perspectives and attempts to provide a comprehensive integrated approach to Responsible AI.

20.2 CONCEPTUALIZING RESPONSIBLE AI

While the term has gained wide currency, its interpretation varies across institutions, governments, and industry stakeholders, as well as perspectives. For policymakers, the task is to integrate these diverse perspectives into coherent and enforceable frameworks that ensure AI advances the public interest.

The idea of Responsible AI has grown alongside the increasing role of AI. Yet the field lacks a unified definition of Responsible AI [17]. For instance, [15] interprets it from a philosophical angle, focusing that with the power of AI comes the responsibility to use it wisely. While [17] describes it as a *human-centered approach*, [18] define it as AI being developed ethically for the good of society and the environment, not just for individuals, focusing more on the intent behind AI systems. On the other hand, [19] takes a broader, systems-level perspective, defining Responsible AI as a blend of values and requirements: ethical principles, philosophical ideas, regulations, and technical standards all working together.

Responsible AI is human-centered and ensures users' trust through ethical decision-making that is fair, accountable, unbiased, nondiscriminatory, and consistent with societal laws and norms. It must also make automated decisions explainable while preserving user privacy through secure implementation [17].

Responsible AI is about developing, evaluating, and using AI systems in ways that are safe, ethical, and trustworthy. Since AI systems are shaped by countless design and deployment decisions, this approach emphasizes keeping people and their goals at the center while upholding enduring values such as fairness, transparency, and reliability [20].

Responsible AI can be seen as a set of guidelines that help organizations use AI in a way that meets legal requirements, stakeholder expectations, and broader ethical standards. When ethical guidelines are integrated into everyday workflows and applications, they help reduce risks, foster trust, and strengthen AI's capacity to deliver positive outcomes for society [21]. In 2019, the OECD released its Principles on Artificial Intelligence, later endorsed by the G20, making them the first international benchmark for AI governance. The framework sets out a vision of AI that promotes innovation while safeguarding human rights, democratic values, and cultural diversity. It highlights transparency, explainability, robustness, and security as essential conditions for accountability, and calls for governance systems that embed these principles throughout the AI lifecycle.

What makes the OECD approach stand out is its strive toward balance: it acknowledges AI's transformative potential while underscoring the risks and ethical duties it brings. By framing AI governance around human dignity and democratic accountability rather than narrow efficiency goals, the principles have become a global reference point, influencing the EU, UNESCO, and national governments. The challenge, however, lies in turning broad ideals like fairness, transparency, and accountability into enforceable mechanisms that work across different contexts.

Responsible AI is not just a collection of principles; it is a practice that needs to function at multiple levels. As [13] puts, effective governance calls for structural, procedural, and relational approaches. This means that principles such as accountability, oversight, robustness, privacy, fairness, and societal well-being should be built into AI systems from the ground up, rather than being added as outcomes.

In the past decade, governments, corporations, NGOs, and researchers alike have all released their own versions of Responsible AI guidelines. [16] describe this as *ethical pluralism*. [9] has identified more than 80 such frameworks worldwide. Though most of these frameworks/guidelines converge around common themes, fairness, transparency, accountability, and privacy, they differ in scope, emphasis, and the mechanisms they propose for enforcement.

This is often described as principle proliferation. On one hand, it signals an important recognition that AI cannot be governed by technical standards or market forces alone; it must also be guided by ethical considerations. On the other, it risks creating fragmentation. When one framework emphasizes inclusiveness, another sustainability, and a third competitiveness, the result can be blurred priorities and weakened accountability.

For policymakers, this presents a dilemma. Adopting widely recognized principles gives AI governance legitimacy and signals a commitment to responsible innovation. But the growing number of overlapping frameworks risks creating confusion and diluting accountability. The real challenge is to move from broad consensus on values to enforceable rules and mechanisms that can stand legal tests and guide real-world practice.

This is where public policy becomes central. While ethical frameworks articulate ideals such as fairness, accountability, and inclusiveness, it is policy that translates these into laws, institutions, and administrative routines. The global nature of AI makes this even more complex, as systems cross borders and values like fairness or privacy vary across cultures. Effective governance therefore requires international alignment while leaving space for local adaptation.

As discussed earlier, the conversation at the global level has already converged on certain baseline principles like fairness, transparency, accountability, and respect for human rights. The challenge for policymakers is now to move from abstract consensus to specific institutional design, crafting laws, regulatory bodies, and oversight mechanisms that can operationalize ethical commitments in diverse social and political contexts.

Responsible AI in policy is not just about including values; it is more about the reflection of those values in practice by building governance systems around them. The next section discusses how corporations and applied frameworks, such as Microsoft's Responsible AI Standard and IBM's guidelines, have attempted to move in this direction, and what lessons policymakers can learn from them.

20.3 CORPORATE PERSPECTIVE ON RESPONSIBLE AI

While international organizations like the OECD and UNESCO have been central in setting normative baselines for Responsible AI, corporations have arguably done more to operationalize these principles into concrete frameworks. This is hardly surprising: technology firms are the primary creators and deployers of AI systems, and they face immediate reputational, financial, and regulatory consequences if their systems are judged to be unsafe, discriminatory, or unethical.

For example, Microsoft's Responsible AI Standard [22] outlines six principles: fairness, reliability and safety, privacy and security, inclusiveness, transparency, and accountability. Each principle is linked to a set of actionable requirements. Fairness requires proactive bias testing in algorithmic models. Transparency involves providing clear documentation of how systems work and what data they rely on. Accountability requires establishing governance mechanisms so that human actors remain responsible for AI outcomes. Microsoft embeds these principles into its product lifecycle, requiring internal reviews and compliance checkpoints before AI technologies are released. For Microsoft, fairness is no longer an aspirational value but a measurable target, achieved through bias detection tools and fairness metrics. Accountability is ensured through internal review boards that assess high-risk AI projects. Transparency is embedded through the creation of model cards and system documentation. These processes show how firms can move from ethical statements to enforceable standards, though they remain largely self-regulatory.

IBM has also been a prominent actor in shaping the Responsible AI agenda. Its framework emphasizes that Responsible AI is a set of guiding principles that govern the design, development, deployment, and use of AI, with the overarching goal of building trust. IBM stresses that AI must align with stakeholder values, legal standards, and ethical principles, while also mitigating risks and maximizing positive societal outcomes [21]. IBM's approach frames Responsible AI as a matter of trust-building. The company recognizes that adoption of AI technologies depends not just on technical performance but on whether users and the public perceive them as legitimate and fair. In other words, the social license to operate is as important as legal compliance. This perspective is crucial for public policy, as it suggests that ethical AI governance cannot be reduced to technical safeguards alone; it must also cultivate public confidence in institutions.

Industry-led education platforms like DataCamp have also attempted to integrate Responsible AI into actionable pillars for practitioners. The framework focuses on four basic values: fairness, efficacy, transparency, and accountability [23]. What stands out in DataCamp's contribution is its candid recognition of the *implementation gap*. As they note, the real challenge lies not in agreeing on ethical ideals but in translating them into "real-world processes and controls." Without clear governance playbooks, organizations risk exposing themselves to bias, inefficiency, and regulatory vulnerabilities.

Corporate frameworks frequently proceed on the assumption that AI systems can be made fully explainable and controllable. However, contemporary systems remain adaptive, opaque, and effectively operate as "black boxes." Principles such as fairness or accountability are difficult to operationalize when the artifact itself is autonomous, dynamic, and heterogeneous [13]. Microsoft's use of review boards and compliance checks offers a strong example of structural and procedural governance, fitting within the framework outlined by [13]. These mechanisms embed oversight and accountability into organizational processes. Yet, these models often reveal a gap when it comes to relational governance, i.e., sustained engagement and collaboration with the communities most affected by AI systems. Without such relational practices, governance risks becoming inward-facing, prioritizing compliance over lived experience.

While corporate frameworks provide valuable models, they also raise critical questions. Most notably, they remain self-regulatory. Companies set their own standards, evaluate compliance internally, and disclose information selectively, creating risks of conflicts of interest, as firms may be incentivized to prioritize market competitiveness over ethical compliance. For instance, while Microsoft and IBM articulate strong principles, they are also commercial players whose primary accountability is to their shareholders. Public policy, thus, cannot rely solely on corporate self-regulation. As noted in [11], binding regulation and independent oversight are necessary to ensure that ethical principles are consistently and equitably enforced.

The corporate engagement with Responsible AI adds another layer to the problem of *principle proliferation*. Alongside OECD principles and national frameworks, we now have a multiplicity of corporate standards: Microsoft's six principles, IBM's trust-based model, DataCamp's four pillars, and countless others from Google, Meta, and Amazon. While these frameworks share significant commonalities, their differences can generate confusion when it comes to implementation across different contexts.

There is a dire need for coherent and enforceable standards, even more so from a policy perspective. Corporate frameworks bring useful insights and innovations, but they cannot take the place of public accountability. Policymakers should recognize and build on these contributions while resisting the temptation to hand over governance entirely to self-regulation. Corporate models can be valuable points of reference, yet they must ultimately feed into binding legal and institutional mechanisms that guarantee AI systems are developed and used responsibly.

The abundance of frameworks risks creating fragmentation and uneven accountability. For policymakers, the challenge is not to reproduce every principle put forward by corporations, but to adopt and harmonize them into standards that are coherent, enforceable, and firmly rooted in democratic values. The next section discusses the regional and national perspectives on how the Asia-Pacific Economic Cooperation (APEC) and India have approached AI governance.

20.4 AI GOVERNANCE IN APEC AND INDIA

At the regional level, the APEC Policy Support Unit has been one of the most active bodies in exploring both the opportunities and risks of AI in public governance. Its 2022 policy brief underscores that AI can be deployed across the entire policy cycle, from agenda setting and formulation to decision-making, implementation, and evaluation [24]. In practice, this could mean detecting unusual health patterns through the analysis of triage data, forecasting the effects of tax reforms, or optimizing urban traffic flows. In this way, APEC highlights AI's potential to increase efficiency, enhance the quality of public services, and reduce administrative burdens.

APEC is clear-eyed about the technology's limitations. AI excels at pattern recognition and data-driven inference, but it struggles with inherently human concepts such as fairness, justice, and equity. It also lacks the ability to grasp causality, cultural nuance, and contextual meaning. AI has the tendency to reinforce bias in public decision-making, widen existing inequalities, and weaken trust in institutions if not governed properly. A striking example comes from the Netherlands childcare benefits scandal, where a self-learning fraud detection algorithm disproportionately flagged families on the basis of ethnicity, leading to severe social repercussions [25].

Another key area that is highlighted by the APEC brief is nonneutrality in AI systems. The underlying design of these systems is indeed framed by human choices and assumptions.

Algorithms mirror the biases, prejudices, and lived experiences present in the societies that produce them. The data they rely on is shaped by structural inequalities, ethical dilemmas, and practical constraints in collection and use. Responsible AI, therefore, is not simply a matter of regulating machines; it is about ensuring human accountability at every stage of design, development, and deployment. To meet this challenge, APEC has stressed the importance of creating supportive conditions that embed ethical safeguards while still allowing space for innovation.

In the Indian context, Responsible AI in the public policy domain presents a somewhat complex scenario. Being an emerging economy aiming at tremendous digital transformation (initiatives like UPI, Aadhar, etc.), India also has a diverse democracy that still grapples with structural inequalities, questionable data governance, and limited capacity.

In 2018, NITI Aayog introduced India's National Strategy on Artificial Intelligence (NSAI), with a roadmap as to how AI be adopted in key public sectors, such as health, agriculture, and education. This strategy also gave the renowned idea of *AI for All*, which has become a guiding mantra to ensure inclusive and citizen-centric AI systems. A crucial part of this vision was the commitment to making AI not just innovative, but also safe, responsible, and trustworthy for society at large.

In 2021, NITI Aayog, in collaboration with the World Economic Forum as the knowledge partner, further released an Approach Document for India, which framed principles of Responsible AI governance within constitutional values. The document considers system-level concerns, pre-deployment, deployment, and post-deployment considerations, as well as accountability, privacy, and security risks. The document also elucidates legal and regulatory approaches for managing AI systems, as well as the technological approach to AI system management. Building on this foundation, it proposes seven guiding principles: safety and reliability, equality, inclusivity, privacy, transparency, accountability, and reinforcement of positive values [26, 27]. Together, these principles are echoed in the *AI for All* vision.

As pointed out by [28, 29], India's approach to Responsible AI has been heavily influenced by the political economy of informational capitalism. Public data, for instance, is treated as both a public resource and an economic asset. Policies encourage AI innovation by granting firms access to large-scale datasets, often without strong privacy safeguards. Although protection mechanisms have now been introduced through the *Digital Personal Data Protection Act* (2023), critics argue it prioritizes state access to data over individual rights.

Similarly, frameworks for nonpersonal data governance explicitly frame data as an economic commons to be harnessed for innovation, rather than primarily as a matter of rights or consent. Despite that a lot is being done in this space, there is still a requirement for a more inclusive and democratized approach to this.

20.5 TOWARD A COHERENT FRAMEWORK

Conceptualizing Responsible AI for policy requires weaving together ethical ideals, economic realities, and institutional mechanisms. At the global level, frameworks such as the OECD principles provide a normative foundation. At the corporate level, companies like Microsoft and IBM illustrate how principles can be translated into operational standards, though largely within self-regulatory regimes. At the regional level, APEC emphasizes the importance of governance frameworks, cross-sectoral partnerships, and trust-building. At the national level, India's experience reveals both the developmental promise of AI and the risks of overreliance on self-regulation and the commodification of citizen data.

Table 20.1 summarizes key definitions and frameworks of Responsible AI across scholarly, corporate, and policy domains.

The economic implications, ranging from job displacement and inequality to competitiveness and public trust, make Responsible AI not only an ethical imperative but also an economic necessity. Policymakers must act decisively to embed Responsible AI into governance structures through enforceable regulation, democratic oversight, capacity-building, and international collaboration.

A review of frameworks from the OECD [12], corporations (Microsoft, IBM, DataCamp), national contexts [26], and academic research [17, 13] highlights both convergence and fragmentation. While most agree on the need for AI systems to be fair, transparent, accountable, and secure, significant gaps remain in implementation, contextual adaptation, and treatment of trust.

This focus on guidelines reflects a widespread recognition that AI cannot be governed solely by technical or market logics. Yet this abundance has produced what [9] call "principle proliferation," an expanding set of guidelines that converge around values such as fairness, accountability, transparency, and privacy, but diverge in emphasis, scope, and enforcement.

Table 20.1 Comparative frameworks of Responsible AI

Source	*Definition/focus*	*Key elements*	*Unique contribution*
Responsible AI: A Structured Literature Review [17]	Responsible AI is human-centered, fosters trust through ethical decision-making that is fair, unbiased, nondiscriminatory, explainable, privacy-preserving, and consistent with societal laws and norms.	Fairness, accountability, explainability, privacy, human rights, trust as *outcome*.	Provides one of the first unified definitions across fragmented literature; frames trust as the outcome, not a principle.
OECD AI Principles [12]	AI should be trustworthy, respect human rights and democratic values, and be robust, secure, and transparent.	Human rights, democratic values, fairness, transparency, accountability, robustness.	First intergovernmental standard on AI ethics; widely endorsed (by G20).
Microsoft Responsible AI Standard [22]	Framework for building AI based on six principles: fairness, reliability and safety, privacy and security, inclusiveness, transparency, accountability.	Corporate governance checkpoints, fairness metrics, transparency documentation, internal audits.	Concrete operationalization of principles across product lifecycle.
IBM [21]	Responsible AI = a set of principles guiding design, development, and use of AI to align with stakeholder values, legal norms, and ethics, while mitigating risks and maximizing positive outcomes.	Trust, stakeholder values, ethics, risk mitigation, compliance.	Frames Responsible AI around trust-building as a condition for adoption.
NITI Aayog [26]—Responsible AI for All (India)	Grounded in constitutional values; distinguishes system-level risks (black-box, exclusion, accountability gaps) from societal risks (jobs, misinformation). Core principles: safety, equality, inclusivity, privacy, transparency, accountability, human values.	Safety and reliability, inclusivity, equality, privacy and security, transparency, accountability, positive human values.	Anchors AI governance in the Indian motto of "AI for All"; situates Responsible AI in developmental state strategy.
AI Governance Framework in Health Care Organizations [30]	Co-designed AI governance framework for safe, responsible use in health care.	Health care-focused governance, stakeholder workshops, iterative refinement.	Shows how governance frameworks can be practically embedded in high-stakes sectors.
Responsible AI [31]	Responsible AI as socio-technical systems spanning design to deployment.	Risk minimization across lifecycle, alignment with societal norms and values.	Emphasizes AI's place within broader social systems, not isolated artifacts.
ESG-AI [32]	Assessment of AI through environmental, social, and governance lenses.	Impact evaluation, corporate commitment, sustainable risk management.	Bridges AI ethics with ESG for investor and organizational accountability.
Hourglass Model [33]	Organizational-level governance framework spanning system lifecycle.	Layers of governance (environmental, organizational, system), lifecycle alignment.	Offers a scalable model for operationalizing AI governance in organizations.
Trust in AI [34]	Multi-level AI governance involving states, firms, citizens, and trust.	Trust dimensions (competence, integrity, benevolence), stakeholder interplay.	Positions social trust as central governance element across levels.
Responsible AI systems [35]	Roadmap for Responsible AI systems highlighting trust infrastructure.	Regulatory context, trustworthy tech, auditability, governance.	Provides a comprehensive societal trust-building roadmap for AI systems.

Source: Authors' compilation.

Across nearly all frameworks examined, four core principles consistently appear:

1. Fairness: Avoidance of bias, discrimination, and unjust exclusion (OECD, Microsoft, NITI Aayog).
2. Transparency: Explainability of decisions, disclosure of AI use, and clarity of processes (OECD, [17]).
3. Accountability: Assignment of responsibility to human actors, with mechanisms for redress when harms occur (IBM, [13]).
4. Privacy and Security: Protection of personal data, prevention of misuse, and maintenance of robust, safe systems (Microsoft, NITI Aayog).

These principles can be seen as the normative baseline for Responsible AI. They represent a global consensus and provide legitimacy for policy initiatives. However, by themselves, they are insufficient, as they risk remaining aspirational.

One of the most important insights from [17] is that trust should not be treated as a principle in itself, but as the *outcome* of successfully embedding fairness, accountability, transparency, and privacy into AI systems.

This reframing has two implications; one, trust becomes measurable. Instead of declaring AI trustworthy, policymakers can evaluate whether systems earn trust through observable outcomes (e.g., reduction in bias incidents, citizen satisfaction surveys, adoption rates). Second, trust aligns with legitimacy. By treating trust as an outcome, Responsible AI becomes linked to the legitimacy of democratic institutions. Citizens trust not just the technology, but also the governance systems that oversee it.

[13] highlight that Responsible AI must move beyond principles to encompass three types of governance practices: structural practices (rules, regulations, allocation of authority e.g., AI Acts, regulatory bodies); procedural practices (audits, monitoring, compliance protocols e.g., Microsoft's bias testing, model cards); and relational practices (stakeholder engagement, collaboration, and trust-building across communities).

Together, these practices ensure that principles are embedded across the AI lifecycle. Ultimately, Responsible AI is about more than minimizing risks. It is about ensuring that AI systems reinforce democracy, equity, and social welfare. As AI becomes ever more pervasive, getting this balance right is not optional. It is essential if the technology is to serve humanity as a whole rather than privileging the interests of a few.

REFERENCES

1. Kaplan, A., & Haenlein, M. (2018). Siri, Siri, in my hand: Who's the fairest in the land? On the interpretations, illustrations, and implications of artificial intelligence. *Business Horizons*, 62(1), 15–25. https://doi.org/10.1016/j.bushor.2018.08.004
2. Russell, S. J., & Norvig, P. (2020). *Artificial intelligence: A modern approach* (4th ed.). Pearson. http://lib.ysu.am/disciplines_bk/efdd4d1d4c2087fe1cbe03d9ced67f34.pdf
3. McAfee, A., & Brynjolfsson, E. (2017). Machine, platform, crowd: Harnessing our digital future. In *Andrew McAfee, Erik Brynjolfsson*. W. W. Norton & Company. https://wwnorton.com/books/Machine-Platform-Crowd/
4. Wirtz, B. W., Weyerer, J. C., & Geyer, C. (2018a). Artificial intelligence and the public sector—Applications and challenges. *International Journal of Public Administration*, 42(7), 596–615. https://doi.org/10.1080/01900692.2018.1498103
5. Wirtz, B. W., Weyerer, J. C., & Geyer, C. (2018b). Artificial intelligence and the public sector—Applications and challenges. *International Journal of Public Administration*, 42(7), 596–615. https://doi.org/10.1080/01900692.2018.1498103

6. Jacobs, J. (2023). The artificial intelligence shock and socio-political polarization. *Technological Forecasting and Social Change*, 199, 123006. https://doi.org/10.1016/j.techfore.2023.123006
7. Georgieva, K. (2024). *AI will transform the global economy*. Let's Make Sure It Benefits Humanity. *IMF Blog*. https://www.imf.org/en/Blogs/Articles/2024/01/14/ai-will-transform-the-global-economy-lets-make-sure-it-benefits-humanity
8. International Monetary Fund. (2024, January 14). *AI will transform the global economy. Let's make sure it benefits humanity*. IMF. Retrieved September 5, 2025, from https://www.imf.org/en/Blogs/Articles/2024/01/14/ai-will-transform-the-global-economy-lets-make-sure-it-benefits-humanity
9. Jobin, A., Ienca, M., & Vayena, E. (2019). The global landscape of AI ethics guidelines. *Nature Machine Intelligence*, 1(9), 389–399. https://doi.org/10.1038/s42256-019-0088-2
10. Ferrara, E. (2023). Fairness and bias in artificial intelligence: A brief survey of sources, impacts, and mitigation strategies. *Science*, 6(1), 3. https://doi.org/10.3390/sci6010003
11. Camilleri, M. A. (2024). Artificial intelligence governance: Ethical considerations and implications for social responsibility. *Expert Systems*, 41(7). https://doi.org/10.1111/exsy.13406
12. Organisation for Economic Co-operation and Development. (2019). *Recommendation of the council on artificial intelligence*. OECD Legal Instruments.
13. Papagiannidis, E., Mikalef, P., & Conboy, K. (2025). Responsible artificial intelligence governance: A review and research framework. *The Journal of Strategic Information Systems*, 34(2), 101885. https://doi.org/10.1016/j.jsis.2024.101885
14. Valle-Cruz, D., Criado, J. I., Sandoval-Almazán, R., & Ruvalcaba-Gomez, E. A. (2020). Assessing the public policy-cycle framework in the age of artificial intelligence: From agenda-setting to policy evaluation. *Government Information Quarterly*, 37(4), 101509. https://doi.org/10.1016/j.giq.2020.101509
15. Dignum, V. (2019). Responsible artificial intelligence. In *Artificial intelligence: Foundations, theory, and algorithms/Artificial intelligence: Foundations, theory, and algorithms*. https://doi.org/10.1007/978-3-030-30371-6
16. Floridi, L., & Cowls, J. (2019). A unified framework of five principles for AI in society. *Harvard Data Science Review*. https://doi.org/10.1162/99608f92.8cd550d1
17. Göllner, S., Tropmann-Frick, M. A. R. I. N. A., & Brumen, B. (2023, June). Responsible artificial intelligence. In *Conference on Information Modelling and Knowledge Bases EJC 2023*. https://doi.org/10.3233/faia231146
18. Lu, Q., Zhu, L., Whittle, J., & Xu, X. (2023). *Responsible AI: Best practices for creating trustworthy AI systems*. Addison-Wesley Professional. https://doi.org/10.1145/3626234, https://legalinstruments.oecd.org/en/instruments/OECD-LEGAL-0449
19. Bennetot, A., Donadello, I., El Qadi El Haouari, A., Dragoni, M., Frossard, T., Wagner, B., ... & Diaz-Rodriguez, N. (2024). A practical tutorial on explainable AI techniques. *ACM Computing Surveys*, 57(2), 1–44. https://doi.org/10.1145/3670685
20. Microsoft. (2024, September 13). *What is responsible AI?* Microsoft Learn. Retrieved September 5, 2025, from https://learn.microsoft.com/en-us/azure/machine-learning/concept-responsible-ai?view=azureml-api-2
21. Stryker, C. (2024, February 6). *What is responsible AI?* IBM. Retrieved September 5, 2025, from https://www.ibm.com/think/topics/responsible-ai
22. Microsoft. (2022). *Responsible AI standard v2: General requirements*. Retrieved September 5, 2025, https://msblogs.thesourcemediaassets.com/sites/5/2022/06/Microsoft-Responsible-AI-Standard-v2-General-Requirements-3.pdf
23. DataCamp. (2025, April). *The 4 pillars of responsible AI* [Webinar description]. *DataCamp*. Retrieved September 5, 2025. https://www.datacamp.com/webinars/the-4-pillars-of-responsible-ai
24. Wirjo, A., Calizo Jr, S., Vasquez, G. N., & Andres, E. A. S. (2022). Artificial intelligence in economic policymaking | APEC. In APEC (Policy Brief No. 52). APEC Policy Support Unit. https://www.apec.org/publications/2022/11/artificial-intelligence-in-economic-policymaking
25. Heikkilä, M. (2022, April 13). Dutch scandal serves as a warning for Europe over risks of using algorithms. *POLITICO*. https://www.politico.eu/article/dutch-scandal-serves-as-a-warning-for-europe-over-risks-of-using-algorithms/

26. NITI Aayog. (2021). *Responsible AI for all: Approach document (Part 1)*. Government of India. Retrieved September 5, 2025, from https://indiaai.gov.in/research-reports/responsible-ai-for-all-approach-document
27. NITI Aayog. (2022). *Responsible AI for all*. Government of India. https://www.niti.gov.in/sites/default/files/2022-11/Ai_for_All_2022_02112022_0.pdf
28. Indian School of Public Policy. (2024). *Ethics of AI in public policy in the Indian context*. Retrieved September 5, 2025, from https://www.ispp.org.in/ethics-of-ai-in-public-policy-in-the-indian-context/
29. Joshi, D. (2024). AI governance in India—law, policy and political economy. *Communication Research and Practice*, 1–12. https://doi.org/10.1080/22041451.2024.2346428
30. Freeman, S., Wang, A., Saraf, S., Potts, E., McKimm, A., Coiera, E., & Magrabi, F. (2025). Developing an AI governance framework for safe and responsible AI in health care organizations: Protocol for a multimethod study. *JMIR Research Protocols*, 14(1), e75702. https://doi.org/10.2196/75702
31. Gunasekara, L., El-Haber, N., Nagpal, S., Moraliyage, H., Issadeen, Z., Manic, M., & De Silva, D. (2025). A systematic review of responsible artificial intelligence principles and practice. *Applied System Innovation*, 8(4), 97. https://doi.org/10.3390/asi8040097
32. Lee, S. U., Perera, H., Liu, Y., Xia, B., Lu, Q., Zhu, L., ... & Nottage, M. (2025). Integrating ESG and AI: A comprehensive responsible AI assessment framework. *AI and Ethics*, 1–28. https://doi.org/10.1007/s43681-025-00741-5
33. Mäntymäki, M., Minkkinen, M., Birkstedt, T., & Viljanen, M. (2022). Putting AI ethics into practice: The hourglass model of organizational AI governance. arXiv preprint arXiv:2206.00335. https://doi.org/10.48550/arXiv.2206.00335
34. Choung, H., David, P., & Ross, A. (2023). Trust in AI and its role in the acceptance of AI technologies. *International Journal of Human–Computer Interaction*, 39(9), 1727–1739. https://doi.org/10.1080/10447318.2022.2050543
35. Herrera-Poyatos, A., Del Ser, J., de Prado, M. L., Wang, F. Y., Herrera-Viedma, E., & Herrera, F. (2025). Responsible artificial intelligence systems: A roadmap to society's trust through trustworthy AI, auditability, accountability, and governance. arXiv preprint arXiv:2503.04739. https://doi.org/10.48550/arXiv.2503.04739

Chapter 21

Responsible AI and the next generation

Policy challenges and governance opportunities

Harmeet Malhotra and Komal Sharma

21.1 INTRODUCTION

The G-20 declaration's "pro-innovation regulatory/governance approach" is consistent with India's strategy as the fourth largest economy in the world with a vibrant artificial intelligence (AI) startup ecosystem, thereby using AI to improve production and efficiency in vital industries such as health, agriculture, education, and defense through resulting increased economic growth. This ecosystem has been positioned by the government, largely through the IndiaAI Mission [1]. The goal of this program has been to build domestic AI innovation skills, to grow AI applications, and to lay the foundation for growth—a computer resource base, development of domestic AI models, and a national dataset platform.

A specific "safe and trusted" pillar of the IndiaAI Mission [1], with MeitY's proposal for an AI Safety Center, reflects growing institutional ambition [2, 3]. This pillar of the IndiaAI Mission has assisted the development of technologies, policy, and framework tools. The Ministry of Electronics and Information Technology (MeitY) has already sought partnerships for a proposal for an AI Safety Center. Governance knowledge also suggestsan integrated, "whole-of-government" approach, including a central AI incident database, technical secretariat, and municipal coordination.

In December 2023, the government amended its intermediary liability regime to expand provisions on deep fakes and misinformation generated by AI [3]. In March 2024, the guidelines were amended to the effect that content associated with AI would need notice and labeling [4, 5] in response to reports of biased outputs among large language models produced elsewhere. On balance, the increasing use of AI raises new governance issues [6, 7], requiring integrated solutions in areas such as data protection, intellectual property, liability, and social equity.

India's approach to AI governance can be viewed as both proactive and reactive and can also be understood as both policy- and action-based. Effectively, the strategy involves improving the domestic AI ecosystem, promoting specific industries to harness the benefits of AI, considering frameworks to mitigate both known and unknown risks, and improving the competitive position of India in the global AI landscape. With the introduction of NITI Aayog's 2018 National Strategy for Artificial Intelligence, India entered the AI space with a focus on five priority areas for AI application: healthcare, agriculture, education, smart cities and infrastructure, and smart mobility.

To reiterate the overall strategy in the country, NITI Aayog published two documents further outlining its original framework in 2021. Ethics were established as a key category, including principles and pillars such as transparency, accountability, inclusion, and privacy in the first document, Principles for Responsible AI. The second document, Operationalizing Principles for Responsible AI, emphasized the need to apply the principles in practice and offered suggestions for implementation, including capacity building, ethics-by-design, and policy initiatives.

DOI: 10.1201/9781003685890-21

21.2 EVOLUTION OF AI GOVERNANCE IN INDIA (2018–2024)

In the past six years, India's AI governance landscape has changed in a manner that reflects India's commitment to leveraging AI for socioeconomic growth alongside emerging ethical, legal, and sociological concerns. India's AI governance journey can be traced from guidance documents, which articulated vision setting, to binding norms and institutional initiatives related to the responsibilities of AI. While there has been substantive work, India appears to be making slow but deliberate and contingent progress in its work toward responsible AI governance.

21.2.1 Early vision: the National Strategy for Artificial Intelligence (2018)

India's official journey into the domain of AI started in 2018, when NITI Aayog released the National Strategy for Artificial Intelligence. As part of this strategy, the authors provided the now common vision of "AI for all," and classified the following five priority sectors in which AI could have the highest socioeconomic impact:

- Education—through adaptive learning platforms
- Healthcare—through early diagnosis and affordable access
- Agriculture—through precision farming and crop management
- Smart Mobility and Smart Transportation—through better traffic management and safety
- Smart Cities and Smart Infrastructure—through enhanced urban mobility and service delivery.

While the 2018 strategy was largely aspirational, it was a plan to enable India to position itself as both a consumer and producer of AI-led solutions. It emphasizedinnovation for social good, to leverage AI to reduce development deficit in a country with a large and heterogeneous population.

21.2.2 Ethical foundations: responsible AI principles (2021)

In pursuit of this vision, NITI Aayog released two significant documents in 2021:

- **Principles for Responsible AI**, with four ethical pillars: transparency, accountability, privacy, and inclusivity
- **Operationalizing Responsible AI**, where the principles were operationalized in practice through ethics-by-design, policy measures, and capacitybuilding.

The reports together constituted the first comprehensive attempt by India to integrate ethics into AI governance. Although they were advisory in nature, their contributions were significant in shaping the discourse on responsible innovation and paving a way for regulatory intervention.

21.2.3 Institutional push: IndiaAI Mission (2022)

In 2022, the Government of India, through its Ministry of Electronics and Information Technology (MeitY), announced the IndiaAI Mission, with the goal of building a strong national AI ecosystem. The mission focused on:

- Building high-quality computing infrastructure.
- Developing indigenous AI models and tools.
- Building a national dataset platform that provides data in a high-quality, easily accessible, and usable way.
- Fostering AI innovation (via startups, research labs, public-private partnerships).

This was a shift from high-level vision documents to real on-the-ground institutional and infrastructural support for AI, grounded in self-reliance and to compete globally.

21.2.4 Legal backbone: Digital Personal Data Protection Act (2023)

One key development was the passage of the Digital Personal Data Protection (DPDP) Act, 2023, India's first standalone data protection statute. AI systems are entirely data-based; therefore, the DPDP Act has serious implications for the governance of AI. It introduces the following provisions:

- Consent-based processing of personal data.
- Purpose limitation to limit misuse of data that is collected.
- Duties of data fiduciaries specifically addressing accountability and transparency.
- A Data Protection Board to enforce compliance.

The legislation starts to form a robust legal framework for ensuring that AI systems respect individuals' rights to privacy and data protection, which is one of the greatest challenges of the present era of digitalization.

21.2.5 Regulatory responses to emerging risks (2023–2024)

The emergence of generative AI models in 2023, and the subsequent proliferation of deepfakes and AI-generated disinformation, led India to enact reactive governance measures:

- Amendments were made to the Intermediary Liability Rules in December 2023 that required platforms to act against AI-generated disinformation and potentially harmful deepfakes.
- In March 2024, the revised guidelines mandated notice and labeling of AI-generated content, which stated a move toward increased transparency and consumer protection.

These actions demonstrate India's responsive capacity to the challenges posed by rapidly evolving AI technologies.

21.2.6 Toward institutional oversight: AI Safety Institute (2024 proposal)

Most recently, the government proposed the establishment of an AI Safety Institute within MeitY [8, 9], which would prioritize safety research, audits, and risk assessments. The suggestion indicates a progressive plan that will evolve from ad-hoc interventions to an institutional strategy for coordinating AI oversight.

21.2.7 Summary of milestones

To better understand the trajectory, Table 21.1 summarizes the key milestones from 2018 to 2024.

Table 21.1 Key Milestones in India's AI governance (2018–2024) [5, 10, 11, 12, 13]

Year	*Milestone/policy document*	*Key features*	*Relevance to AI governance*
2018	*National Strategy for Artificial Intelligence* (NITI Aayog)	Introduced **"AI for All"** vision; identified five priority sectors: HealthCare, Agriculture, Education, Smart Cities andInfrastructure, and Smart Mobility.	First comprehensive roadmap; positioned India as an AI innovator for social good.
2021	*Principles for Responsible AI* (NITI Aayog)	Ethical pillars: **Transparency, Accountability, Privacy, and Inclusivity**.	Established normative foundation for responsible AI.
2021	*Operationalizing Responsible AI* (NITI Aayog)	Proposed mechanisms: **ethics-by-design, capacity building, and policy interventions**.	Shifted focus from theory → practice, though advisory in nature.
2022	*IndiaAI Mission (MeitY draft proposal)*	Announced plan for a national AI innovation ecosystem with **computing infrastructure, indigenous models, and national datasets**.	Institutional push toward building AI capacity and competitiveness.
2023	*Digital Personal Data Protection Act (DPDP)*	India's **first standalone data protection law**; introduced **consent-based processing, purpose limitation, data fiduciary obligations, and Data Protection Board**.	Legal backbone for AI governance focusing on data rights and privacy.
2023	*Amendments to Intermediary Guidelines (Dec 2023)*	Clauses added for **AI-generated misinformation, deepfakes**.	First regulatory recognition of generative AI risks.
2024	*Revised AI Guidelines (March 2024)*	Mandatory **notice and labeling of AI-generated content**; push for **safe and trusted AI**.	Proactive step toward transparency in generative AI outputs.
2024	*AI Safety Institute (proposal under MeitY)*	Dedicated body for **AI risk research, safety standards, and auditing tools**.	Institutional mechanism to ensure safe deployment of AI in India.

21.2.8 Reflection

Figure 21.1 demonstrates the gradual development of policy, ethics, legislation, and institutional development in the history of AI governance in India. Starting from the orientation toward a vision of AI for good, India has layered moral values, proposed objectives to develop infrastructure, protected citizens' data through legislation, and has begun to address issueslike bias and misinformation, which have become tangible threats. The trajectory of India's governance indicates a commitment to create and maintain a balance between innovation and accountability, albeit with unresolved issues, especially concerning the capacity to enforce across sectors and coherence among sectors.

21.3 LEGAL AND POLICY FRAMEWORK FOR AI IN INDIA

In India, navigating AI management involves negotiating a complex legal landscape with multiple, interfacing laws, core constitutional principles, and policy frameworks. India has adopted a multi-faceted approach, relying on its constitutional protections, existing laws,

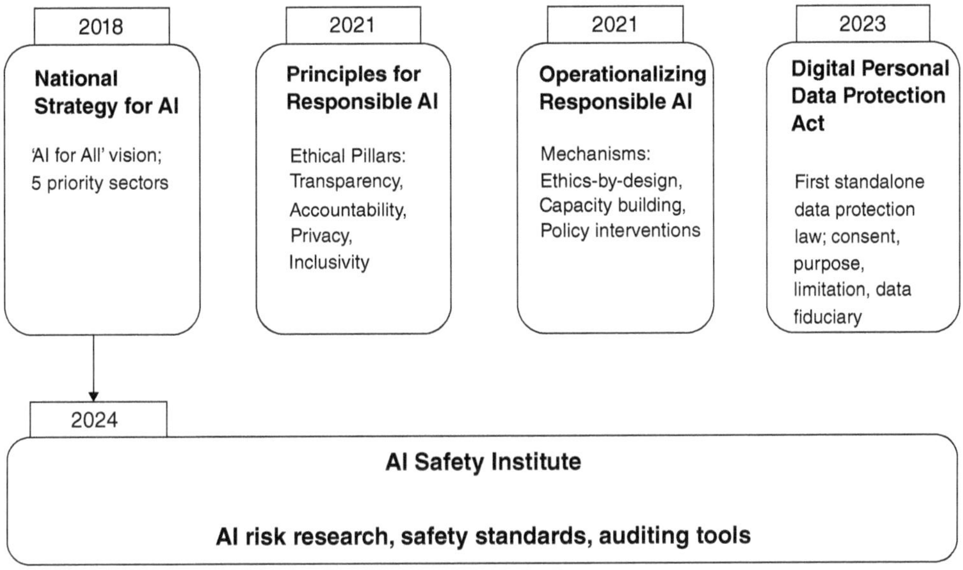

Figure 21.1 Evolution of AI governance in India (2018–2024).

and new digital laws (the European Union has proposed a wholesale AI Act). As Figure 21.1 illustrates,this is an evolving framework that promotes the need to balance innovative practices with accountability.

21.3.1 Constitutional principles and AI

Even in the absence of an AI-specific legal framework, indicating how AI applications could be configured is the Constitution of India:

- **Right to Privacy (Article 21)**: This was established in Justice K.S. Puttaswamy v. Union of India (2017). Privacy is now a fundamental right. Any AI application that is processing personal information must review this standard.
- **Equality and Nondiscrimination (Articles 14–16)**: These clauses could be pursued in respect of biased or discriminatory algorithms that are used to employ, lend, or grant access to welfare provision.
- **Freedom of Speech (Article 19)**: AI moderation and AI selection in a content filter environment must balance the competing concepts of free expression and limitation on hate speech, misinformation, or threats to public order.

These constitutional safeguards together mean that AI governance in India remains consistent with human dignity and democracy.

21.3.2 Sectoral laws and AI relevance

India is presently employing a sectoral approach to manage AI-based risks:

- **Information Technology Act, 2000**—Governs cybercrime, intermediary liability, and digital security. Amendments of the Act tend to cover AI-generated content, increasingly including misinformation and deepfakes.

- **Copyright Act, 1957**—Provides the framework for intellectual property; however, uncertainty still exists on whether AI-generated creative works can be copyright protected and whether ownership rests withthe developer, user, or AI system.
- **Competition Act, 2002**—Relevant to AI-powered markets, especially with respect to algorithmic collusion, data monopolies, and anti-competitive behavior of digital platforms.
- **Consumer Protection Act, 2019**—Provides remedial means for unfair trade practices in the context of AI, e.g., manipulative recommendation algorithms and pricing algorithms which are opaque intheir determination due to AI algorithms.

This yields an incomplete set of potential protections for the use of AI technology without the comprehensiveness of a separate AI statute.

21.3.3 Digital Personal Data Protection Act, 2023

The DPDP Act, 2023 is India's first comprehensive data protection law, with significant implications for AI governance. It regulates:

- **Consent-based processing** of personal data.
- **Purpose limitation** to prevent misuse.
- **Data fiduciary obligations**, including transparency and accountability.
- A **Data Protection Board** to enforce compliance.

For AI systems, which are data-driven by nature, the DPDP Act ensures that development and deployment respect individual privacy rights. However, critics point out gaps, such as wide government exemptions and limited safeguards for cross-border data flows.

21.3.4 Emerging regulatory measures (2023–2024)

In response to risks from generative AI, India has begun implementing **targeted rules:**

- **Intermediary Guidelines Amendment** (**Dec 2023**): Platforms are required to monitor and act against AI-generated misinformation and deepfakes.
- **Revised Guidelines** (**March 2024**): Introduced mandatory **labeling and notices** for AI-generated content, aiming to enhance transparency and trust.

These interventions demonstrate India's agile response to fast-evolving risks but also highlight the reactive nature of regulation in the absence of a dedicated AI law.

21.3.5 Institutional mechanisms

To strengthen governance capacity, India has proposed and initiated several institutional mechanisms:

- **AI Safety Institute** (**proposed, 2024**):It is a national body for AI risk research, safety audits, and regulatory guidance.
- **Bureau of Indian Standards** (**BIS**): It is engaged in developing AI-related standards.
- **Inter-Ministerial Coordination** (**proposed**):It is a "whole-of-government" approach involving MeitY, NITI Aayog, and sectoral regulators to ensure coherence across domains.

Such institutions are critical for building technical expertise and monitoring compliance in a rapidly changing technological landscape.

21.3.6 Challenges in the current framework

Despite progress, several gaps persist:

- **Fragmentation:** AI is currently governed through multiple, uncoordinated laws.
- **Accountability:** Lack of clarity on liability when AI causes harm.
- **Bias and Fairness:** Absence of statutory safeguards to prevent discrimination.
- **Enforcement:** Regulators often lack the technical expertise to oversee complex AI systems.

21.3.7 Reflection

India displays a hybrid AI legal and policy framework that fuses the current body of law dealing with digitaltechnologies, constitutional principles, and further protections such as the DPDP Act. This gradual approach permits flexibility, but it could potentially create gaps in the law as AI is more quickly integrated into key sectors. For a more forward-looking framework, India would benefit from harmonization, capacity building, and possibly an AI-specific law adapted for India's socioeconomic context. Figure 21.2 shows the AI governance ecosystem in India.

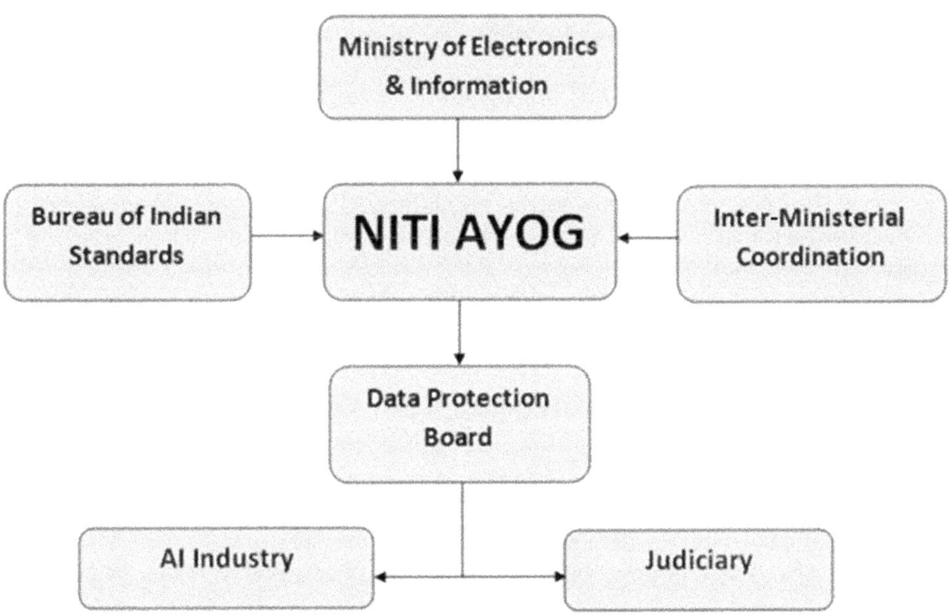

Figure 21.2 AI governance ecosystem in India.

21.4 DIGITAL PERSONAL DATA PROTECTION ACT, 2023

TheDPDP Act, 2023 is India's first standalone data protection law and is fundamental to regulating AI. This article summarizes relevant provisions of the Act and their implications forAI, indicating that this legislation is an important vehicle for regulating the technology.

Radiating from this central point are four key features of the Act, each with a brief explanation of its relevance to AI:

1. Consent-Based Processing:
   a. **Meaning:** Personal data can only be collected and processed with the clear, informed, and explicit consent of the individual (the Data Principal).
   b. **Relevance to AI:** AI systems, which are inherently data-hungry, are now legally required to obtain consent for every piece of personal data they use. This directly controls how AI models are trained and how they operate, preventing the unconsented use of personal information.
2. Purpose Limitation:
   a. **Meaning:** Data collected for a specific, stated purpose cannot be used for any other purpose without obtaining fresh consent.
   b. **Relevance to AI:** This feature is critical for preventing the "scope creep" of AI. For example, a company cannot collect data for "service improvement" and then use it to train a new AI model for an entirely different purpose, like personalized advertising, without first seeking new consent.
3. Data Protection Board:
   a. **Meaning:** The Act establishes a new regulatory authority to oversee its enforcement. This board is empowered to investigate complaints, address data breaches, and impose penalties for noncompliance.
   b. **Relevance to AI:** This creates a formal, institutional mechanism for regulating the data aspects of AI. If an AI system causes harm due to a data breach or privacy violation, the Data Protection Board has the authority to intervene, investigate, and hold the responsible organization accountable.

The image basically demonstrates that, by regulating AI's most essential input, data, the DPDP Act provides a robust legal framework for AI regulation. The Act ensures that AI is developed and used in a way that respects basic privacy rights, through requirements for consent, limitation of purpose, and accountability.

21.5 CHALLENGES AND GAPS IN INDIA'S AI LEGAL FRAMEWORK

While we have made continual improvements to build an AI governance ecosystem, there are still clear gaps, as summarized in Table 21.2. These issues are not confined to India; they are international challenges, as demonstrated by some overseas case studies. The following caselets suggest that a lack of appropriate legal protection can lead to real-world damage.

21.5.1 No standalone AI law

India relies on a patchwork of existing statutes (IT Act, DPDP Act, Competition Law) rather than a unified AI law.

Table 21.2 Description of challenges and gaps

Sr. No.	*Challenge/gap*	*Description*
1	No Standalone AI Law	AI does not have a unified legal framework, and instead, it relies on some combination of existing laws (such as the IT Act and the DPDP Act).
2	Algorithmic Accountability	There are no explicit legal provisions defining who is liable when an AI system malfunctions or causes harm, leaving a major gap in accountability.
3	Bias and Discrimination	The legal system lacks specific statutory safeguards to address and prevent algorithmic bias.
4	AI in Criminal Justice	The use of AI technologies such as predictive policing and surveillance lacks clear legal oversight, raising significant concerns about civil liberties and due process.
5	Intellectual Property Rights	There is legal ambiguity under the Copyright Act, 1957, regarding the ownership of creative works that are generated by AI.
6	Cross-Border Data Flow	India's data localization requirements can clash with the global nature of AI development and training, potentially hindering innovation that relies on international data processing.
7	Enforcement Capacity	Regulatory authorities, including the Data Protection Board and the Bureau of Indian Standards (BIS), may lack the requisite level of technical knowledge that would enable them to properly regulate and manage AI-driven challenges.

Caselet: The European Union's AI Act (expected 2025) demonstrates the value of a comprehensive, risk-based approach. By comparison, India's fragmented model risks regulatory blind spots where harmful AI applications may escape oversight.

21.5.2 Algorithmic accountability

Liability when AI systems produce errors or cause harm is still ambiguous.

Caselet - COMPAS Algorithm (US Courts): Black defendants with real profiles were often given greater "recidivism risk" scores than White defendants by the COMPAS risk assessment tool used by the criminal justice system. Courts have struggled with who should be held responsible—the algorithm, the state that used the tool, or the software developer. This highlights how urgently clarity around accountability measures is needed.

Table 21.2 outlines key legal and regulatory challenges related to artificial intelligence in India.

21.6 INTERNATIONAL PERSPECTIVES AND LESSONS FOR INDIA

21.6.1 The European Union: risk-based and rights-oriented

The **European Union (EU AI Act)** is the world's first comprehensive AI law, expected to come into force in 2025. Its central principle is a **risk-based approach:**

- **Unacceptable risk AI** (e.g., social scoring) is banned outright.
- **High-risk AI** (e.g., in healthcare, hiring, law enforcement) must comply with strict requirements such as transparency, data quality, human oversight, and conformity assessments.

- **Low/minimal risk AI** (e.g., chatbots, video games) face lighter obligations such as disclosure requirements.

The EU's framework emphasizes fundamental rights, consumer safety, and ethical safeguards over speeding up and scaling fast. The EU's framework is clear legally, has strong enforcement mechanisms, and conforms to the EU's rule of law, but critics argue that it could potentially slow startups and add additional regulatory burden on small actors.

21.6.2 The United States: sectoral and innovation-first

The US has resisted a single federal AI law, instead adopting a **sectoral approach** combined with voluntary guidelines. Examples include:

- **NIST AI Risk Management Framework** (2023)[14, 15]: Provides standards for responsible AI development, focusing on trustworthiness, bias mitigation, and robustness.
- **White House AI Bill of Rights** (2022)[16]: Outlines principles for safe and transparent AI, including protections against algorithmic discrimination.
- Regulation at the **state level** (e.g., Illinois' Biometric Privacy Law, California Consumer Privacy Act).

This approach reflects the US priority of **fostering innovation**, leaving regulation to evolve alongside industry practices. The model encourages experimentation but has been criticized for **lack of coherence and weak enforcement.**

Lesson for India: India's AI startup ecosystem may thrive from a sector-specific, innovation-centric approach. However, in the absence of robust accountability mechanisms, it may create unintended consequences such as misinformation and bias. India may also need a hybrid approach to accountability and innovation.

21.6.3 China: state-centric and security-oriented

China has taken a **top-down regulatory approach**, prioritizing state control, data sovereignty, and national security:

Personal Information Protection Law (PIPL, 2021): Strict rules on data consent, processing, and cross-border transfers.

Regulations on Recommendation Algorithms (2022): Require platforms to ensure algorithms promote "positive values" and avoid harmful content.

Generative AI Regulation (2023) [17]: Mandates labeling of AI-generated content and government approval for foundation models.

China's strategy is in line with its larger government model, which emphasizes national security, censorship, and strict control over digital platforms. It limits openness, freedom of innovation, and international cooperation even while it guarantees stringent state monitoring.

Lesson for India: India can learn from China's emphasis on **data protection, labeling, and content moderation**, but must avoid excessive state control that undermines free expression, innovation, and democratic values.

21.6.4 Comparative summary

Table 21.3 shows the comparison between global AI governance models—highlighting the EU's rights-based clarity, the US's innovation-driven flexibility, and China's security-focused control—offering India a balanced path combining risk-based regulation, accountability, and innovation freedom.

Table 21.3 Comparative analysis of AI approaches followed by developing nations

Region	*Approach*	*Key Strength*	*Key Weakness*	*Lesson for India*
EU	Risk-based, rights-oriented	Strong legal clarity, safeguards for citizens	May over-regulate and stifle startups	Adapt a **risk-based framework** but keep safeguards affordable.
US	Sectoral, innovation-first	Encourages rapid innovation and startup growth	Lacks coherence and enforcement	Combine **sectoral flexibility** with stronger accountability.
China	State-centric, security-first	Strong oversight, strict data sovereignty	Limits openness, freedom, and global cooperation	Balance **data sovereignty** with democratic values and innovation freedom.

21.6.5 Reflection

India is at a crossroads in global AI governance. It cannot afford overregulation, as the EU has, that could limit economic growth, and it cannot rely on self-regulation based on an assumption of shared interests, as the US does, due to its considerable size and diversity. Finally, it cannot give up political freedoms in the way that China has done. The Indian model, therefore, needs to be a **hybrid approach**—

- risk-based where harms are high,
- sectoral where innovation is critical,
- rights-oriented to protect citizens, and
- innovation-friendly to empower startups and local ecosystems.

This **contextualized governance model** will allow India to carve out a distinct position in the global AI landscape, balancing innovation, rights, and accountability.

21.7 STRENGTHENING INDIA'S AI ECOSYSTEM

Figure 21.3 shows the five-pillar model of strengthening India's AI ecosystem. To enhance India's AI ecosystem, a more holistic and capacity-building approach is required than just regulatory. While academia must support by training qualified professionals, being funded to carry out research in ethics and Indian (governance) AI solutions, and working on public open datasets for innovators, it is the startup ecosystem creating entirely home-grown solutions that needs concentrated support in funding, incubators, and guidelines, while reducing redtape.

And lastly, PPPs will act as facilitators in centralizing resources that designers can use without hassle. This may include cloud solutions, safety audit programs, or the national AI compute facility for cloud computing, etc.

21.8 FUTURE ROADMAP FOR RESPONSIBLE AI IN INDIA

India's roadmap for a collaborative AI future suggests a practice-driven, principles-based approach to navigate duty and innovation. In addition to having engaged consideration of an AI Act for the long term, India might consider that a transformative AI governance

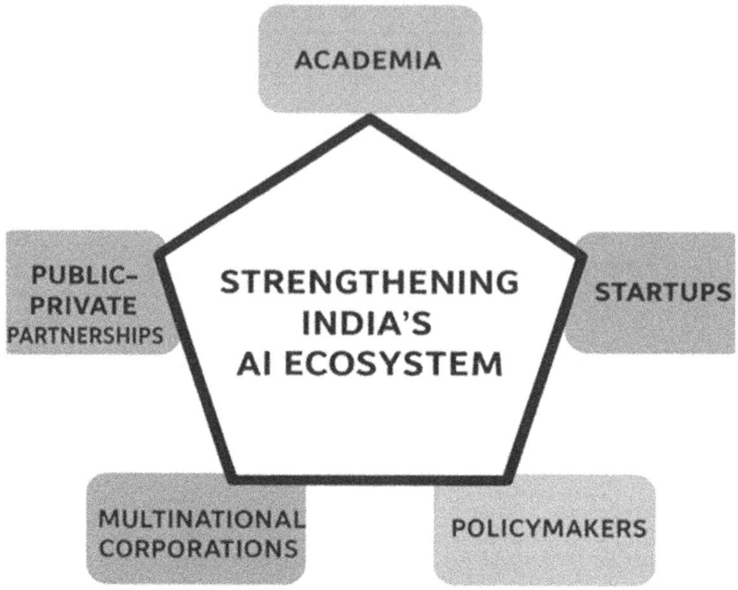

Figure 21.3 Five-pillar model of strengthening India's AI ecosystem.

framework that pulls together existing laws and sectoral guidance will enhance the clarity of all use cases. In this regard, the prototype AI Safety Institute will be an important center for the generation of knowledge and capacity building (across research, technology audits, risk assessments, and techcounsel to regulators), which will help form the basis for a responsible AI ecosystem. In implementing this, it will also be critical to continue to build capacity for the AI workforce, who will need to be reskilled in the transformations of the workplace due to AI-enabled systems, as well as building the capacity of regulators with the technical skill sets to monitor and observe complex systems. Figure 21.4 shows the future roadmap for responsible AI in India.

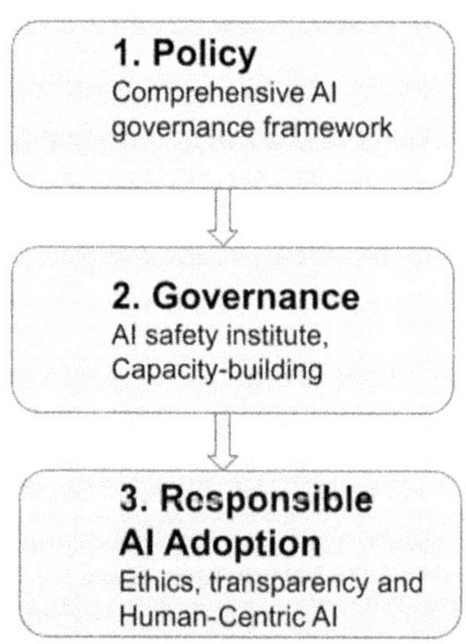

Figure 21.4 Future roadmap for responsible AI in India.

21.9 CONCLUSION

India's path to AI governance demonstrates both the benefits and difficulties of deploying a new, and profoundly altering, technology within a complex socioeconomic context. In the last few years, India has moved from lofty initiatives, such as the National Strategy for AI (2018), to clearer institutional and regulatory instruments, including the DPDP Act (2023) and recommendations for AI safety. These initiatives suggest a willingness on the part of Indian policy makers to protect rights, develop public trust, and promote innovation.

India has seen fruitful developments; however, our study demonstrates that there remain considerable gaps related to intellectual property rights, accountability, bias mitigation, and enforcement capacity [18].

Many of the global case studies point to the potential for actual harm that can occur from inadequate protections, stressing the importance of creating transparent, principle-based governance. Still, we noted in the comparative analysis of the US, China, and the EU that there is no single best approach; India must create a hybrid, unique approach that considers India's development aspirations, democratic principles, and innovative capacity.

India's success in the future will rely on fortifying its AI ecosystem, which includes universities, startups, policymakers, international companies, and public-private partnerships, while integrating ethics, accountability, and transparency into all tiers of policy. Crucial actions will include creating a specialized AI Safety Institute, encouraging global cooperation, and providing technical know-how to regulators.

In conclusion, responsible AI in India presents both a historic opportunity and a policy dilemma. India can create an AI governance model that is both internationally relevant and locally responsive if it is led by human dignity, inclusivity, and foresight. This will guarantee that the next generation of AI contributes to fair growth and social trust.

REFERENCES

1. NITI Aayog. (2021a). *Principles for responsible AI*. Government of India.
2. Mohanty, A., &Sahu, S. (2024, November 21). *India's advance on AI regulation*. Carnegie Endowment for International Peace.
3. Elbashir, P., &Desikachari, K. B. (2025, March 13). *India's path to AI autonomy*. Atlantic Council.
4. Parliamentary committee calls for stronger laws on deepfakes. (2025, August). *MediaNama*.
5. New AI rules in India may force creators to label their content as AI-generated. (2025). *Hindustan Times*.
6. Batool, A., Zowghi, D., &Bano, M. (2023, December). *Responsible AI governance: A systematic literature review* (arXiv:2401.10896). arXiv. https://arxiv.org/abs/2401.10896
7. Takshashila Institution. (2025). *State of AI governance, 2024*.Takshashila Institution.
8. International Association of Privacy Professionals (IAPP). (2025, July). *Global AI governance law and policy: India*.IAPP.
9. AI Governance in India—Law, policy and political economy. (2024). *Taylor & Francis Online*.
10. India's cyber laws face deepfake reckoning. (2025). *Policy Circle*.
11. Government of India. (2023). *Digital personal data protection act, 2023*. Ministry of Law and Justice.
12. NITI Aayog. (2018). *National strategy for artificial intelligence: #AIforAll*. Government of India.
13. NITI Aayog. (2021b). *Operationalising responsible AI*. Government of India.
14. *Thaler v. Perlmutter*, 43 F.4th 1203 (D.C. Cir. 2023).
15. National Institute of Standards and Technology. (2023). *AI risk management framework*. U.S. Department of Commerce.
16. White House, Office of Science and Technology Policy. (2022). *Blueprint for an AI bill of rights*. Washington, DC.

17. Cyberspace Administration of China. (2023). *Interim measures for the management of generative artificial intelligence services*. Government of China.
18. Ali, S. A., Ansari, M., Alam, M., &Rakshit, S. (2023). Federated learning for business intelligence: Predictive maintenance in Industry 4.0. In *AI-based data analytics*.

Chapter 22

Challenges and future trends in next-generation AI

Integrating neuroscience, edge computing, and sustainability

Aditi Aggarwal, Preeti Sharma, and Pankaj Kumar Varshney

22.1 INTRODUCTION

With uses in health care, transportation, finance [1], agriculture, education, and defense, artificial intelligence (AI) has become one of the 21st century's most revolutionary technologies. Advances in deep learning, reinforcement learning, and natural language processing over the last ten years have made it possible for machines to carry out tasks that were previously thought to be exclusive to humans, such as medical image diagnosis and real-time language translation. But these advancements have also brought to light important drawbacks. Present-day AI systems are energy-hungry, computationally costly, and frequently fragile in the face of hostile inputs or unfamiliar surroundings. Furthermore, concerns about privacy, latency, energy consumption, and accessibility are raised by the centralization of AI processing in massive cloud data centers [2].

The need for effective, flexible, and sustainable AI has never been higher as these systems transition from research labs to real-world, ubiquitous applications—integrated into wearable health devices, smartphones, autonomous cars, smart homes, and industrial Internet of Things (IoT) systems. Because of this paradigm shift, AI must work closer to the data source (edge computing), learn and adapt more like biological systems (neuroscience-inspired AI), and operate within resource efficiency and sustainability constraints (Green AI) [3].

Next-generation AI is greatly inspired by neuroscience. The human brain uses only about 20 watts of power, which is about the same as a dim lightbulb, to achieve remarkable intelligence, in contrast to modern deep neural networks. Biological systems are energy-efficient and adaptive due to characteristics like sparse connectivity, spiking communication, event-driven processing, and local plasticity rules. A promising route toward scalable, effective, and resilient AI that can flourish in resource-constrained environments is provided by translating these mechanisms into neuromorphic computing architectures and algorithms [4].

In contrast, edge computing disperses computational intelligence among local devices, such as smartphones, microcontrollers, smart sensors, and self-governing robots, instead of relying on centralized cloud servers [5]. By minimizing data transfers, this model improves privacy, lowers communication latency, and permits context-specific adaptation. However, the direct deployment of large deep learning models [6] is impractical because of the severe hardware limitations that edge devices frequently face in terms of processing power, memory, and battery life [2].

Simultaneously, sustainability in AI design and distribution has emerged as a global necessity. In comparison to entire homes or small industries, advanced deep learning model training frequently uses current in the megawatt-hour range and generates carbon emissions. With expectations of billions of devices coming online over the coming decade, the environmental impact of AI can be a significant obstacle. Thus, the next era of AI should not only be intelligent and optimal, but also environmentally responsible.

DOI: 10.1201/9781003685890-22

A special chance to overcome these obstacles is presented by the convergence of neuroscience, edge computing, and sustainability. Effective, biologically installed computations can be produced by neuroscience-inspired algorithms; intelligence is guaranteed by edge computing and privacy protection; and ongoing practices guarantee that the advancement and application of AI align with international objectives for environmental preservation and energy conservation. When combined, these domains have the chance to create a new generation of AI systems that are not only more potent but also more sensible, ethical, and long-lasting [7].

22.2 NEUROSCIENCE-INSPIRED ARTIFICIAL INTELLIGENCE

The human brain has long served as an inspiration for AI. The fundamentals of biological neurons were simplified in early neural network models, such as the perceptron. Modern deep learning architecture, however, is limited to actual neurobiological processes, despite its name. Because of this discrepancy, AI that is inspired by neuroscience has emerged. Its goal is to bring artificial computations closer to the principles found in the brain, allowing for greater flexibility, power, and efficiency [8].

22.2.1 Brain-inspired computational models

Numerous key aspects of biological intelligence that are readily transferred to artificial systems are revealed by neuroscience research:

1. **Spiking Neural Network (SNN):** The SNN brain's spike behavior mimics how neurons transmit information through discrete electric pulses, in contrast to traditional artificial neural networks (ANNs), which rely on continuous activation values. This method allows for energy efficiency, compatibility with neuromorphic hardware, and temporal coding [9].
2. **Computer processing of neuromorphorfish:** Neuromorphic processors (like Intel Loihi and IBM TrueNorth) are inspired by the structural and functional features of the brain and are made to operate in parallel using observable techniques that use less power than GPUs [10].
3. **Cognitive architecture:** Models like Soar and ACT-R attempt to replicate complex cognitive functions like memory, reasoning, and judgment. Deep learning offers more effective interpretation and human-like logic than pattern recognition [11].
4. **Adaptation and plasticity:** Through processes like synaptic plasticity, biological brains exhibit lifelong learning. In order to create systems that optimize without making terrible mistakes, AI researchers are searching for analogous techniques like meta-learning, reinforcement learning, and self-organizing neural networks.

22.2.2 Advantages of neuroscience-inspired AI

1. **Energy efficiency:**Many orders of brain-inspired model size can lower computation costs by utilizing phenomenon-driven architecture and pig-based processing**Strength and adaptability:** Neuroscience-based models are better able to handle noise, incomplete or dynamic data, making them suitable for real-**world environments.**
2. **Interpretability:** One of the main issues in contemporary deep learning is addressed by certain cognitive models, which shed light on the reasons behind an AI's decisions.
3. **Compatibility with hardware:** Neuromorphic components can be incorporated into laptops and other edge devices, opening the door to long-lasting, location-based intelligence.

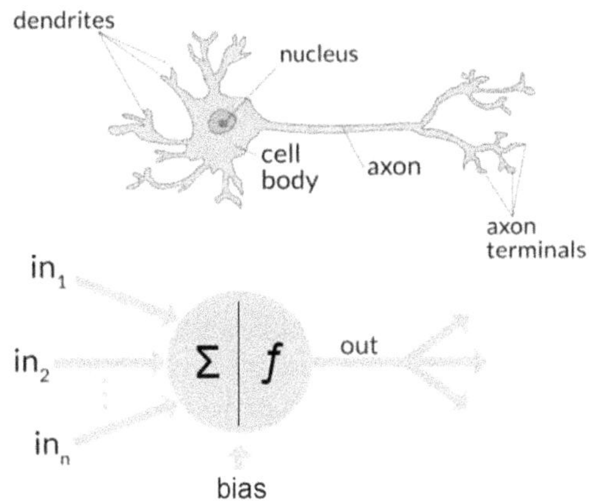

Figure 22.1 Comparison between biological brain and AI neural networks [23].

22.2.3 Challenges in neuroscience-inspired AI

Despite these advantages, there are still a lot of difficulties:

- Standardized frameworks for creating biologically inspired models are lacking.
- For big datasets, it can be challenging to scale mechanisms like the brain.
- Premature copying and a limited comprehension of human cognition.
- Special hardware is needed, but it is not yet widely accessible and is still undergoing development.

22.2.4 Future prospects

Neuroscience-inspired AI is expected to converge with edge data processing and sustainability goals. As neuromorphic processors mature, they may be distributed in mobile devices, IoT sensors, and medical implants. In addition, efficient training algorithms can revolutionize the energy profile of spiking model AI combined with algorithms, making it more compatible with global sustainability goals (Figure 22.1)[22].

- Left side: illustration of biological neurons with dendrites, axons, and synaptic spikes.
- Right side: diagram of artificial neural network layers (input, hidden, output).
- Labels: highlight differences (event-driven spikes vs. continuous activations, synaptic plasticity vs. static weights, energy efficiency vs. high computational cost).

22.3 AI AT THE EDGE: EDGE AND FOG COMPUTING INTEGRATION

The exponential growth of the connected equipment fabric of the IoT and 5G network has created a great demand for real-time, decentralized AI processing. Traditionally, the AI models are trained and deployed in the cloud, where calculation resources with large and large calculations are available. However, cloud-centered architecture is on challenges, bandwidth, privacy and energy consumption. To address these limitations, edge and fog computing for the next-generation AI have emerged as an important paradigm [12].

22.3.1 Edge AI

Edge AI refers to the deployment of AI algorithms directly on devices such as smartphones, sensors, drones, or autonomous vehicles, where data is generated. By processing the information locally, edge AI reduces dependence on centralized servers and ensures real-time responsiveness.

Key benefits:

- **Low latency:** Decisions can be made within milliseconds, critical for autonomous driving and health-care monitoring.
- **Privacy and security:** Sensitive data need not be transmitted to the cloud, reducing the risk of breaches.
- **Energy efficiency:** Reduces continuous communication overhead with remote servers.

22.3.2 Fog computing

Fog computing acts as a middle layer between the cloud and the edge. While edge devices train the model before sending data to localized estimates, fog knots (e.g. local server or gateway) overall, preprocesses and sometimes clouds. Efficiency and scalability are balanced in this method.

Key applications:

- Smart health-care systems where patient monitoring data is aggregated at hospital servers before cloud analysis [13].
- Industrial IoT, where predictive maintenance data is analyzed at the factory floor before long-term storage (Figure 22.2).

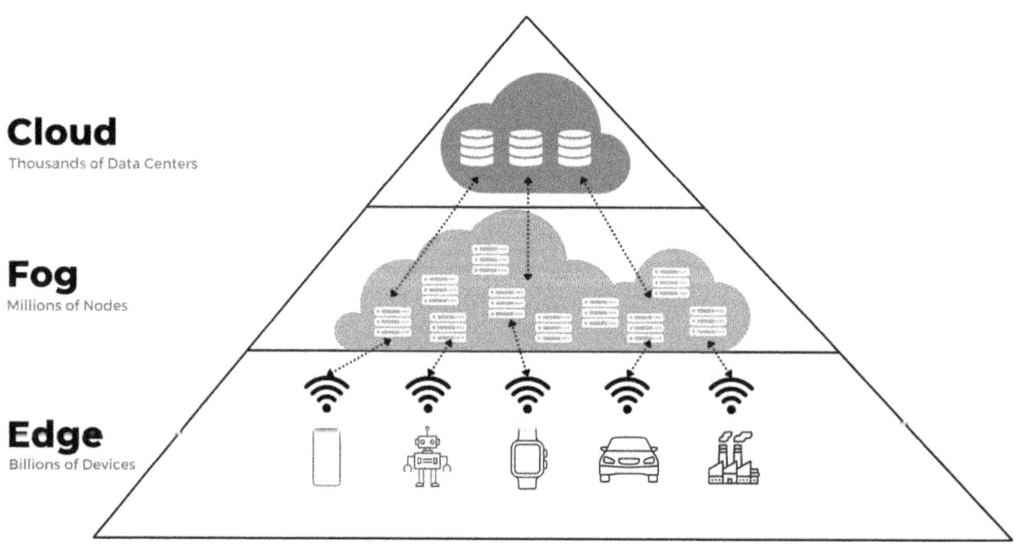

Figure 22.2 Cloud vs. fog vs. edge computing architecture [24].

22.3.3 Challenges of edge and fog AI

Despite its promise, several barriers remain:

- **Hardware constraints:** Edge devices have limited processing power and memory, requiring lightweight AI models.
- **Model compression:** Techniques such as pruning, quantization, and knowledge distillation are essential but may compromise accuracy.
- **Security threats:** Decentralized devices are vulnerable to adversarial attacks and physical tampering.
- **Interoperability:** Lack of standardized frameworks across diverse devices and vendors.

22.3.4 Case studies

1. **Health care:** Wearables with AI capabilities, like ECG sensors, can identify irregularities instantly and independently of the cloud.
2. **Autonomous vehicles:** Safety in dynamic environments is ensured by edge-based decision systems that lessen dependency on connectivity.
3. **Smart grids:** By instantly analyzing data from millions of sensors, edge AI optimizes energy distribution.

22.3.5 Future directions

Cloud computing, edge, and fog will all be integrated into the next generation of AI. Local decision-making and cloud-powered analysis will be balanced by hybrid architectures. With advancements in neuromorphic components, federated learning, and low-power products, edge AI's role in sustainable deployment will only grow (Table 22.1).

Table 22.1 Comparison of cloud, edge, and fog computing for AI workloads [3, 7, 9, 13–14]

Feature	*Cloud computing*	*Edge computing*	*Fog computing*
Location of processing	Centralized data centers	Directly on end devices (e.g., sensors, phones)	Intermediate nodes (e.g., gateways, local servers)
Latency	High (depends on network connectivity)	Very low (milliseconds)	Moderate (faster than cloud, slower than edge)
Bandwidth usage	High (continuous data transfer)	Low (local processing)	Medium (aggregates before cloud transfer)
Privacy	Vulnerable to breaches during transmission	High (data stays local)	Moderate (some data stays local, some sent upward)
Energy efficiency	Low (energy-intensive server farms)	High (minimal transmission overhead)	Balanced efficiency
Use cases	Large-scale model training, big data analytics	Autonomous vehicles, wearables, smart sensors	Industrial IoT, smart cities, health-care systems

22.4 SUSTAINABILITY IN AI

The environmental and ecological effects have grown in importance as AI advances. The computational power needed to run modern deep learning models is enormous and frequently equals the energy used by small businesses or entire households. For instance, hundreds of tons of carbon emissions may result from training a large transformer-based model, raising questions about whether AI is compatible with global climate targets. As a result, the sustainability of AI has emerged as a clear obstacle for the upcoming generation of intelligent systems [14].

22.4.1 Environmental costs of AI

AI's environmental burden arises primarily from:

1. **Model training:** Large neural networks require thousands of GPU hours. For example, GPT-class models demand petaflop-days of computation.
2. **Data storage:** Massive datasets (petabytes in size) consume significant storage energy.
3. **Inference at scale:** Billions of daily queries to AI models also contribute to ongoing energy consumption.
4. **Hardware lifecycle:** Manufacturing, deploying, and eventually discarding AI hardware adds to the carbon footprint.

22.4.2 Approaches to Green AI

To mitigate these issues, the emerging paradigm of **Green AI** emphasizes efficiency alongside accuracy. Strategies include:

- **Model optimization:** Techniques such as pruning, quantization, and low-rank factorization reduce memory and computation.
- **Transfer learning and reuse:** Instead of training from scratch, models can be fine-tuned for specific tasks, lowering energy usage.
- **Energy-aware training:** Using renewable energy-powered data centers (e.g., Google's carbon-neutral AI infrastructure).
- **Hardware innovation:** Neuromorphic chips and low-power accelerators designed for AI inference.

22.4.3 AI for sustainability

Ironically, while AI contributes to environmental challenges, it also provides **tools for sustainability**. Examples include:

- **Smart energy grids:** AI optimizes renewable energy integration, balancing supply and demand.
- **Climate modeling:** Machine learning improves predictive accuracy in weather forecasting and climate change simulations [15].
- **Agriculture:** AI enables precision farming, reducing fertilizer and water usage.
- **Waste management:** Computer vision systems support automated recycling.

22.4.4 Challenges and limitations

Despite progress, achieving sustainable AI is not straightforward:

- Trade-off between model performance and efficiency.
- Lack of standardized metrics for measuring carbon footprint across models.
- Limited accessibility of green computing hardware for researchers in developing regions.
- Policy gaps in regulating AI's ecological impact [16].

22.4.5 Future directions

Next-generation AI will need to prioritize energy-efficient algorithms, carbon-aware scheduling of workloads, and global standards for reporting AI's environmental impact. Interdisciplinary efforts combining AI research, environmental science, and policy will be vital to aligning AI development with the United Nations' Sustainable Development Goals (SDGs) (Figure 22.3 and Table 22.2) [17].

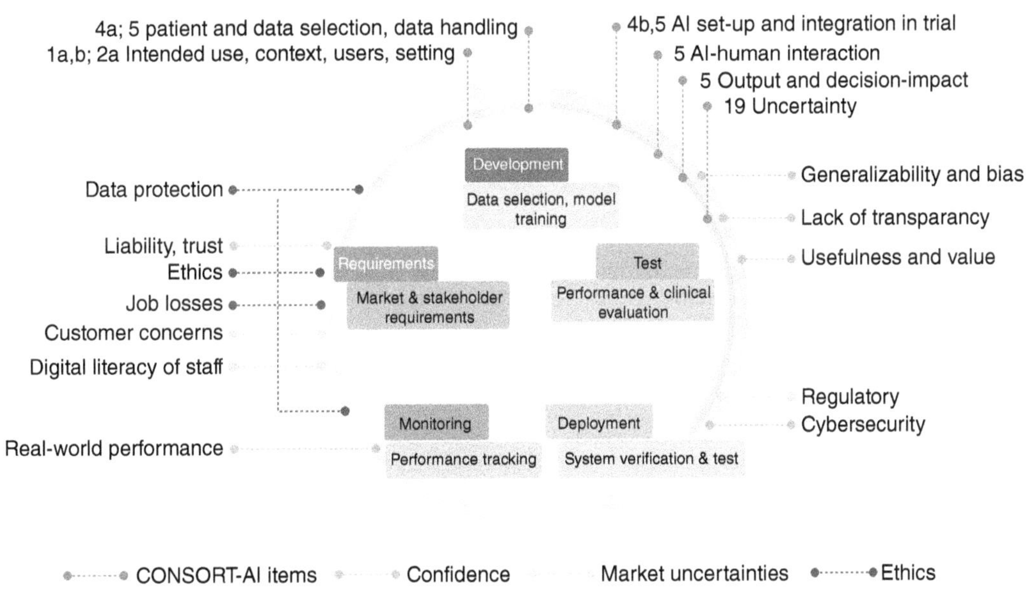

Figure 22.3 AI lifecycle and sustainability impact [25].

Table 22.2 Estimated carbon footprint of major AI models [2, 17, 19, 21]

AI model (year)	*Parameters (approx.)*	*Compute (FLOPs)*	*CO_2 emissions (metric tons)*	*Notes*
BERT (2019)	110 million	3.3×10^{16}	~0.3	Widely used NLP model
GPT-3 (2020)	175 billion	3.1×10^{23}	~552	Large-scale generative model
AlphaGo Zero (2017)	N/A	1.3×10^{21}	~96	Reinforcement learning model
BLOOM (2022)	176 billion	3.5×10^{23}	~400	Open-access LLM
PaLM (2022)	540 billion	8.9×10^{23}	~1,000+	Google's large-scale LLM

22.5 INTERDISCIPLINARY INTEGRATION: NEUROSCIENCE + EDGE + SUSTAINABILITY

While neuroscience, edge computing, and sustainability represent significant advances individually, the true potential of next-generation AI lies in their integration. The combination of biologically inspired intelligence with decentralized architecture and green principles provides a blueprint to create AI systems that are not only powerful but also responsible and ready for the future.

22.5.1 Synergy between neuroscience and edge AI

- **Neuromorphic chips at the edge:** Neuroscience-inspired processors, such as Intel's *Loihi*, are optimized for event-driven computation [18]. When deployed at the edge, they enable energy-efficient, low-latency intelligence for real-time decision-making.
- **Spiking neural networks for IoT:** Unlike traditional deep models, spiking networks process data asynchronously, making them ideal for IoT sensors and wearable devices with limited power supply.
- **Cognitive edge devices:** Cognitive architectures, when deployed locally, can enable devices that reason, plan, and adapt in real-world settings without cloud dependence.

22.5.2 The triangular integration framework

The convergence of neuroscience, edge computing, and sustainability creates a **holistic paradigm** for AI:

- **Neuroscience** provides the *algorithmic inspiration* (e.g., plasticity, spiking, cognition).
- **Edge computing** provides the *deployment infrastructure* (localized, real-time, distributed).
- **Sustainability** provides the *ethical and ecological compass* (Green AI, carbon-aware optimization) [4].

Together, they form a resilient, efficient, and human-centered AI ecosystem capable of addressing 21st-century global challenges (Figure 22.4).

22.6 CHALLENGES IN NEXT-GENERATION AI

Despite the promising convergence of neuroscience, edge computing, and sustainability, the pathway toward next-generation AI is not without obstacles. These challenges span technical, ethical, and policy domains, requiring interdisciplinary collaboration to address them effectively.

22.6.1 Ethical challenges

- **Privacy and security:** Decentralized edge systems, while reducing cloud dependence, are vulnerable to adversarial attacks and physical tampering.

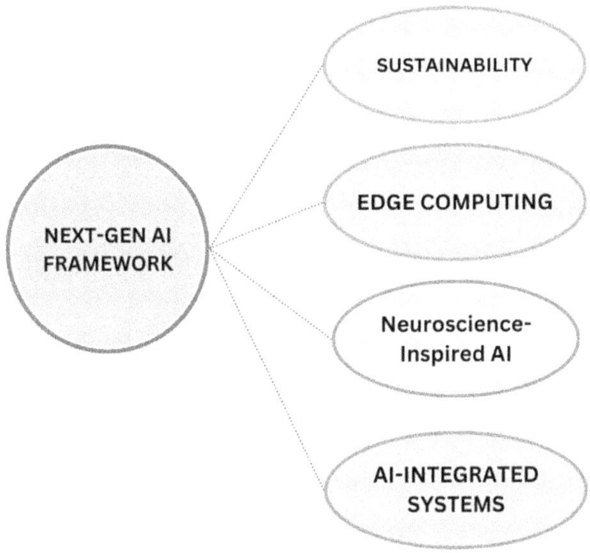

Figure 22.4 Framework of next-gen AI.

- **Explainability and transparency:** Neuroscience-inspired models may be more interpretable, but neuromorphic and deep networks still struggle with black-box decision-making.
- **Bias and fairness:** Sustainable AI frameworks must ensure that efficiency-driven decisions do not reinforce algorithmic bias or marginalize vulnerable populations.

22.6.2 Governance and policy challenges

- **Lack of policies:** No existing frameworks are available for measuring carbon footprint, energy usage, or sustainability compliance in AI.
- **Certain gaps:** Policies often lag behind technological advancement, leaving AI applications underregulated in terms of ecological and ethical impact.
- **Resource inequality:** Different countries face challenges due to resource inequality (Table 22.3).

Table 22.3 Key challenges and proposed mitigation strategies [5, 7]

Category	*Challenge*	*Proposed mitigation strategy*
Technical	Scalability of spiking neural networks	Develop standardized frameworks and scalable training algorithms
	Hardware limitations at the edge	Invest in low-power neuromorphic chips and model compression
	Accuracy–efficiency trade-off	Hybrid models combining lightweight inference with cloud-based refinement

(*Continued*)

Table 22.3 (Continued)

Category	*Challenge*	*Proposed mitigation strategy*
Ethical	Privacy and security risks in edge AI	Federated learning, encryption, and blockchain-based trust mechanisms
	Lack of interpretability	Incorporate explainable AI techniques into neuromorphic and edge models
	Algorithmic bias	Diverse datasets and fairness-aware learning algorithms
Governance	Absence of carbon accounting standards	Establish global metrics for AI sustainability reporting
	Policy lag	Develop adaptive, forward-looking AI governance frameworks
	Resource inequality	Promote open-source neuromorphic tools and subsidize access in developing regions

22.7 FUTURE TRENDS IN NEXT-GENERATION AI

Numerous technical disciplines study machines that interact, teach, and reshape the world as AI enters a transformative phase. The next generation of AI is primarily driven by developments in quantum technologies, edge computing, neuroscience, and sustainability. Although this convergence is still in its early stages, it is expected to result in AI systems that are more efficient, adaptable, and reference-free, thereby better meeting human needs. These trends reflect two imperatives: leveraging technological advancements while ensuring AI development is energy-efficient, socially responsible, and adaptable to different applications. In the following sections, we find new trends that will shape the future of AI.

22.7.1 Neuromorphic hardware at the edge

Data processing from neuromorphic computing signifies a paradigm shift in the architecture and deployment of AI systems. Neuromorphic processors, as opposed to conventional von Neumann architecture, are designed to mimic the composition and functions of biological neural networks. These systems process information by using spikes, which communicate by sending electrical impulses that mimic brain activity. This approach enables real-time, highly energy-efficient computation and is particularly suitable for edge devices such as drones and smart systems. The use of neuromorphic hardware has many transformative benefits. First, these devices can reduce latency and dependency on cloud infrastructure by performing computations locally, which may allow for quick decision-making in crucial applications such as robotics and autonomous navigation. Second, these systems can support distributed AI due to their intrinsic low-power consumption, which is in line with sustainability goals.

22.7.2 Self-adaptive and context-aware AI

Conventional AI models' capacity to adjust to changing circumstances is constrained by their frequent reliance on static architectures and task-specific training. The next generation of AI systems, on the other hand, is anticipated to be independent and reference-fishing, constantly learning from their surroundings to modify their behavior. These systems draw inspiration from human cognition, and will integrate elements of memory, logic, and perception so that they can demonstrate lifelong learning without experiencing scary to forget them—an event where newly acquired knowledge interferes with previously learned information.

Self-adaptive AI has far-reaching applications, from personal health care—where systems adjust treatment schemes based on developing patient data—to industrial automation, where machine adapters operate in real-time parameters. Integration of reference awareness will also increase the interaction between human meters, so that the AI system can explain the signs of the fine position, understand the intentions and react with appropriate functions [20].

22.7.3 Federated and decentralized sustainable AI

With the growth of distributed AI systems, privacy and energy efficiency are becoming more significant. One promising method in this area is federated learning, which enables multiple edge devices to train machine learning models without collaborating on raw data. AI aligns with ethical and legal requirements by protecting privacy and minimizing centralized data collection [16].

The decentralized AI system can be modified to use less energy and leave a smaller carbon footprint when paired with sustainability considerations. By using strategies like carbon-aware planning, adaptive composition, and model compaction, federated AI can become both environmentally responsible and privacy-preserving. These methods are especially applicable in large industrial networks, IoT ecosystems, and smart cities where distributed intelligence must be successfully used in odd infrastructure and equipment.

22.7.4 Deal with quantum computing

Quantum computing promises a leap in computational capacity and an exponential acceleration of optimization, cryptography, and simulation functions. "Quantumaugmented AI," or the application of quantum principles to AI, has the potential to create new problem-solving opportunities that are currently safe for use with conventional computers. For example, quantum algorithms are capable of accurately simulating climate models, optimizing complex neural networks, and accelerating drug discovery.

Furthermore, an AI system that is both scalable and energy-efficient can be achieved by combining quantum computation with neuromorphic architecture. Research in hybrid architectures and quantum-inspired algorithms is advancing quickly, despite the fact that practical quantum integration is still in its infancy. This field has long-term implications for both theoretical and applied AI.

22.7.5 Bio-AI hybrids and cognitive AI

Advances in bio-AI hybrids—a system that combines artificial processors and organic neurons—are made possible by research into neural and brain-computer interfaces. AI can interpret and communicate directly with neural signals thanks to this hybrid architecture's capacity to combine human cognition and machine intelligence. According to preliminary cognitive AI experiments, these systems can improve human capabilities, enable neuroprosthetics, and assist medical intervention for neurological conditions.

In the long run, cognitive AI envisions an unbroken human-machine symbiosis in which people and artificial systems work together to solve problems, be creative, and make decisions. These systems can effectively alter the interface between intelligent machines and humans by utilizing brain insights to achieve high efficiency, relevant logic, and spontaneous understanding (Figure 22.5).

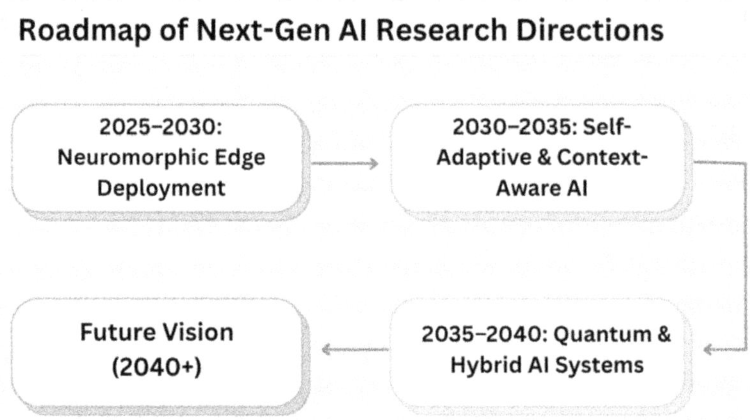

Figure 22.5 Roadmap of next-gen AI research directions.

22.8 CONCLUSION

The development of AI enters a significant stage. Even though automation, large models, and deep learning have advanced significantly, the next generation of AI should go beyond current boundaries to become more adaptable, efficient, and durable. This chapter has identified three key areas—sustainability, edge computing, and neuroscience—that will have an impact on this future. Hardware limitations, accuracy trade-offs, and scalability are examples of technical issues that need to be addressed. Ethical concerns that need to be addressed include bias, fairness, transparency, and privacy. Government decision-makers ought to establish global standards for carbon accounting, sustainable AI reporting, and fair access to edge and neuromorphic technologies.

Developments such as bio-AI-hybrids, federated learning, quantum-enhanced models, neuromorphic hardware, and self-adaptive AI will also influence the direction of AI in the future. All things considered, this development suggests a future in which AI is inherently responsible, decentralized, and supervised by humans—a transformative tool to address some of the most important problems facing the planet [21, 22]. Finally, rather than distinct technological advancement, the future of AI lies in the integration of neuroscience, computation, and sustainability principles. Through interdisciplinary cooperation in the domains of informatics, neuroscience, environmental studies, and policy, the next generation of AI can be a force for innovation, inclusivity, and the welfare of the planet.

REFERENCES

[1] Ansari, M., Chhabra, A., Hasan, N., & Alam, M. (2025). *Quantum finance: Revolutionizing financial markets with quantum computing*. In *AI-driven finance in the VUCA world* (pp. 96–105). Auerbach Publications.

[2] Strubell, E., Ganesh, A., & McCallum, A. (2019). Energy and policy considerations for deep learning in NLP. In *Proceedings of ACL* (pp. 3645–3650).

[3] Ansari, M., Ali, S. A., & Alam, M. (2022). Internet of things (IoT) fusion with cloud computing: Current research and future direction. *International Journal of Advanced Technology and Engineering Exploration*, *9*(97), 1812.

[4] Upadhyay, A. K., Varshney, M., & Varshney, P. K. (2024). Techniques for accurate forecasting of household energy consumption. *Journal of Computational Analysis & Applications*, *33*(6).

[5] Roy, K., Jaiswal, A., & Panda, P. (2019). Towards spike-based machine intelligence with neuromorphic computing. *Nature*, *575*(7784), 607–617.

[6] Varshney, P. K., Bharara, V., Sharma, N., & Mehta, H. K. (2024). Critical evaluation of deep learning models for heart disease detection. In *OTCON* (pp. 1–7). IEEE.

[7] Shi, W., Cao, J., Zhang, Q., Li, Y., & Xu, L. (2016). Edge computing: Vision and challenges. *IEEE Internet of Things Journal*, *3*(5), 637–646.

[8] Ali, S. A., Ansari, M., Alam, M., & Rakshit, S. (2023). Federated learning for business intelligence: Predictive maintenance in Industry 4.0. In *AI-based data analytics* (pp. 125–140). Auerbach Publications.

[9] Varshney, P. K., Agrawal, G. S., & Sharma, S. K. (2016). Energy consumption analysis of wireless ad hoc networks routing protocols. *INROADS*, *5*(1s), 250–253.

[10] Marković, D., Mizrahi, A., Querlioz, D., & Grollier, J. (2020). Physics for neuromorphic computing. *Nature Reviews Physics*, *2*(9), 499–510.

[11] Ansari, M., & Alam, M. (2024). IoT-cloud-based air pollution forecasting model. *Arabian Journal for Science and Engineering*, *49*(3), 3135–3162.

[12] Alam, M., Ara, K., Javed, M. S., & Ansari, M. (2014). Traffic attack detection using neural networks. In *ICDMKE* (pp. 2–4).

[13] Ansari, M., & Alam, M. (2021). IoT-cloud analysis of air pollution data. In *ICDAM 2021* (pp. 125–139). Springer.

[14] Ansari, M., & Alam, M. (2023). IoT-cloud-enabled smart framework. In *Intelligent systems and applications* (pp. 87–105). Springer.

[15] Rana, C. (2025). Road object detection using ML. In *Machine learning in multimedia* (pp. 50–60). CRC Press.

[16] Zhang, Y., Zheng, Y., & Yang, Q. (2022). Federated learning for sustainable AI. *Nature Machine Intelligence*, *4*, 123–132.

[17] Benaich, I., & Hogarth, N. (2022). *State of AI report 2022*.

[18] Davies, M., Srinivasa, N., Lin, T. H., et al. (2018). Loihi neuromorphic processor. *IEEE Micro*, *38*(1), 82–99.

[19] Verdecchia, R., Cardellini, V., & Cortellessa, V. (2021). Green AI: A survey. *ACM Computing Surveys*, *54*(6), 1–36.

[20] Rana, C. (2022). IoT-based health monitoring system. In *IC3I* (pp. 1727–1732). IEEE.

[21] Schwartz, R., Dodge, J., Smith, N. A., & Etzioni, O. (2020). Green AI. *Communications of the ACM*, *63*(12), 54–63.

[22] Kairouz, P., McMahan, H. B., Avent, B., et al. (2021). Advances in federated learning. *Foundations and Trends in Machine Learning*, *14*(1–2), 1–210.

[23] Applied Go. (n.d.). *Perceptron tutorial*. https://appliedgo.net/perceptron/

[24] Share Google Images. (n.d.). *Image resource*. https://share.google/images/JHDTBW6d5K1znbwKP

[25] ResearchGate. (n.d.). *AI lifecycle figure*. https://www.researchgate.net/figure/The-Life-Cycle-of-Artificial-Intelligence-Applications-Schwendicke-and-Krois-2021_fig1_381466522

Index

Pages in *italics* refer to figures and pages in **bold** refer to tables.

For Product Safety Concerns and Information please contact our EU
representative GPSR@taylorandfrancis.com
Taylor & Francis Verlag GmbH, Kaufingerstraße 24, 80331 München, Germany

www.ingramcontent.com/pod-product-compliance
Ingram Content Group UK Ltd.
Pitfield, Milton Keynes, MK11 3LW, UK
UKHW052218190926
475702UK00006B/1089

* 9 7 8 1 0 4 1 1 6 5 4 6 0 *